S0-DUT-745

The Chemical Constituents of Citrus Fruits

ADVANCES IN FOOD RESEARCH

Edited by

C. O. CHICHESTER E. M. MRAK G. F. STEWART

University of California
Davis, California

Supplement 1

Phenolic Substances in Grapes and Wine, and Their Significance. 1969
VERNON L. SINGLETON AND PAUL ESAU

Supplement 2

The Chemical Constituents of Citrus Fruits. 1970
J. F. KEFFORD AND B. V. CHANDLER

The Chemical Constituents of Citrus Fruits

J. F. KEFFORD and B. V. CHANDLER

Commonwealth Scientific and Industrial Research Organization
Division of Food Preservation
Ryde, New South Wales, Australia

1970

ACADEMIC PRESS New York and London

ACADEMIC PRESS, INC.
111 Fifth Avenue, New York, New York 10003

United Kingdom Edition published by
ACADEMIC PRESS, INC. (LONDON) LTD.
Berkeley Square House, London W1X 6BA

LIBRARY OF CONGRESS CATALOG CARD NUMBER: 75-84257

PRINTED IN THE UNITED STATES OF AMERICA

CONTENTS

Preface *vii*

Chapter 1 **Introduction** 1

Chapter 2 **General Composition of Citrus Fruits** 5
- I. Effects of Rootstock 8
- II. Effects of Maturity and Storage 10
- III. Effects of Fruit Size 13
- IV. Effects of the Nutrient Status of the Tree 14
- V. Effects of Horticultural Sprays 17
- VI. Effects of Climatic Factors 20

Chapter 3 **Carbohydrates** 23
- I. Soluble Carbohydrates 24
- II. Structural Carbohydrates 24

Chapter 4 **Acids** 31
- I. Trends with Maturation 33
- II. Sites and Mechanisms of Acid Biosynthesis 35
- III. Analytical Aspects 36

Chapter 5 **Vitamins** 39
- I. Ascorbic Acid 39
- II. Analytical Aspects 41

Chapter 6 **Inorganic Constituents** 43
- I. Analytical Interest 43
- II. Horticultural Aspects 47

Chapter 7 **Nitrogen Compounds** 49
- I. Amino Acids 52
- II. Analytical Applications 53
- III. Nitrogen Bases 54
- IV. Proteins 56
- V. Nucleic Acids 57

Chapter 8 **Enzymes** 59
- I. Pectolyzing Enzymes 59
- II. Esterases 60
- III. Phosphatases 61
- IV. Oxidoreductases 61

V. Nitrate Reductase 63
VI. Carboxypeptidase 64
VII. Enzymes of the Tricarboxylic Acid Cycle 64

Chapter 9 **Pigments** 67
I. Changes during Maturation 67
II. Oranges 68
III. Grapefruit 69
IV. Lemons 74
V. Mandarins and Tangerines 74
VI. Hybrids 75
VII. Analytical Aspects 77

Chapter 10 **Lipids** 81

Chapter 11 **Volatile Flavoring Constituents** 85
I. Oranges 90
II. Mandarins 95
III. Grapefruit 96
IV. Lemons 97
V. Limes 99
VI. Other Citrus and Related Fruits 99
VII. Biogenetic Aspects 100
VIII. Analytical Aspects 101

Chapter 12 **Simple Polyphenolic Compounds** 105
I. Phloroglucinol and Derivatives 105
II. Phenolic Acids 105
III. Coumarins 106

Chapter 13 **Flavonoids** 113
I. The Isomeric Rhamnoglucosides 116
II. Analytical Methods 118
III. Lemons 122
IV. Grapefruit 125
V. Oranges 126
VI. Other Citrus Species 129
VII. Biochemical Aspects 133
VIII. Technological Aspects 138
IX. Flavonoids as By-Products 145

Chapter 14 **Triterpenoids and Derivatives** 149
I. Steroids and Triterpenoids 149
II. Limonoids 150

Chapter 15 **Research Needs** 165

References 169

Author Index *213*
Subject Index *225*

PREFACE

The chemistry of citrus fruits was reviewed in *Advances in Food Research* Volume 9 (1959), and at that time the senior author made the rash claim that the catalog of constituents was approaching completion. No abatement, however, in the published work on this subject has been apparent—new compounds continue to be found in citrus fruits, and knowledge about constituents already known has advanced greatly. Accordingly, this supplementary volume is presented to bring the 1959 review up to date.

In some areas, notably the volatile flavoring constituents and the limonoid bitter principles, the new knowledge almost entirely supersedes the earlier accounts, but in other areas it will be necessary to consult both reviews for a complete record.

The original intention was to review the ten-year period 1958 to 1967, but in fact the literature to the end of 1968 has also been reviewed, with substantial coverage of papers appearing in 1969; in all about 1000 references. The sources are diverse, ranging from chemical and biochemical journals through food science and technology to plant physiology and horticulture. It is hoped that the bringing together of widely scattered information about the composition of citrus fruits will meet a real need among all scientists and technologists concerned with these fruits as foods in fresh and processed forms. The technology of citrus products, however, is considered to be a subject in itself and has not been reviewed, but in many sections of the text the technological significance of specific constituents has been indicated.

We record our gratitude to Mrs. Joan Anderson for her painstaking care in the preparation of the manuscript and the tables.

Chapter 1 Introduction

World production of citrus fruits is now about 30 million tons (Anonymous, 1967a); the proportion of the crop consumed in the form of processed citrus products continues to increase and there has been an attendant expansion in international trade in citrus juices and concentrates. A particular trend apparent in the last 10 years is the rapid growth in the market for ready-to-drink citrus beverages, mainly canned, with juice contents generally less than 50%.

Out of these commercial developments has arisen a pressing need for more reliable methods for determining juice content and for detecting adulteration in single-strength and concentrated citrus juices and compounded citrus products. A large volume of published work reports on attempts to meet this need. Estimations of the better-known constituents of citrus fruits—the sugars, acids, and minerals—are of limited use for establishing authenticity because of the relative ease of adulteration with these materials. So the minor constituents—nitrogen compounds, pigments, flavonoids, and vitamins—have been intensively investigated as index compounds, together with empirical chemical indexes such as the formol value and the chloramine value. Further, the interest in many countries in comminuted citrus beverages which incorporate major proportions of the whole fruit has encouraged the collection of analytical data for whole citrus fruits and citrus peels as well as for citrus juices.

Major accounts are available of the biochemistry and physiology of the lemon (Bartholomew and Sinclair, 1951) and the orange (Sinclair, 1961), and the nutritive properties of citrus fruits have been reviewed elsewhere (Kefford, 1966).

Throughout this book, wherever possible, citrus fruits are described by the botanical or horticultural names favored by Swingle and Reece (1967) and Hodgson (1967). To assist readers in identifying particular fruits and in following the taxonomic relationships between them, summarized schemes of classification of citrus fruits and hybrids are set out in Tables I and II. It should be realized, however, that even within named varieties there may be wide variation in composition under the influence of many factors: genetic factors giving rise to clones and horticultural selections, climatic and regional factors, and orchard procedures. Moreover, between fruit on a single tree and within individual fruits there is further variability in composition (Kefford, 1959).

TABLE I CLASSIFICATION OF CITRUS FRUITS[a]

General Name	Botanical Name	Varietal Groups
Sweet orange[b]	*Citrus sinensis*	Normal oranges Navel oranges Blood oranges Low-acid oranges
Bitter or sour orange	*Citrus aurantium*	Bitter (Seville) oranges Bittersweet oranges
Mandarin	*Citrus reticulata*	Common mandarins Tangerines Satsuma mandarins (*C. unshiu*) Mediterranean mandarins (*C. deliciosa*) Small-fruited mandarins
Grapefruit	*Citrus paradisi*	Pale-fleshed grapefruits Red-fleshed grapefruits
Pummelo or Shaddock	*Citrus grandis*	Common pummelos Pigmented pummelos Low-acid pummelos
Lemon[b]	*Citrus limon*	Acid lemons Low-acid lemons
Lime	*Citrus aurantifolia*	Small-fruited acid limes Large-fruited acid limes Low-acid limes
Citron	*Citrus medica*	Acid citrons Low-acid citrons
Kumquat	*Fortunella* sp.	–
Trifoliate Orange	*Poncirus trifoliata*	–

[a]After Swingle and Reece (1967) and Hodgson (1967).

[b]Throughout this book the name "orange" used without qualification should be understood to mean the sweet orange, and the name "lemon" without qualification to mean the true lemon, *Citrus limon*.

TABLE II CLASSIFICATION OF CITRUS HYBRIDS[a]

General Name	Parentage			Varieties mentioned in text
Bergamot	*C. aurantium* hybrid?			–
Calamondin	*Citrus?*	x	*Fortunella?*	–
Citrange	*P. trifoliata*	x	*C. sinensis*	Rusk Troyer
Citrangequat	Citrange	x	*Fortunella* sp.	Sinton
Citrangor	Citrange	x	*C. sinensis*	–
Citrumelo	*P. trifoliata*	x	*C. paradisi*	Sacaton
Hybrid lemon	*C. limon*	x	*C. sinensis?*	Meyer
	C. limon	x	*C. medica* or *C. reticulata?*	Borneo lemon Ponderosa Rough lemon (*C. jambhiri*)
Hybrid papeda	*C. ichangensis*	x	*C. grandis?*	*Citrus sudachi* Ichang lemon
Orangelo	*C. sinensis*	x	*C. paradisi*	Sukega
Rangpur	*C. reticulata* hybrid?			East Indian Kusaie Otaheite
Tangelo	*C. reticulata*	x	*C. paradisi* or *C. grandis*	Hyuganatsu (*C. tamurama*) Minneola Natsudaidai Satsumelo
Tangor	*C. reticulata*	x	*C. sinensis*	Murcott Ortanique Temple

[a]After Swingle and Reece (1967) and Hodgson (1967).

Frequent reference will be made to component parts of citrus fruits, some of which are basic morphological structures and other technological fractions. These components are defined in the following terms and the approximate proportions in which they occur in citrus fruits are indicated:

(1) The *flavedo* (10%), or outer peel, is a layer of tissue underlying the epidermis and containing chromoplasts and oil sacs.

(2) The *albedo* (12-30%), or inner peel, is a layer of spongy white tissue beneath the flavedo. The albedo and the *core*, or central axis, contain the vascular system which supplies the fruit with water and nutrients.

(3) The *peel* (20-50%) is the flavedo and albedo together. The *peel juice* is the free fluid from the peel, extracted mechanically.

(4) The *endocarp* (50-80%) within the peel is sometimes called the pulp and is the principal edible portion of citrus fruits. It consists of a series of *segments*, carpels, or locules, each of which contains a compact mass of elongated, thin-walled *vesicles*.

(5) The *juice* (35-55%) is the cell contents from the vesicles, extracted by means of a variety of hand or mechanical devices, and usually screened. In addition to constituents in solution, the juice contains suspended chromoplasts and tissue fragments in amounts dependent on the fineness of screening.

(6) The *rag and pulp* (20-25%) is the fraction screened from the juice and consists of fragments of the core, the segment walls, and the juice vesicles.

Chapter 2 General Composition of Citrus Fruits

Much of the information that has been reported on the general composition of citrus fruits has mainly local relevance in the growing region concerned. Sources of routine information of this kind are listed in Table III; juice yield, soluble solids, acidity, and ascorbic acid are commonly reported, and in a number of cases more complete analyses including sugars, ash constituents, nitrogen, and formol and chloramine values (see Chapter 7).

Different parties in the chain from the primary producer to the consumer are interested in the composition of citrus fruits from different points of view: the grower in the yield and size of the fruit and the yield of soluble solids in pounds per tree or pounds per acre, the processor in the yield of juice, the yield of soluble solids in pounds per ton of fruit, and in the character of the fruit; and the consumer in palatability, in terms of sugar-acid balance and flavor, and in nutritive value. It is commendable to find more widespread reporting of yields of soluble solids per tree or per ton of fruit. An attempt has been made by Royo (1962) to calculate an index of quality for oranges–from the consumers' viewpoint–based on yield of juice, soluble solids, acidity, ascorbic acid, formol number, and organoleptic scores.

The composition of citrus juices may be influenced by the method of extraction from the fruit. Mechanical extractors are now almost universally used for the commercial extraction of citrus juices and are capable of adjustment to vary the pressure on the fruit. With the Shamouti orange, which has an ovoid shape and a thick peel, Danziger and Mannheim (1967) found that increasing the extraction pressure increased the yield of juice but decreased the soluble solids and acidity because some of the juice was derived from the peel rather than from the juice vesicles. Extraction pressure had less effect on the composition of Valencia juice because this orange is spherical and has a thinner peel. Proximate analyses of "juices" from the peels of Florida oranges are given by Swift

TABLE III SOURCES OF INFORMATION ON THE COMPOSITION OF CITRUS FRUITS[a]

Region	Varieties	References
Algeria	Orange (8 var.), mandarin (3 var.), Clementine, grapefruit	Buffa and Bellenot (1962a)
Australia	Orange (21 var.)	Cox (1969)
Brazil	Orange, tangerine, lemon, lime (33 var. in all)	Cardinali and Seiler (1958)
California	Orange (2 var.)	Beisel and Kitchel (1967)
	Orange (2 var.), lemon	Joseph *et al.* (1961); Birdsall *et al.* (1961)
	Mandarin, hybrids (13 var. in all)	Cameron *et al.* (1966)
	Rangpur (16 var.)	D. Singh and Schroeder (1962)
Colombia	Orange	Rojas and Zambiano (1969)
China	Orange, tangerine (23 var. in all)	Chu *et al.* (1957)
Egypt	Orange, low-acid lemon	Ragab (1962a,b)
	Orange, mandarin	El-Zhorkani (1968)
Florida	Navel orange (28 var.)	Gardner and Reece (1960)
	Orange	Westbrook and Stenstrom (1964); Floyd *et al.* (1969)
	Tangerine	Westbrook and Stenstrom (1959)
	Grapefruit	Long *et al.* (1961); Westbrook and Stenstrom (1963)
	Tangelo (4 var.)	Harding *et al.* (1959)
	Hybrids (tangerine-grapefruit, 5 var.)	Scott and Hearn (1966)
	Murcott tangor	Deszyck and Ting (1960a)
	Lemon (13 var.)	Grierson and Ting (1960); Rouse and Knorr (1969).
German market	Orange (6 countries)	Koch and Haase-Sajak (1964)
	Tangerine (2 countries)	Benk and Bergmann (1963b)
	Clementine	Benk (1961a)
	Deformed oranges	Benk (1961b)
Greece	Orange, mandarin, lemon	Exarchos and Aspridis (1962, 1965, 1968)
	Bitter orange	Zaganiaris (1959)
India	Sour orange, lemon, pummelo	Sarkar (1958)
	Grapefruit (2 var.)	Randhawa *et al.* (1958)
	Tangelo (5 var.)	Singh (1964)
	Jambhiri	Chaliha *et al.* (1964b)

TABLE III *continued*

Israel	Orange, lemon, grapefruit	Ludin and Samish (1962)
	Orange, lemon, grapefruit (juices and comminuted)	D. Wagner and Monselise (1963)
Italy	Orange	La Face (1960)
	Orange, lemon	Cananzi (1958a,b); I. Calvarano (1960, 1963a); Di Giacomo and Rispoli (1966)
	Sour orange	Kunkar (1964a)
	Mandarin	Di Giacomo *et al.* (1968d,e)
	Bergamot	Benk and Wildfeuer (1961)
	Sweet lime (2 var.)	Benk and Bergmann (1963a)
Jamaica	Ortanique tangor	Blazquez (1967)
Japan	Satsuma mandarin (4 var.)	Nishiura (1964)
	Natsudaidai	Inoue *et al.* (1968); Inoue and Yamamoto (1968)
	Hyuganatsu hybrid	Kadota and Nakamura (1963)
	Kumquat (*Fortunella*)	Shigeyama and Murakami (1962)
Lebanon	Orange (5 var.)	Maleki and Sarkissian (1967)
Mexico	Orange (2 var.)	S. H. Hart (1961)
Morocco	Orange (5 var.)	Krummel (1963)
	Orange (3 var.), grapefruit	Huet (1962b)
New Zealand	Lemon (5 var.)	Dawes (1969)
Pakistan (East)	Citrus (12 var.)	Fazl-i-Rubbi *et al.* (1959)
	Lime (9 var.)	B. D. Mukherjee *et al.* (1958)
Russia	Lemon (2 var.)	Sholakhova (1962)
	Grapefruit, pummelo	Fishman (1969)
South Africa	Navel orange (4 var.)	Marloth and Basson (1959)
Spain	Orange (5 var.)	Krummel (1963)
	Navel orange (2 var.)	Primo *et al.* (1962)
Texas	Orange (7 var.)	S. H. Hart (1961)
	Valencia orange	Lime and Tucker (1961)

[a]References not cited elsewhere in Section II.

(1965b). Primo *et al.* (1963a, 1964b) had earlier drawn attention to differences in the yield of juice from oranges, and in the pectin content and pectinesterase activity of the juice, when extracted by hand and mechanical methods. With lemons, Vandercook *et al.* (1966) reported that extraction and finishing procedures affected the composition of the juice only slightly, pectin being the only constituent showing a marked effect (see Chapter 3). The effects of extraction procedures may be particularly significant with the citrus fruits containing bitter flavonoids (see Chapter 13). The amount of pulp incorporated into grapefruit, pummelo, and sour orange juices has an important influence on juice quality and may be controlled by adjustment of the extraction equipment (Hendrickson *et al.*, 1958a; Huet, 1961b; Buffa and Bellenot, 1962b).

The major factors affecting the general composition of citrus fruits are, however, the horticultural and climatic conditions under which they are grown. A whimsical comment by Monselise and Turrell (1959) suggests that there are at least 3.67×10^{11} permutations and combinations of variables in citrus culture. It is not surprising, therefore, that it is difficult to draw general conclusions about the influence of particular factors on the composition of citrus fruits.

I. Effects of Rootstock

Ten further years of rootstock trials have confirmed in general the effects on citrus composition and quality reported in the previous review (Kefford, 1959).

With all citrus scions, rough lemon rootstocks tend to give high yields of fruit with a low juice content, while the juice is low in soluble solids content and acidity; at the other extreme, trifoliate orange and its hybrids tend to give low yields of fruit with a high juice content, while the juice is high in soluble solids and acidity.

These opposite tendencies are well illustrated by Gardner and Horanic (1967) and Gardner *et al.* (1967) with Valencia scions in Florida, where rough lemon stocks gave the highest yield of fruit but the lowest soluble solids content, about 1° Brix below the next lowest, while trifoliate orange hybrids (citrangors) were highest in soluble solids and acidity but among the lowest in yield of fruit.

Because rough lemon stocks give high yields of fruit from an early stage in the life of the tree, they generally surpass other stocks in yields of soluble solids per tree, while the reverse is true for trifoliate orange and its hybrids. Thus in a 17-year trial with Valencia and Parson Brown scions grown on six stocks in Florida (Gardner and Horanic, 1961), rough lemon stocks gave the lowest yield of soluble solids per box but the highest yield of soluble solids per tree, although they were overtaken after 13 years by Cleopatra mandarin, Duncan grapefruit, and sweet orange stocks.

In a study of Washington Navel oranges on eight stocks, and Valencia oranges on three stocks, during six seasons in Australia, Kefford and Chandler (1961) confirmed the general experience with rough lemon and trifoliate orange stocks, the effects being similar with both scion varieties but more marked with Washington Navel. Tangelo and Cleopatra mandarin stocks resembled trifoliate orange in their effects on orange composition; Kusaie and East Indian limes (Rangpurs) and sweet lime resembled rough lemon; and sweet orange stock occupied an intermediate position. Recently, Bowden (1968) has reported similar observations on Valencia and Joppa oranges grown in Queensland. With both varieties, trifoliate orange stock was consistently superior to rough lemon in juice yield, soluble solids content, soluble solids per ton of fruit, and total carotenoids in the juice, and with Valencias only, in acidity and ascorbic acid content. With Joppas the order of ascorbic acid contents was reversed, rough lemon giving the highest values. In all these characteristics fruit from trees on sweet orange stock generally occupied the middle position.

In South Africa, the experience has also been very similar (Marloth, 1959; Marloth and Basson, 1960); in addition, Empress mandarin has performed well there as a stock, being similar to sweet orange in its effects on orange composition.

In New Zealand, trifoliate orange was consistently superior to rough lemon and sweet orange as a rootstock for several varieties of orange, giving higher juice content and soluble solids content (Fletcher and Hollies, 1965).

The adverse effects of rough lemon stocks on quality and composition are shown not only with oranges but also with Murcott tangors (W. G. Long *et al.*, 1962), tangelos (Harding *et al.*, 1959), grapefruit (M. Cohen and Reitz, 1963; Gardner and Horanic, 1966), and limes (Colburn *et al.*, 1963).

In a rootstock trial with Lisbon lemons grown on 7 stocks in two regions in Arizona, rough lemon and Troyer citrange stocks were surprisingly grouped together in giving the juices of lowest acidity (Hilgeman *et al.*, 1966). When acid production was calculated as pounds of acid per ton of fruit, the results illustrated the specific interactions between rootstock and region (soil, climate) that often make it difficult to generalize about rootstock influences; in the Salt River Valley, sour orange, Rangpur, and sweet orange stocks gave the highest values of pounds of acid per ton, and Troyer citrange and rough lemon stocks the lowest values; but, at Yuma, Sacaton citrumelo, Ponkan mandarin, and Troyer citrange stocks showed the highest, and sweet and sour orange stocks the lowest yields in pounds of acid per ton.

In Texas, Columbia sweet lime stock gave low soluble solids contents and low acidities in lemons, oranges, tangerines, and tangelos (Woodruff and Olsen, 1960).

Specific effects of rootstocks on bitterness in oranges are discussed in Chapter 14.

II. Effects of Maturity and Storage

Changes in the composition of citrus fruits during maturation on the tree have been recorded in the course of a number of investigations, and it is appropriate also to review here changes that occur when citrus fruits are held after removal from the tree.

While citrus fruits do not ripen off the tree in the same way as pome fruits, some of the changes characteristic of maturation, notably decreasing acidity (see Chapter 4) in oranges, mandarins, and grapefruits, continue during storage. The fruit begins to lose weight as soon as it leaves the tree and there is generally a loss of soluble solids and ascorbic acid, on the basis of the original fresh weight of the fruit, although sometimes these constituents increase in concentration because of the overall loss in weight.

The losses occur because transpiration, respiration, and other metabolic processes continue in the stored fruit, until eventually the storage life is terminated by physiological or fungal breakdown. The storage life may however be extended by holding at moderately low temperatures and by coating the skin with materials to restrict loss of water and to increase its resistance to fungal attack.

When lemons and limes are stored at moderate temperatures (60°-70°F) and controlled humidities, there is movement of water and solubles from the peel to the endocarp so that the peel becomes thinner and the endocarp increases in juice content; this is the procedure known as "curing."

Gradients in the distribution of soluble constituents within individual citrus fruits have again been demonstrated in a painstaking analytical study of oranges and grapefruit by Ting (1969). The ripening process in these fruits appears to advance from the peripheral regions of the endocarp towards the core. Thus soluble solids contents were higher and acidity and total nitrogen contents were lower at the periphery than at the core. Axial gradients were also observed with soluble solids and total nitrogen contents increasing from the stem end to the stylar end of the fruit. Ascorbic acid was more uniformly distributed.

A. Oranges, Mandarins, and Hybrids Long experience with Florida oranges, involving analysis of more than 10,000 individual fruits, has been summarized by Harding and Sunday (1964) in a series of curves for early-, mid-, and late-season oranges which illustrate that soluble solids content in the juice, and pounds of solids per box increase with maturity, while acidity declines. Similar experience with oranges in Queensland was reported by Bowden (1968) who found that pounds of solids per ton increased more steeply with maturity in the Joppa, a midseason variety, than in the Valencia.

Workers under Dr. Paul Harding in the U.S. Department of Agriculture, Agricultural Marketing Service, Orlando, Florida, have extended their comprehensive studies of the composition, maturation, and seasonal variability of Florida citrus fruits to include tangelos (Harding *et al.*, 1959) and the Murcott tangor (W. G. Long *et al.*, 1962; cf. Deszyck and Ting, 1960a). During a period of 4 months around normal commercial maturity, all of these fruits showed typical maturation patterns with rising soluble solids content, falling acidity, and little change in ascorbic acid and juice content.

Hilgeman *et al* (1967a,b) followed the maturation of Valencia oranges in Arizona during the period from the 10th to the 15th month after blossoming. The fruit increased in weight and in juice content, while the juice increased in total soluble solids content and in net soluble solids (total soluble solids minus acid), but decreased in acidity. The production of net soluble solids per tree was calculated and also found to increase. In an earlier study of Valencia oranges grown in Arizona, Khalifah and Kuykendall (1965) had found that the acidity was about halved during ripening from the 11th to the 14th month from blossoming, and continued to decrease after harvest, during storage up to 20 weeks at 33° and 48°F. The effects of storage on soluble solids and ascorbic acid contents were not consistent, but generally ascorbic acid content declined, in some treatments by as much as 25%.

When Valencias and some other late varieties of oranges hang on the tree beyond normal maturity they are subject to "regreening" of the rind. Miller and Glass (1957) examined the effect of regreening on the composition of oranges and concluded that there was a tendency for full-colored fruits to be higher in soluble solids and ascorbic acid than regreened fruits, although statistical differences were established only in a few cases.

Maturation curves for mandarins, tangelos, and tangors, showing a general increase in Brix:acid ratio but considerable variation between growing areas, are presented by McCarty *et al.* (1965).

Under conditions simulating commercial distribution of Valencia oranges in America (2 weeks in an iced car at 48°- 50°F and 90 - 95% relative humidity, followed by 2 weeks at room temperature), Pritchett (1962) found that the weight and juice content of the fruit decreased by 5-8% of the initial values. When the composition of the juice was calculated on the basis of the original fresh weight of the fruit, about 3% of the soluble solids, 10% of the acid, and 5-14% of the ascorbic acid had been lost.

When Sathgudi oranges and Coorg mandarins were stored 3-4 weeks at 80°- 90°F (Dala *et al.*, 1962a,b), weight losses of 20 - 25% occurred but were partially controlled to about 16% by wax emulsion skin coatings; the juice content decreased in a corresponding way, while the concentrations of soluble solids and sugars (both reducing and nonreducing) in the juice increased, and the acidity and ascorbic acid content decreased. These changes were less marked when the fruit was precooled and stored at lower temperatures (Moorthy *et al.*, 1962). In other studies of the storage of Indian oranges and mandarins, which included treatments with 2,4-D, 2,4,5-T, and α-naphthaleneacetic acid to give storage lives of 35 - 50 days at 50°- 85°F or 75 - 90 days at 40°F, the trends in composition followed the expected pattern of increasing sugar content and decreasing acidity and ascorbic acid content (Lodh *et al.*, 1963; Kohli and Bhambota, 1965; Rodriques *et al.*, 1963).

Substantially the same pattern appeared in the maturation and storage of Russian oranges, according to Gershtein (1962), who also reported storage changes in the peel: there was a small decrease in sugar content, but ascorbic acid, hesperidin, pectin, and oil contents increased, probably because of loss of moisture.

In *Citrus tamurana*, a presumed pummelo hybrid grown in Japan, the acidity fell in typical fashion during maturation but there was little change in total sugar and reducing sugar contents, while the ascorbic acid content fluctuated (Ueyanagi and Morioka, 1965; Morioka and Ueyanagi, 1966).

B. Lemons and Limes When the maturation of lemons in Florida was followed through a 2-month period by Oberbacher *et al.* (1961), it was found that the juice content increased both on weight/weight and volume/volume bases, but the acidity and soluble solids content varied in an irregular way, although in general they tended to decrease. Florida lemons held for 1 to 2 weeks at 60°F until fully colored increased in juice content by about 12% and in acidity by about 5% of the initial values (Grierson, 1968).

The effects of storage on lemons are well illustrated by observations on California lemons held for 3 months at three temperatures: 40°, 55°, and 75°F (Eaks, 1961). There was an overall weight loss of about 10%; the peel content decreased and the juice content increased; and the soluble solids and acidity increased both in concentration and in percentage of original fresh weight. The changes were greatest at 75°F and least at 40°F. Ascorbic acid in the juice was well retained in storage at 55°F but declined at 75°F, and also at 40°F owing to chilling injury. In the peel the ascorbic acid actually increased in storage at 40°F and 55°F, both in concentration and in percentage of original fresh weight. Analogous studies on limes revealed very similar effects of storage on composition (Eaks and Masias, 1965).

When Vandercook *et al.* (1966) made regular analyses of California lemons stored at 58°F and 85% relative humidity, they failed to find a consistent increase in acidity, but the acidity during the last 8 weeks of storage was higher than in the first 7 weeks. They reported, however, as the most striking change in composition, a loss of malic acid amounting to about 60% in 15 weeks.

The growth and maturation of the Indian lime known as "kagzi nimbu" was followed from 1 to 6 months after blossoming by Subramanyam *et al.* (1965). During this period the weight and volume of the fruit increased according to a typical sigmoid growth curve. In the first 3 months the moisture content of the fruit increased but from then on it remained constant. Accumulation of acid, reducing sugars, and ascorbic acid became marked in the fourth month; thereafter, the acidity increased continuously, but the sugar content leveled off, and the amount of ascorbic acid in the whole fruit declined. The polyphenol and pectin concentrations on a fresh-weight basis decreased throughout growth.

In the Meyer lemon, which is probably a hybrid and is less acid than true lemons, the pattern of maturation is again one of increasing reducing sugar content in the juice, and decreasing acidity and ascorbic acid content (Kennedy and Schelstraete, 1965).

III. Effects of Fruit Size

It is well known that small oranges and grapefruit tend to be sweeter than large ones. A number of workers have explored further the relations between the size and the composition of citrus fruits.

Accumulated experience with Florida oranges indicates that the pounds of solids per box decreases with increasing size (Harding and Sunday, 1964). While the volume of juice per fruit may increase with size (W. G. Long, 1962), the juice content as percentage by weight generally decreases. In addition, an inverse relation between size and soluble solids content has been established for Valencia oranges (Smith, 1967), Navel oranges (Marloth and Basson, 1959), Marsh grapefruit (W. G. Long *et al.*, 1959), and Duncan grapefruit (W. G. Long *et al.*, 1961). In some cases, the inverse relation holds for acidity and ascorbic acid content also.

A study by Kretchman and Jutras (1962) of the influence of pruning on grapefruit showed that the more severe the pruning the greater the size of the grapefruit and the juice content, but the soluble solids content and the acidity were depressed, according to the general trend. However, in the measurements of Smith (1963a) on grapefruit at a single picking, one observation was out of line in that the acidity increased with size, although the juice content, and the soluble solids and ascorbic acid contents decreased.

In the Japanese Natsudaidai no appreciable differences in the contents of sugars, ascorbic acid, or naringin were found between large and small fruit but the small fruit were higher in acidity, while the large fruit were higher in total and amino nitrogen (Shimba and Nakayama, 1963). Like other citrus fruits, the Natsudaidai had a higher concentration of soluble solids–sugars, acids, ascorbic acid, and amino acids–in the calyx half than in the stylar half of the fruit, and fruit growing on the south side of the tree had higher acidity than fruit on the other three sides (Hattori *et al.*, 1959).

The age of the tree influences both the size of the fruit and its composition. In studies on Navel oranges in South Africa, Marloth and Basson (1959) found that as the trees advanced in age the fruit decreased in size and in the thickness of the peel but increased in soluble solids content. Significant positive correlations of soluble solids content, pounds of solids per ton, acidity, and total carotenoids in the juice, and a negative correlation of fruit weight with the age of trees of Valencia and Joppa oranges in Queensland were established by Bowden (1968).

Comparative studies of Florida oranges at 3 and 5 years suggested that the fruit on the older trees matured more slowly since the Brix:acid ratio was lower on equivalent picking dates (Rouse *et al.*, 1965a; Rouse, 1967b).

IV. Effects of the Nutrient Status of the Tree

Extensive empirical studies continue to be reported on the nutrition of citrus trees. Many of these include observations on effects on fruit quality and composition. Usually, however, the experiments are designed to guide horticultural practice in particular regions and it is difficult therefore to extract general findings that cut across local conditions of soil, season, and climate.

In a few cases, sufficient information has accumulated to permit the optimum nutritional status of the tree to be defined in terms of leaf analysis for specific nutrients, while the level of nutrients in the fruit itself may also be a useful index of the nutritional status of the tree (Koo, 1962).

A. Nitrogen Status There has been some clarification of the effects of nitrogen fertilization on the yield and composition of citrus fruits, although experience in different growing regions is not consistent.

Studies on oranges in Florida have shown that increasing rates of nitrogen fertilization consistently give increasing yields of smaller fruit; the yield of juice from the fruit is little affected but the juice shows increasing soluble solids content and increasing acidity; the Brix:acid ratio decreases, so that commercial maturity is delayed, as is shown also by delayed disappearance of chlorophyll from the peel (Reitz and Koo, 1960; Calvert and Reitz, 1963, 1964; Stewart and Wheaton, 1965). Tangerines also show high acidity and delayed maturity under high nitrogen fertilization (Koo and McCornack, 1965).

The chemical form of the nitrogen fertilizer, whether organic or inorganic nitrogen in several forms, had little effect on the yield of oranges or the yield of juice from the fruit, but did influence the composition of the juice. In general, ammonium nitrogen, applied as ammonium salts or anhydrous ammonia, produced fruit higher in soluble solids content than did other sources of nitrogen; sodium nitrate had the opposite effect and at high rates produced symptoms of sodium toxicity in the tree (Leonard *et al.*, 1961; Stewart *et al.*, 1961; Calvert *et al.*, 1962; Smith, 1967).

In Florida (Smith 1963a), the effects of nitrogen status on composition of grapefruit were generally similar to those reviewed for oranges. Linear increases in soluble solids content and acidity with increasing levels of nitrogen fertilization were demonstrated by Sites *et al.* (1961), but there was an accompanying decrease in fruit yield and tree health. The yield and composition of grapefruit juice were little affected by the chemical form of the nitrogen fertilizer (Smith and Rasmussen, 1961a).

Limes grown in Florida showed increasing yields of fruit with increased juice, soluble solids, and acidity with increasing nitrogen status (T. W. Young and Koo, 1967).

Workers in California have recorded somewhat different observations. Having defined an optimum nitrogen status in terms of a leaf nitrogen content of 2.4% in spring cycle leaves from nonfruiting terminals, W. W. Jones and Embleton (1967) found that raising the nitrogen level affected the quality of Washington Navel oranges adversely, causing the peel to thicken and the yield of juice and the ascorbic acid content to fall; at the same time there were no significant effects on soluble solids content or acidity.

Australian experience confirms the adverse effects of heavy nitrogen fertilization on the quality of Valencia and Washington Navel oranges as indicated by increased peel thickness, decreased juice content, and increased acidity in the juice (Bouma, 1961; Bouma and McIntyre, 1963; Cary, 1968). The production of total soluble solids per ton of fruit was affected only slightly by nitrogen status, but when calculated per tree it was highest at intermediate nitrogen levels and very low with no applied nitrogen (De Fossard and Lenz, 1967).

The effects of nitrogen nutrition were further elucidated by observations suggesting that high nitrogen status encouraged renewed growth of the peel of oranges during the maturation stage when peel growth normally has virtually ceased. The nitrogen content of the endocarp was little affected by nitrogen status, but at high nitrogen levels the nitrogen content of the peel increased greatly (Bouma, 1959b). Since the increase in peel nitrogen was accounted for largely as protein nitrogen, while the peel increased in thickness and dry weight, continued or renewed cell division and protein synthesis late in the maturation of the fruit were indicated.

B. Phosphorus Status The unfavorable effects of both deficiency and excess of phosphorus in the tree on the quality and composition of citrus fruits have been confirmed.

In a Florida study of Pineapple oranges planted in virgin soil and fertilized with four different levels of phosphate for 20 years (Smith *et al.*, 1963), there was little effect on the yield or quality of the fruit for the first 10 to 12 crops; but, thereafter, at the high phosphate levels there was a progressive decrease in the soluble solids content, acidity, and ascorbic acid content of the juice. Substantially similar observations were made on Valencia oranges by Anderson (1966).

A long-term fertilizer trial was also conducted in Australia where Valencia and Washington Navel oranges were grown in a soil with an inherent phosphorus deficiency aggravated by ammonium sulfate applications which had made the soil highly acid. Under these conditions, the fruit quality was greatly improved by application of superphosphate; the peel thickness decreased, the juice content increased, and the acidity decreased (Bouma, 1959a; Cary, 1968). The depressing effect of high phosphorus status on soluble solids content and acidity was confirmed in Washington Navel oranges grown on rooted cuttings (Bouma, 1961).

In a study on mandarin oranges grown in Russia, Gamkrelidze (1965) reported also that the acidity decreased, and the sugar content increased then decreased, with increasing applications of phosphorus.

Massive applications of phosphorus increased the yield of lemons in California but had no effect on the composition of the fruit (Embleton *et al.*, 1967). This was in contrast to what was found for oranges.

C. Potassium Status In Florida, orange trees maintained at a high potassium status have commonly yielded large fruit with thick peel, low soluble solids, and high acidity (Reitz and Koo, 1960; Smith and Rasmussen, 1959), the effect of high potassium levels being more marked on Valencia than on Hamlin oranges. In a 5-year trial, Koo (1962) observed maximum yields of soluble solids (as pounds per tree) from Hamlin oranges when the potassium content in spring flush leaves was about 1.5%, and from Valencia oranges at a leaf potassium level of about 1.2%.

Observations on grapefruit grown in Florida showed that increasing potassium status decreased the juice yield and soluble solids content but increased the acidity and the ascorbic acid content so that, as with oranges, commercial maturity was delayed (Smith and Rasmussen, 1960; Smith, 1963a). Increasing acidity was also the principal effect of increasing potassium fertilization of limes in Florida (T. W. Young and Koo, 1967). It is surprising, therefore, that Dancy tangerines responded to potassium fertilization by producing fruit of lower acidity (Koo and McCornack, 1965).

It is noteworthy also that the general effects of heavy potassium applications to oranges and grapefruit were substantially reversed with lemons grown in California (Embleton *et al.*, 1967). Massive applications of potassium increased the yield of fruit, the juice content of the fruit, the soluble solids content, acidity, and ascorbic acid content of the juice, and the pounds of acid and pounds soluble solids per ton of the crop. Similar favorable effects on the juice yield, soluble solids content, and acidity in mandarin oranges following application of potassium as glauconite were reported from Russia (Lezhava, 1967).

D. Minor Elements Foliar application of manganese to Valencia orange trees moderately deficient in manganese increased the yield of fruit and the soluble solids content of the juice, and hence the yield of solids per ton of fruit; foliar sprays of zinc on moderately deficient trees decreased the yield of juice but increased the ascorbic acid content (Labanauskas *et al.*, 1963).

Application of boron to mandarin trees in Russia increased the yield of fruit and the ascorbic acid content, and improved the sugar-acid balance (Menagarishvili, 1962).

V. Effects of Horticultural Sprays

A. Oil Sprays The usefulness of petroleum oil sprays for controlling pests in citrus orchards is offset by their detrimental effects on tree health and fruit quality. Much work has been done with the object of relating the specifications of oils, in terms of molecular size, hydrocarbon composition, and extent of refinement, to pesticidal and phytotoxic properties (Riehl, 1967).

In trials on Hamlin, Pineapple, and Valencia oranges in Florida, Trammel and Simanton (1966) found that the depressing effect of oil sprays on soluble solids content increased with molecular size, as indicated by distillation temperature, and was significant only with oils of which 50% distilled above 440°F at 10 mm Hg. There were few significant effects on juice content or acidity and again these were observed only with the heavier oils. The extent of refinement as indicated by unsulfonated residue had no effect on fruit composition but was relevant to other phytotoxic symptoms. These workers also found no effect of hydrocarbon composition. The results of Dean and Hoelscher (1967), from work on Pineapple oranges in Texas, do not conform entirely with Florida experience since they found that a light naphthenic oil had a slightly greater effect in depressing soluble solids and increasing acidity than heavier naphthenic and paraffinic oils and a "re-formed" oil.

B. Arsenate Sprays Workers in Florida have explored further the effects of lead arsenate sprays on the composition of grapefruit. When the spray was applied post-bloom and the fruit was sampled 5-10 months later, sprayed fruit decreased in acidity by 30 to 40% over this period but unsprayed fruit by only 25% of the initial values (Deszyck and Ting, 1958). The same workers (Deszyck and Ting, 1960b) established that arsenate, in addition to decreasing acidity, increased the nonreducing and total sugar contents of the juice as compared with untreated fruit; this was an absolute increase in sugar content since the juice content was not affected. Arsenate sprays also increased the total flavonoid content of the grapefruit juice.

The magnitude of the effects of arsenate varied with rate of application, variety, season, and growing area (Reitz and Hunziker, 1961); for instance, the effects were more marked with Ruby Red than with Marsh and Duncan varieties, and were greater in wet seasons than in dry seasons.

Decreased acidity in Valencia oranges sprayed with calcium arsenate has been reported from South Africa (Pretorius, 1966) and, in line with previous experience, the effect was counteracted by subsequent spraying with copper oxychloride or by the addition of copper, zinc, and manganese to the arsenate spray.

Attempting to elucidate the biochemical basis for the action of arsenate, Vines and Oberbacher (1965) postulated that arsenate competes with phosphate and interferes with citric acid accumulation in citrus fruits. In a study of oxidative phosphorylation by mitochondrial preparations from orange juice vesicles, they found that arsenate uncoupled phosphorylation (with citrate as substrate) without affecting oxygen uptake.

In the juice of Natsudaidai sprayed with lead arsenate, Ito and Izumi (1966) found an arsenic content of 0.08 ppm.

C. Plant Growth Regulators Gibberellic acid (gibberellin A3), the naturally occurring plant growth regulator, is recommended in California as a horticultural spray for lemons and Navel oranges since it has desirable effects when growers wish to hold fruit on the tree beyond normal maturity.

Studies on lemons have shown that the principal effect of gibberellic acid is to delay maturation of the fruit; when applied to nearly mature but still green fruit, it retards disappearance of chlorophyll from the peel. When Lisbon lemons of the same age were analyzed after storage for 1½ months, fruit treated with gibberellic acid had a lower content of juice, with a higher soluble solids content and markedly higher ascorbic acid content, than untreated fruit, but the acidity was not affected (Coggins *et al.*, 1960b).

When applied to Navel oranges, gibberellic acid appeared to have a specific effect on the maturation and senescence of the peel, but no effect on the juice content or composition (Coggins and Hield, 1962). The gross effects of gibberellic acid on the peel are to delay coloring by retarding both the disappearance of chlorophyll and the accumulation of carotenoid pigments (see Chapter 9), to increase the mechanical firmness of the peel, and to decrease the incidence of a sticky exudate (the chemistry of which has not been elucidated) that becomes a nuisance in the packing house (Coggins and Eaks, 1967).

An attempt to explore the biochemical mechanism of the effects of gibberellic acid was made by L. N. Lewis *et al.* (1967). When Navel oranges were treated about 6 months after anthesis, there was little difference between treated and control fruit until about 1 month after the normal harvesting date in California, i.e., until about 11 months after anthesis. From this time on the marked accumulation of sugars which characterizes senescence in the albedo was retarded in the treated fruit, while there was a simultaneous increase in glucose respiration and oxygen uptake. In addition, there was a lower ratio of monovalent to divalent cations and a higher level of phosphorus in the treated fruit. On the basis of these findings an effect of gibberellic acid on the integrity of mitochondrial membranes was postulated.

Effects of gibberellin on some enzymes and other components of the flavedo of Shamouti oranges were reported by Monselise and Goren (1965), and compared with effects of 2-chloroethyltrimethylammonium chloride (CCC). Gibberellin increased peroxidase and catalase activity in the flavedo while CCC decreased indoleacetic acid oxidase activity. As a result of gibberellin treatment both reducing sugar and hesperidin contents were lower than in fruit treated by CCC; there was little effect on ascorbic acid or nitrogen contents from either treatment.

When Valencia orange trees were sprayed with gibberellic acid (Coggins *et al.*, 1960a) at a stage when they were carrying both mature fruit and very small fruit of the succeeding crop, the mature fruit showed more regreening and decreasing soluble solids content with increasing gibberellic acid dosage; the subsequent crop showed symptoms of delayed maturation as indicated by retarded loss of chlorophyll, increased peel thickness, decreased juice content, and increased acidity; soluble solids and ascorbic acid were not affected.

Grapefruit reacted to gibberellic acid treatments in the same general manner as Valencia oranges (Coggins *et al.*, 1962); mature fruit showed increased regreening with no difference in juice composition. The subsequent crop was retarded in maturation with slower loss of chlorophyll and decreased juice content, but with higher soluble solids and ascorbic acid contents and lower acidity.

In Russian studies, gibberellic acid sprays were found to increase the yield, seediness, and ascorbic acid content of lemons and oranges (Sholokhava and Domanskaya, 1962).

Spraying of Pineapple oranges in Florida with 2,4-dichlorophenoxyacetic acid and 2,4,5-trichlorophenoxypropionic acid to reduce fruit drop retarded maturation by 3 to 4 weeks so that accumulation of soluble solids and loss of acidity in the treated fruits consistently lagged behind the controls (Phillips and Meagher, 1966).

Application of α-naphthaleneacetic acid to Wilking mandarins for thinning purposes had no effect on juice content, or soluble solids and acid in the juice (Hield *et al.*, 1962). Ascorbic acid applied to assist abscission of oranges also had no effect on the juice content, soluble solids or acidity, but possibly increased the ascorbic acid content of the juice (Cooper and Henry, 1967).

Preharvest sprays of chlorophenoxyacetic acid, 2,4,5-trichlorophenoxyacetic acid, and β-naphthoxyacetic acid were applied to mandarin oranges in Coorg, India, by Rodrigues and Subramanyam (1966); the results of fruit analyses were not statistically analyzed but they suggest that there was little effect on composition although there were favorable effects on storage quality.

Fluoride sprays have been applied to citrus trees not for any beneficial purpose but to simulate the effects of fluorides in polluted atmospheres. However, fluoride treatments that produced in the trees characteristic symptoms of fluorine toxicity, together with significant decreases in yield of fruit, had no effect on the juice yield and the soluble solids, acid, and ascorbic acid contents of Washington Navel oranges (Brewer *et al.*, 1967). Leonard and Graves (1966), studying Valencia oranges, reported a trend towards higher acidity and lower soluble solids content with increasing fluorine in the leaves, derived from airborne fluorides, but they could not establish significance.

VI Effects of Climatic Factors

A. Moisture Status A number of investigators have confirmed earlier observations that abundant soil moisture increases the juice content of citrus fruits but decreases soluble solids content and acidity; however, under conditions of heat or moisture stress the fruit loses moisture and increases in solids content.

In a comparative study of the effect of climate on the quality of grapefruit, Hilgeman (1966) found that the consistently high soil moisture in Florida in the summer produced fruit with a higher juice content and thinner peel, but lower in soluble solids and acid. This was in contrast to Arizona where internal water deficits in the trees were much greater.

Irrigation of oranges and grapefruit in Florida (Koo, 1963) increased the yield of fruit, and also the production in pounds of solids per acre, but not in proportion to the increased yield of fruit because the soluble solids content was lowered. The acidity was also depressed by irrigation, in line with the finding by Deszyck and Ting (1958) of an inverse relation between rainfall and acidity in grapefruit over six seasons.

Irrigation of tangerines also depressed the soluble solids content and acidity (Koo and McCornack, 1965), while in a wet year the soluble solids content was consistently lower throughout the season by 0.5° Brix than in the 4 following years (Westbrook and Stenstrom, 1959). Similar effects of wet and dry seasons on tangelos were reported by Harding *et al.* (1959).

B. Temperature The properties of oranges, lemons, and grapefruit grown in four climatic districts in Southern California were compared by Monselise and Turrell (1959). The moisture content of the peel decreased while the dry weight of peel per unit volume and the pressure to puncture the peel increased from the coast toward the desert, i. e., in the direction of increasing summer temperatures and increasing evaporation, and also from the base to the top of the tree.

A more extensive study of the gross effects of climate on the growth of Valencia orange trees and the quality of the fruit was made by Cooper *et al.* (1963) who observed commercial plantings in Florida, Texas, Arizona, and California. Oranges grown in moderate to high relative humidities had higher juice contents and thinner peels than those grown in drier climates. High day and night temperatures encouraged early accumulation of soluble solids and accelerated the loss of acid (see Chapter 4). Thus the time from blossoming for Valencia oranges to reach a Brix:acid ratio of 9 was found by Scora and Newman (1967) to be closely related to the climatic conditions in the growing region as indicated broadly by the mean annual temperature: in the Rio Grande Valley of Texas (23°C), 9 months; in Florida (21°C), 11 months; in the inland desert of California and Arizona (19°-22°C), 11-12 months; and on the California coast (15°C), 17-18 months.

These climatic influences were generally confirmed in a comprehensive study by Reuther *et al.* (1969) of Valencia oranges grown in 6 regions in California, Arizona, Texas, and Florida. The times from blossoming for the fruit to reach a Brix:acid ratio of 9 ranged from 7½ to 8½ months in Weslaco, Texas to 14 to 15 months in Santa Paula, California. These workers were unable, however, to discover any simple relation between rate of ripening and regional climatic parameters, and they suggest that the many metabolic processes involved in the maturation of citrus fruits are controlled by independent mechanisms which are influenced in different ways by the external environment. They agree with other workers that high temperatures, particularly night temperatures in the fall and winter, encourage a rapid decline in juice acidity, but they were unable to establish consistent effects of climate on soluble solids and ascorbic acid contents.

Within the California-Arizona region Tucker and Reuther (1967) defined four zones of decreasing mean maximum temperature in the order: desert, central interior, southern interior, and coast, and found that this was also the order in which Valencia and Navel oranges first reached commercial maturity. For each zone, the seasonal trends in soluble solids, acidity, and juice content were sufficiently consistent to be fitted by equations which might be used to predict the fruit composition in a particular season from values measured early in that season.

There is further evidence that the composition of oranges is particularly influenced by the amount of heat received during certain critical periods. Using observations on Washington Navels harvested around the same date in 8 successive years in California, W. W. Jones *et al.* (1962) were able to establish a positive correlation between the Brix:acid ratio and the heat units above 55°F for April-May in the previous year, which was the time of flowering and fruit set. Earlier flowering leads to more advanced maturity at a given date in the next year. These workers also derived negative correlations between the acidity and the heat units for two periods—for the same April-May period, and also for August, which is the time of acid production in the fruit. High availability of heat units in August is therefore considered to be conducive to low acidity.

Marked differences in quality between oranges in different regions of New

Zealand are probably due to differences in available heat units above 55°F; regions receiving around 1500 day degrees per season gave fruit much lower in juice and soluble solids content than regions receiving around 2000 day degrees (Fletcher and Hollies, 1965).

C. Freezing Several times in the last 10 years, citrus areas in different parts of the world have been subjected to weather conditions under which crops were frozen on the trees; notably the severe freeze in Florida in December, 1962.

The formation of ice crystals in citrus fruits causes disorganization of the cell structure to an extent depending upon the duration of freezing conditions. When the fruit is harvested shortly after freezing the composition is little affected, but fruit remaining on the tree may show shrinkage and desiccation of the juice vesicles as well as a loss of weight and reduced juice content (Bissett, 1958; Rouse *et al.*, 1958; W. G. Long *et al.*, 1962).

The damage to cell structures may also lead to some disturbance of normal metabolic processes. In oranges that have been frozen the acidity declines and the Brix:acid ratio increases more rapidly than in normal fruit, but the contents of flavonoids and limonoid bitter principles, and the pectinesterase activity may be higher. An increase in the protopectin content in the peel of oranges 6 to 8 weeks after freezing, suggesting renewed synthesis of pectin, was regarded by Rouse *et al.* (1963) as evidence of "recuperation" of structural elements.

Chapter 3 Carbohydrates

The pioneering work of W. B. Sinclair and his school at the University of California Citrus Experiment Station, Riverside, established the nature of the carbohydrate constituents of citrus fruits by the use of methods which were codified by Sinclair and Jolliffe (1960). The peel, pulp, and juice components of cit rus fruits were first separated into constituents soluble and insoluble in 80% ethanol. The alcohol-soluble fraction contained soluble mono- and disaccharides, nonvolatile organic acids, amino acids, and other minor constituents, while the alcohol-insoluble fraction contained the carbohydrates contributing to cell structure, and some proteins.

The alcohol-insoluble solids were further extracted with water and dilute acid to give water- and acid-soluble pectic substances, then with 2% hydrochloric acid to give "hemicellulose," and finally, with zinc chloride-hydrochloric acid to give cellulose. The pectic substances were determined as calcium pectate, as total uronide carbon dioxide, or as anhydrogalacturonic acid by colorimetric reaction with carbazole.

When seven methods for the determination of pectin were critically evaluated by Koch and Hess (1964), using citrus and apple pectins and orange and apple juices as test materials, they favored the colorimetric estimation of galacturonic acid with carbazole. This method was superior to titrimetric and gravimetric methods in accuracy and reproducibility, and, moreover, permitted the determination of pectin, calcium pectate, and protopectin in the presence of each other.

Workers in Florida have used an alternative method of fractionating pectic substances by progressive extraction with water, 0.75% ammonium oxalate, and 0.05 *N* sodium hydroxide (Rouse and Atkins, 1955). Water extraction removes the pectins with relatively high methoxyl content; ammonium oxalate sequesters calcium and magnesium ions and so permits the extraction of pectic and low-methoxyl pectinic acids present as water-insoluble calcium and magnesium salts; and the fraction soluble in sodium hydroxide is regarded as protopectin, the water-insoluble parent material that yields pectin by acid hydrolysis.

I. Soluble Carbohydrates

Application of ion-exchange chromatography by Minutilli and Albanese (1958) to the analysis of the sugars in Sicilian orange and lemon juices confirmed the presence of sucrose, fructose, and glucose as the common sugars. The same three sugars were identified by Chaliha *et al.* (1964b) in the juice of the Assam rough lemon.

Bean (1960) set out to explore the metabolic basis for the different composition of oranges and lemons (see also Chapter 4) and found that the difference in sugar storage was apparently not due to lack of sucrose-synthesizing mechanisms in the lemon since active systems were present at all stages of growth. The most likely pathway for the formation of sucrose in oranges and lemons was condensation of fructose diphosphate with uridine diphosphate glucose.

The distribution of sugars in different parts of the juice vesicle of Natsudaidai was examined by Kodama *et al.* (1964) but no marked differences were found.

Examination of the sugars in Spanish and American processed orange juices by gas chromatography of the trimethylsilyl derivatives (Alberola *et al.*, 1967, 1968) showed major peaks due to α-glucose, β-glucose, fructose, and sucrose, while among several minor peaks was one attributed to galactose. This is the first report of the occurrence of galactose in orange juice.

The alcohol-soluble solids in the peels of oranges and grapefruit were examined by Ting and Deszyck (1961). This fraction made up of 30 to 50% of the dry weight of the peels, the proportion increasing with increasing maturity. The total sugar content was about 80% and the sugars present were glucose, fructose, and sucrose, with traces of xylose and rhamnose. In general, the sucrose content was less than the total content of reducing sugars.

II. Structural Carbohydrates

A. Changes during Maturation Using the methods already mentioned, Sinclair and Jolliffe (1958a, 1961a) followed changes in the pectic substances of Californian Valencia orange peel and pulp throughout the growth of the fruit from the size of a green pea. The rapid early growth demands structural carbohydrates so that there was a steep initial rise in the content of pectic substances, reaching a peak after about 4 months and then declining slowly (on a total solids basis) owing to the accumulation of soluble carbohydrates. In the peel the acid-soluble pectic substances first predominated but were later overtaken by the water-soluble fraction. In the pulp, however, the acid-soluble pectic substances remained the major fraction throughout growth. The water- and acid-soluble fractions accounted for 70 to 80% of the total pectin while the residue (protopectin) decreased with advancing maturity. There was little change in the pectic substances in the juice throughout maturation.

The extent of esterification of the carboxyl groups of Valencia peel pectin increased during maturation, rising rapidly to about 80% methylation.

Valencia oranges are subject to "granulation," a physiological disorder characterized by thickening and hardening of the walls of the juice vesicles. As might be expected, the affected vesicles contain increased amounts of structural carbohydrates as indicated by higher contents of alcohol-insoluble solids and pectic substances, which, in turn, show a higher percentage methylation (Sinclair and Jolliffe, 1961b).

Changes in Florida citrus fruits during maturation have been extensively studied by A. H. Rouse, C. D. Atkins, and E. L. Moore at the Citrus Experiment Station, Lake Alfred. Valencia and Pineapple oranges and Silver Cluster grapefruit from mature trees were examined during two seasons; the results of this work are summarized in Table IV. The fruits were divided by hand into five component parts which did not correspond precisely with components separated in commercial citrus processing. The "peel" fraction comprised flavedo, albedo, and core; the "membrane" fraction was made up of segment walls; the "juice" was extracted from the juice vesicles by hydraulic pressing, and the residue was the "juice sacs"; the seeds made up the fifth component.

In all three fruits the membrane fraction was highest in total pectin and protopectin contents and in the purity, methoxyl content, and jelly grade of the extracted pectin. The grapefruit pectins were higher than the orange pectins in jelly grade, an index of the extent of polymerization of the polygalacturonic acid.

In these Florida studies of citrus fruits from the time when they were approaching full size, there were few marked or consistent changes with maturity in the distribution of the pectic substances between the water-soluble, ammonium oxalate-soluble, and sodium hydroxide-soluble fractions, and in the commercially important characteristics of jelly grade and methoxyl content. Subsequent reports from the same workers on the characteristics of oranges from 3- and 5-year-old trees (Rouse *et al.*, 1965a; Rouse, 1967b) indicated that the pectin content in the component parts was generally higher in the fruit from the young trees. A further recent contribution from the Florida laboratory reports on the distribution of the pectic substances in lemons (Rouse and Knorr, 1969).

Included in Table IV are data from Rouse (1967a) relating to the pectic substances in some by-products of citrus processing. As sources of pectin, orange

TABLE IV PECTIN CONTENT AND CHARACTERISTICS IN FLORIDA CITRUS FRUITS[a]

Fruit and component part	Percent of whole fruit	Pectin content[b] (% AIS)	Pectin purity[c] (%)	Methoxyl groups[d] (%)	Jelly grade
Valencia orange[e] (December-June)					
Peel	20	25	91	10	206
Membrane	10	34	93	11	314
Juice sacs	20	21	83	10.5	222
Seeds	1	–	58	13	0
Pineapple orange[f] (September-May)					
Peel	21	23	89	9	178
Membrane	14	30	95	10	300
Juice sacs	23	18	86	9	185
Seeds	3.5	–	64	11	0
Silver Cluster grapefruit[g] (August-May)					
Peel	28	22	87	11	229
Membrane	10	34	91	12	323
Juice sacs	20.5	21	85	10	222
Seeds	4	–	55	–	0
Commercial by-products[h]					
Lime peel	–	31	92	11	351
Lemon peel[i]	–	28	97	11	343
Orange cores[j]	–	31	83	12	344
Grapefruit cores[k]	–	31	94	12	366

[a]Mean values for samples taken monthly during the seasons shown, in two successive years.

[b]Percentage by weight of dried alcohol precipitate extracted from alcohol-insoluble solids (AIS).

[c]Percentage of anhydrogalacturonic acid in the alcohol precipitate.

[d]Calculated as percentage of the anhydrogalacturonic acid.

[e]From Rouse *et al.* (1962a,b).

[f]From Rouse *et al.* (1964a,b).

[g]From Rouse *et al.* (1964c, 1965b).

[h]From Rouse (1967a).

[i]Pectin extracted from water-insoluble solids.

[j]Central axis and segment walls from commercial juice extractors.

[k]Central axis and segment walls from commercial sectioning machines.

and grapefruit cores were comparable in yield and quality with lime and lemon peels which are generally regarded as the best sources of citrus pectin.

Information on the yield and quality of pectin from Assam mandarins and rough lemons was reported by Chaliha *et al.* (1963b, 1964b).

Indian limes contained about 30% pectin in fully-developed green fruit and about 24% in yellow fruit on a dry basis (Srivas *et al.*, 1963). During storage at 75°- 86°F, 70 - 80% of the recoverable pectin was lost, the loss being greater in the more mature fruit; the water- and oxalate-soluble pectin fractions increased in proportion and the acid-soluble pectin decreased; the methoxyl content and the jelly grade of the pectin also declined. All of these changes were less marked in storage at 46°-50°F.

In the pummelo, Khan *et al.* (1959) found 14% pectin in the flavedo, 13% in the albedo, and 4% in the pulp.

Tarutani and Manabe (1963) found the Japanese mandarin and Natsudaidai to have similar pectin contents, while the mandarin pectin had a methoxyl content greater than 8%. In both species, the total pectin content of the peel and the pulp decreased during maturation (Miura *et al.*, 1965). The content of water-soluble pectin was higher in the peel than in the pulp, and higher in Natsudaidai than in the mandarin.

An examination of the juice vesicles of Natsudaidai by Kodama *et al.* (1964) revealed more cellulose in the stalk of the vesicle than in the vesicle wall.

B. Composition of Orange Juice Cloud Citrus juices are generally opaque because of the presence of "a heterogeneous mixture of cellular materials and perhaps emulsoids held in suspension by pectin," to quote Scott *et al.* (1965). These workers investigated the composition of this cloud in orange juices freshly extracted on commercial juice extractors from a range of varieties grown in Florida during several seasons. After a primary screening through a 16-mesh screen, the juices were centrifuged (fraction A) under conditions (10 min at 600 *g*) equivalent to those commonly used in the quality control determination of "free and suspended pulp." The effluents were then recentrifuged (3 min at 60,000 *g*) immediately to give fraction B, and again after storage overnight to throw down the fine cloud (fraction C). In addition, oranges were divided into the components: albedo, rag, pulp, and pulp-free juice, in an attempt to identify the source of the fragments making up the cloud. Analysis of the several fractions and components gave the results summarized in Table V.

Fraction A, the "free and suspended pulp," resembled the albedo, rag, and pulp components in being high in cellulosic constituents and relatively low in pectin, lipids, and phosphorus. This fraction was evidently made up of fragments from mechanical disintegration of structural tissues. Fractions B and C, the true cloud, were high in pectin, lipids, nitrogen, and phosphorus indicating that they were derived from the contents of the juice vesicles. (See Chapter 7, Section IV).

TABLE V COMPOSITION OF ORANGE JUICE CLOUD AND COMPONENT PARTS OF ORANGES[a]

Fraction	Amount in juice (% dry wt.)	Light-scattering components		Percentage composition of AIS[b]					
		Lipids (%)	AIS[b] (%)	Pectin (%)	Hemi-cellulose (%)	Cellulose (%)	N (%)	Ash (%)	P_2O_5 (%)
A (Free and suspended pulp)	0.19	10	90	63	19	11	4.2	3.5	0.4
B (Cloud)	0.34	25	75	83[c]	3	2	7.6	2.6	1.3
C (Cloud)	0.14	27	73	77[c]	1	1	7.1	3.0	1.6
Albedo	–	7	93	44	13	32	0.7	3.1	Trace
Rag	–	3	97	52	16	32	0.7	3.5	0.1
Pulp	–	11	89	57	19	19	2.8	4.1	0.3
Juice (pulp-free)	–	35	65	80	1	2	6.0	9.4	2.0

[a]From Scott *et al.* (1965).

[b]AIS, alcohol-insoluble solids.

[c]About half of this fraction found to be protein by Baker and Bruemmer (1970).

The composition of the cloud was similar in different varieties of orange and also in fruit damaged by freezing.

A marked effect of extraction and finishing procedures on the soluble pectin content of lemon juice was reported by Vandercook *et al.* (1966)–the pectin content increased with increasing extraction pressure. Finisher pressure increased the pectin content only when the fruit was extracted under heavy pressure, which suggests that albedo particles contributed the additional pectin in this case.

C. Chemistry of Citrus Polysaccharides Application of chromatographic techniques to the separation and identification of the products of hydrolysis of the structural carbohydrates of citrus fruits opened the way to the elucidation of the structures of these polysaccharides.

The structural polysaccharides of the peels of oranges and grapefruit were studied by Ting and Deszyck (1961) using a procedure which involved hydrolysis and chromatographic separation of the carbohydrate monomers. The "pectin" fraction yielded arabinose, galactose, and galacturonic acid; the "hemicellulose" fraction, xylose, arabinose, galactose, glucose, and traces of rhamnose and uronic acids; and the "cellulose" fraction, mainly glucose with xylose, arabinose, uronic acids, and traces of galactose and mannose. Calculations based on the assumption that each hydrolysis product was derived from a single polysaccharide indicated that orange and grapefruit peels may contain: araban 9%, galactose 5%, xylose 2.5%, cellulose glucosan 15-20% in orange peel and 28% in grapefruit peel, and polygalacturonic acid 23%. In total, however, these calculated constituents accounted for only 53-70% of the alcohol-insoluble solids in the peels.

McCready and Gee (1960) posed the question: Is pectin a pure galacturonan or is it a more complex polysaccharide containing some nonuronide sugars as part of the molecule? Among pectins from a number of sources, they examined orange, grapefruit, and lemon pectins, purified by precipitation from aqueous solution by cupric ions, and found them to contain 92.1, 91.7, and 90.4% anhydrouronic acid, respectively. Since the nonuronide sugars (arabinose, galactose, and rhamnose) were detected in each of the purified pectins, McCready and Gee concluded that citrus pectins contain polymerized nonuronide sugars.

Pectins from several sources, including a commercial citrus pectin, were also examined by Zitko and Bishop (1965). Using a procedure for fractionating the pectic acids with sodium acetate, they concluded that there were two principal acidic components: a linear galacturonan free from neutral sugars, and another linear galacturonan to which the neutral sugars, rhamnose, arabinose, and galactose, were attached as side chains.

Subsequently, however, Aspinall *et al.* (1968) presented evidence for the presence of L-rhamnose in the main chains of a citrus pectin. These authors prepared from dried lemon peel the pectin extractable with cold water, which amounted to about one-third of the total pectin. By precipitation as the copper salt and chromatography on diethylaminoethyl-Sephadex they obtained a purified pectin that was probably homogeneous except for differences in extent of esterification. The uronic acid content (76%) of this pectin indicated that about 79% of the sugar residues were galacturonic acid. There was a trace of glucuronic acid, probably less than 1% of the sugar residues. Characterization of oligosaccharides formed by partial fragmentation of the pectin revealed that it consisted of galacturonorhamnan chains in which rhamnose residues were unevenly distributed between blocks of 4-*O*-substituted α-D-galacturonic acid residues. The evidence for the location of the other neutral sugars (arabinose, galactose, and traces of xylose and fucose) generally supported the view of Zitko and Bishop (1965) that they occur in outer chains. In a similar study of Satsuma mandarin pectin, Hatanaka and Ozawa (1968) also concluded that neutral sugars were distributed along the polygalacturonide chain of the pectin.

Among the products of acid hydrolysis of pectin prepared from the peel of the Satsuma mandarin by precipitation as the iron salt, Hirota (1962a) identified galacturonic acid, digalacturonic acid, arabinose, galactose, rhamnose, xylose, glucose, and fructose. Suyama and Tsusaka (1958) reported 93% anhydrogalacturonic acid in the pectin from the peel of the Natsudaidai.

Adulteration of citrus juices with extracts of rag or peels increases the content of structural carbohydrates. One method of detecting such adulteration is to determine the "pentose equivalent" of the juice by destructive distillation with 12% hydrochloric acid, the furfural produced being measured colorimetrically and calculated as xylose. For instance, orange juices gave values equivalent to 0.05-0.1 g xylose per 100 ml, while rag extracts gave values at least three times as great (Sawyer, 1963; Benk, 1968a).

Chapter 4 Acids

It is claimed that the first practical commercial pH meter was developed by Dr. A. O. Beckman in 1935 to measure the acidity of citrus products (Beckman Instruments Inc., 1967). Acidity is an important attribute because the sour taste is a major factor in the acceptability of citrus fruits and their juices. Thus oranges (about 1% acid, pH 3.5) are generally acceptable, grapefruit (1.5-2.5% acid, pH 3.0) are sometimes sour, and lemons and limes (5-6% acid, pH 2.2) are notably sour. Sourness greatly influenced the acceptance of California mandarins, tangelos, and tangors—there was only 10% acceptance by tasters of mandarins with an acidity of 3%, whereas there was 100% acceptance of mandarins with an acidity around 0.8% (Lombard, 1963; Lombard and Brunk, 1963).

A linear relation between the pH of grapefruit juice and the log of the free acid concentration expressed as milliequivalents per gram of soluble solids was demonstrated by Kilburn (1958) and Kilburn and Davis (1959), and it appeared that a common regression line could be drawn for all the citrus fruits examined: grapefruit, orange, tangerine, lemon, lime, and Calamondin. Expressed in another way this means that at a uniform soluble solids content (Brix) the pH of citrus juices is proportional to the log of the titratable acidity. The free acid concentration (as meq/g soluble solids) in grapefruit juice was approximately constant in fruit 7 to 10 months from blossom. Thereafter it declined in accord with the normal maturity trend, but the concentration of combined acids (salts) on the same basis remained relatively constant throughout the period 7 to 16 months from blossom.

Spanish workers under Dr. Eduardo Primo Yufera in the Department de Quimica Vegetal, Valencia, made a very thorough study of the acids in oranges, by thin-layer chromatography of the free acids and their methyl esters, and by gas chromatography of the methyl esters, and extended the already long list of acids known to occur (Table VI; see also Table XII for volatile acids).

TABLE VI ACIDS IN ORANGE JUICE[a]

Acid	Spanish oranges[b,c]				American oranges[d]			
	Comuna	Sanguina	Cadenera	Valencia	Valencia	Hamlin	Pineaple	Parson Brown
Aconitic	tr	tr						
Adipic	+	+	+	+	+	+	+	tr
Benzoic	+	+	+	+	+	+	+	+
Chlorogenic[e]					+			
Citramalic[f]						+		
Citric	+	+	+	+	+	+	+	+
Galacturonic[c]	+	+	+	+				
Isocitric		tr	tr	tr	tr		+	+
Lactic	+	+		+	+	+		
Malic	+	+	+	+	+	tr	+	+
Malonic	+	+	+	+	+	+	+	+
Oxalic	+	+	+	+	+	+	tr	
Phosphoric	+	+	+	+	tr	+	+	+
Quinic[f,g]						+	+	
Succinic	+	+	+	+	tr	tr	+	+
Tartaric	tr	tr	tr	tr	tr	tr	+	+

[a]+, Acid present; tr, trace amount only.
[b]From Primo *et al.* (1963c).
[c]From Sanchez *et al.* (1964).
[d]From Primo *et al.* (1965a).
[e]Siddiqi and Freedman (1963).
[f]Ting and Vines (1966).
[g]Ting and Deszyck (1959).

Among the acids listed in Table VI, malonic, adipic, isocitric, and aconitic acids were reported for the first time in oranges. In addition, four unidentified acidic substances were present in appreciable amounts and these may have included quinic, chlorogenic, and citramalic acids, which had been found in oranges by other workers.

Free galacturonic acid was also reported in Spanish orange juices by Sanchez *et al.* (1964) although Sinclair and Jolliffe (1958b) had failed to find it in American samples.

Citric acid was the dominant acid in all the orange varieties examined but the Spanish workers were at variance with other workers in reporting lactic acid, rather than malic acid, as the next most abundant acid.

In some citrus varieties of low acidity, the order of dominance is reversed. For example, the Palestine low-acid lime had 0.20 g/100 ml malic acid but only 0.08 g/100 ml citric acid (R. B. Clark and Wallace, 1963; Clements, 1964a) while the Faris low-acid lemon (total acidity about 0.5%) also contained more malic than citric acid (Bogin and Erickson, 1965).

The free phosphoric acid content in orange and grapefruit juices, calculated from the pH, the anion and cation concentrations, and the activity coefficients, was about 0.15 moles/liter and represented about 5% of the total inorganic phosphate content (Diemair and Pfeifer, 1962), while lemon juice had 0.5-1.0 moles/liter or about 30% of the total phosphate (Langendorf and Lang, 1961).

The organic acids in the peels as well as the juices of oranges, grapefruit, lemons, tangerines, and the Palestine sweet lime were examined by Clements (1964a,b), by freeze-drying followed by silicic acid column chromatography. In this procedure the peel samples were acidified by means of a cation-exchange resin before extraction with methanol. The peel extracts showed at least 15 peaks in the column eluates and 25-30 components on paper chromatograms. Oxalate, malate, malonate, and citrate were the predominant anions in all the peels, generally in that order of concentration. For example, these four anions accounted for 30-50% of the total anions in Navel orange peel. Malonic acid was always accompanied by another acid—probably chlorogenic acid.

In Navel orange peel the concentrations of total anions and individual anions were about twice as high in the flavedo as in the albedo.

I. Trends with Maturation

As maturity advances, a steady decline in acidity is the most consistent change in the composition of the "sweet" citrus fruits. This change, which has already been discussed in Chapter 2, has been examined in some detail with respect to individual acids in oranges in parallel studies by Clements (1964b) in California and by Rasmussen (1964) in Florida. Valencia oranges on the tree were sampled for a period of 16 months, from fruit set to commercial maturity as well as several months beyond this stage.

In both the pulp and peel of oranges, the total water-soluble organic acids on a dry weight basis increased up to about 6 months from fruit set and then declined. Citric and malic acids made up about 95% of the total water-soluble acids in the pulp throughout the growth cycle. Citric became the major acid from about 2 months after fruit set, reached peak concentration about 5 months after fruit set, and then declined. Since both the citric acid concentration and also the amount of citric acid per fruit decreased (eventually to a level less than half that at maturity), there is evidently not only a dilution of acidity with advancing maturity but also a loss of citric acid by translocation or metabolic conversion.

Malic acid was next to citric in concentration in the pulp. The amount of malic acid per fruit remained relatively constant, so that in late-hanging fruit it amounted to about 20% of the total acids.

Malic acid was the dominant water-soluble acid in the peel where it reached maximum concentration at about the time that citric acid reached maximum concentration in the pulp. The total amount of malic acid in the peel was slightly less than that in the pulp.

The oxalic acid content of the pulp was very low, less than 5 mg/g dry weight, but in the peel oxalate was the major anion, again reaching maximum concentration (16 mg/g dry wt.) 4 to 5 months after fruit set; it is present in the form of insoluble calcium oxalate. In Navel orange peel, while the oxalate content declined, the malonate content increased greatly in the late-hanging fruit.

To examine the effects of climatic factors, these investigations were extended by Rasmussen *et al.* (1966) to Valencia oranges grown in six regions in four states of the United States. The maximum concentration of citric acid, in the range 160-200 mg/g dry weight of pulp, was reached earliest in Texas when the fruit was about 18 weeks from fruit set, and latest in Riverside, California at about 22 weeks from fruit set. These observations suggest that accumulation of citric acid is accelerated by warm nights and spring rains, and delayed by lower night temperatures and low rainfall. The subsequent rate of decline in citric acid concentration was subject to similar climatic influences, the decrease being much more rapid in Texas than in Riverside where autumn and winter nights are 10°-15°F cooler.

In the Hamlin orange the seasonal rise and fall of citric acid content produced a curve similar to that for Valencia oranges but advanced by about 2 months, in line with its earlier maturing character (Ting and Vines, 1966). Succinic acid was also determined in this variety and its seasonal variation ran parallel with that of citric acid. The malic acid concentration fell to a pronounced minimum 5 to 6 months from fruit set, then rose again as the fruit matured. Quinic acid declined steadily in concentration from the first sampling of young fruit.

The Marsh grapefruit appeared to be different from oranges in showing a continuing rise in citric acid content per fruit throughout the growth cycle; the proportion of citric acid in the total acids increased from about 50% in young fruit to more than 90% in mature fruit. The concentrations of malic, quinic, and succinic acids declined up to about 5 months from fruit set and then remained roughly constant (Ting and Vines, 1966).

Genetic effects on acidity are illustrated by the hybridization studies of Soost and Cameron (1961) who found that a low-acid (about 0.1%) pummelo crossed with several mandarin and orange varieties (acidities 1-2%) produced hybrids of moderate acidity (0.6-2.6%), while acid (1-2%) pummelos with the same scions produced hybrids of high acidity (1.4-5.2%).

II. Sites and Mechanisms of Acid Biosynthesis

During the last 10 years, evidence has accumulated which suggests that the organic acids stored in the vacuole of citrus fruits are synthesized within the fruits rather than translocated from sites of formation in the leaves. Soluble enzyme systems and mitochondria capable of acid synthesis have now been isolated from citrus fruits (see Chapter 8).

Erickson (1957) had found that the Faris sweet lemon, a variety of low acidity, remained low in acid (about 0.5%) when grafted on a sour lemon plant, while the Eureka lemon grafted on the sweet lemon plant remained high in acid (about 5%). From these results it seemed unlikely that the leaves were the principal source of acids in citrus fruits.

The juice vesicles were favored by Huffaker and Wallace (1959) as a major site of acid synthesis because they contained active phosphoenolpyruvate carboxylase and carboxykinase enzyme systems capable of fixing carbon dioxide into organic acids. Then Bean and Todd (1960) found the juice vesicles in young oranges to be highly active in incorporating carbon dioxide into citric and malic acids, and also into serine and aspartic acid.

By studying carbon dioxide fixation in the Eureka lemon, the Valencia orange, and the Palestine sweet lime, R. B. Clark and Wallace (1963) hoped to gain further insight into mechanisms of acid synthesis because these three fruits covered a wide range of acidity (7.4, 2.2, and 0.2%; relative concentrations 37:11:1). The amounts of carbon dioxide fixed by vesicle homogenates and intact vesicles were, however, in the reverse order at all stages of growth of the fruits. It appeared that the synthesis and accumulation of organic acids were governed by factors other than the ability of the fruits to fix carbon dioxide.

Further comparative studies were made by Bogin and Wallace (1966a,b) using the Eureka lemon (pH 2.5) and the Tunisian sweet lemon (pH 5.5). The high pH of the sweet lemon is due not only to low acidity (less than 1%) but also to a higher content (1.5 mg/ml) of amino acids than the sour lemon (0.5 mg/ml). In each fruit, organic acids were synthesized by carbon dioxide fixation involving phosphoenolpyruvate carboxylase, nicotinamide-adenine dinucleotide phosphate malic enzyme, and isocitric dehydrogenase. Both the capacity for carbon dioxide fixation and the malic acid content decreased as the fruits matured.

To explain the greater accumulation of citric acid in the sour lemon than in the sweet lemon, these authors advanced a hypothesis based on a number of differences in the biochemical mechanisms in the two fruits. The breakdown of citric acid by the enzyme aconitase was competitively inhibited by citramalate which was synthesized in greater amounts by the sour lemon than by the sweet lemon. The lower production of citramalate in the sweet lemon was possibly due to restricted availability of its precursor, pyruvate, because in this fruit there was a higher level of carbon dioxide fixation by phosphoenolpyruvate carboxylase, there was more amination of pyruvate to alanine, and there was higher activity of catalase contributing to oxidative carboxylation of pyruvate. Overall, therefore, the greater inhibition of aconitase by citramalate in the sour lemon might be expected to encourage the accumulation of citric acid.

From grapefruit, sampled throughout the growing cycle, Vines and Metcalf (1967) were able to isolate mitochondrial preparations capable of oxidative phosphorylation of intermediates in the tricarboxylic acid cycle. The activity of preparations from immature fruit was low but it increased rapidly as the fruit approached maturity, and thereafter generally declined. The seasonal rise and fall in the acidity of grapefruit therefore appears to be correlated with oxidative phosphorylation activity. An explanation of the effect of arsenate in reducing the acidity of grapefruit may lie in its interference with phosphate esterification in the mitochondria (Vines and Oberbacher, 1965) (see Chapter 2, Section V).

III. Analytical Aspects

Workers at the Fruit and Vegetable Chemistry Laboratory of the U.S. Department of Agriculture at Pasadena, California, investigated the acids of lemons in a search for analytical procedures to establish the authenticity of lemon juices. In analyses of the juices of lemons from coastal, inland valley, and desert regions in California and Arizona, they found a highly significant correlation ($r = 0.467, p < 0.1$) between the citric acid content and the L-malic acid content (Vandercook *et al.*, 1963). Although this relation was established for lemons of uncontrolled storage history, it is known that malic acid is lost in storage (Vandercook *et al.*, 1966) and for this reason it is not the most reliable index of the integrity of lemon juice.

Subsequently, the same group (Rolle and Vandercook, 1963) reported a more reliable relation incorporating two other variables: the amino acid content *(A)* (see Chapter 7), and the polyphenolic content *(P)* (see Chapter 13), as well as the L-malic acid content *(M)*, which they expressed in the form:

$$C = 36.54 + 12.04A + 30.06P + 2.71M$$

where C is the calculated citric acid content and all the concentrations are expressed in milliequivalents per 100 ml of juice. If the titrated acidity of a lemon juice sample was more than 20% greater than the acidity calculated from this relation, it lay outside the 99% confidence limits and the juice might reasonably be regarded as abnormal or adulterated by the addition of acid. The validity of the relation was not affected by storage of the fruit, nor by different methods of juice extraction, nor by the presence of chemical preservatives and storage of the juice (Vandercook *et al.*, 1966; Vandercook and Guerrero, 1968).

The analytical method used in these investigations for the determination of L-malic acid is based on the measurement of the optical rotation of the uranyl acetate complex and has been adopted as official, first action, by the Association of Official Analytical Chemists for malic acid in lemon juice (H. Yokoyama, 1965, 1966) as well as for malic acid in other fruit juices, including orange juice (Fernandez-Flores *et al.*, 1968).

Potentiometric neutralization curves of orange and lemon juices were investigated by Primo and Royo (1968) as a means for detecting adulteration with citric acid. If the maximum slope of the curves exceeded 2.5 or 6.0 pH units per milliliter of 0.1 N sodium hydroxide, respectively, for orange juice or diluted lemon juice (one part in five), then adulteration of the juice might be suspected.

Chapter 5 Vitamins

A comprehensive survey of the nutrients, including vitamins, in mature Valencia and Navel oranges and Eureka lemons, representatively sampled in California during two seasons, was reported by Joseph *et al.* (1961) and Birdsall *et al.* (1961). The results of this work are summarized in Table VII.

In line with previous experience, the peels of all three fruits contained higher concentrations of vitamins than the edible portion (pulp) and juice, while the edible portion had higher levels of nicotinic acid, biotin, and pantothenic acid than the juice. Sawyer (1963) also found a higher content of nicotinic acid in the rag than in the juice of Mediterranean oranges. Ascorbic acid was the only vitamin present in amounts of major nutritional usefulness. Navel orange juice contained more ascorbic acid by about 15 mg/100 g than the juices of Valencia oranges and lemons.

I. Ascorbic Acid

General information on the ascorbic acid content of citrus fruits and the influence of horticultural factors has been reviewed in Chapter 2.

Changes in the ascorbic acid content of oranges and lemons throughout growth and development were followed by Eaks (1964). In both Valencia oranges and Eureka lemons the ascorbic acid content of the *whole fruit* increased rapidly to a sharp peak when the fruit weight was around 10 g, then declined and eventually leveled off when the fruits were approaching maturity. The amount of ascorbic acid in the whole fruit rose steeply to the stage of the peak concentration, then increased more slowly throughout the growth of the fruit. Lemons grown on the Californian coast had more ascorbic acid per fruit

TABLE VII VITAMINS IN CITRUS FRUITS[a]

Vitamin	Unit (per 100 g)	Valencia orange			Navel orange			Eureka lemon[b]		
		Peel	Edible portion	Juice	Peel	Edible portion	Juice	Peel	Edible portion	Juice
(Percentage of Whole Fruit)	–	25	75	49	31	68	41	32	67	43
Ascorbic acid	mg	137	40	44	222	57	59	129	53	44
Biotin	μg	5	1	0.8	3	0.7	0.6	2	0.6	0.3
Carotenoids[c]										
Total	mg	9.9	3.4	2.8	12	2.1	1.4	0.3	0.06	0.04
β-Carotene	mg	0.3	0.2	0.2	0.2	0.05	0.03	0.03	0	0
Choline	mg	23	12	8	23	13	6	11	10	6
Folic acid	μg	12	4	3	9	3	2	5	2	1
Inositol	mg	257	204	159	185	187	156	216	109	66
Nicotinic acid	μg	888	491	376	665	477	429	356	129	71
Pantothenic acid	μg	490	276	207	303	221	187	319	194	104
Pyridoxine	μg	176	65	57	102	49	48	172	80	51
Riboflavin	μg	91	33	27	95	45	34	79	21	12
Thiamine	μg	120	130	100	90	100	100	58	42	31

[a]Average values for 2 seasons, California fruit, from Joseph *et al.* (1961) and Birdsall *et al.* (1961).

[b]Cured 60 days at 56°-60°F, 86-88% relative humidity.

[c]Cf. Chapter 9.

than lemons grown inland because of the greater proportion of peel. The peel of lemons contained about three times as much ascorbic acid in concentration and in total amount as the juice, while the flavedo contained about twice as much as the albedo.

Similar trends in ascorbic acid content from fruit set to maturity were observed by Goren and Monselise (1965a) in the peel of Shamouti oranges. Significant correlations between the ascorbic acid content and the contents of reducing sugars and hesperidin suggested that these constituents may be linked in biosynthetic processes in the peel of oranges (see p. 134).

With the object of specifying statistically valid sampling procedures, Primo *et al.* (1963b) examined the distribution of ascorbic acid content in the juice between oranges on individual trees of several Spanish varieties, in particular zones of the trees, and between trees in an orchard. They found, for instance, that to determine ascorbic acid content with 99% confidence limits of ± 1 mg/100 ml a sample of 326 oranges should be analyzed, while 8 oranges would give a result with 95% confidence limits of ± 5 mg/100 ml.

In agreement with common experience in the northern hemisphere, the highest values of ascorbic acid content in the juice were found in oranges on the southern aspect of the tree where they received most sunlight, while the lowest values were found in the interior of the tree where the fruit was most shaded.

When a valid sampling method was used, no significant differences in ascorbic acid content were found between oranges taken at different times on the same day, or with an interval of 24 hours, or from different trees in the same orchard.

Using samples of six oranges taken from each of six Shamouti trees at 4-hour intervals for 24 hours, Goren and Monselise (1965a) did find a significant diurnal fluctuation in the ascorbic acid content of the green *peel* and *rag*, with daytime and nighttime peaks, while the dehydroascorbic acid content showed a broadly antithetic fluctuation.

II. Analytical Aspects

Interesting applications of some vitamin constituents of citrus fruits as index compounds for the detection of adulteration and the estimation of juice content in orange juices and concentrates have been made by workers in the Laboratory of the Government Chemist, London. Nicotinic acid, determined by microbiological assay, was suggested by Sawyer (1963) for this purpose, and subsequently Lisle (1965) proposed a procedure involving the estimation of both nicotinic acid and inositol. Similar coefficients of variation of about 12% were found for the inositol and nicotinic acid contents of orange juice samples from Israel, South Africa, British Honduras, and Southern Rhodesia, and statistical combination of the analytical values led to an equation for the calculation of the orange juice content of concentrates in the form:

$$C = 2.20x + 0.0025y$$

where C is the concentration in units of 10° Brix juice, x is the nicotinic acid content in mg/100 ml, and y is the inositol content in mg/100 ml.

Chapter 6 Inorganic Constituents

I. Analytical Interest

Classical methods for the estimation of fruit content in foods and beverages are based upon determinations of the inorganic constituents, for instance, the total ash and the alkalinity of the ash. The procedure for estimating citrus juice content in beverages on the basis of the alkalinity of the ash has been refined by Morgan (1963) by providing corrections for inorganic sulfites, benzoates, and phosphates that may have been added as preservatives or adulterants. The increasing speed and convenience of analytical methods has, however, encouraged the use of procedures based on the determination of individual elements.

Information on the inorganic constituents of citrus fruits that has become available in the last decade is condensed in Table VIII. In view of improvements in analytical techniques and sampling procedures this table should be regarded as superseding Table XII in the previous review (Kefford, 1959). Information on the mineral constituents of citrus fruits has also been tabulated by Chatt (1966) and Holeman (1963).

Regional differences in the inorganic constituents of oranges were found by Hulme *et al.* (1965) in a survey of fruit from four regions: South Africa, the Eastern Mediterranean (Israel and Cyprus), Spain, and Brazil. The mean potassium contents of whole oranges from South Africa, the Eastern Mediterranean, and Spain were very similar (1710, 1700, 1710 mg/kg, respectively), but the mean for Brazilian oranges (2310 mg/kg) was significantly higher. The phosphorus contents for the four regions were, however, consistent (210, 220, 200, 220 mg/kg, respectively). No significant difference was found between South African Valencia and Navel oranges in potassium, phosphorus, or nitrogen content. To estimate the fruit content of beverages made from whole oranges, Hulme *et al.* (1965) recommend the combined expression $0.05\,(7\,k + 10p + 3n)$,

TABLE VIII INORGANIC CONSTITUENTS OF CITRUS FRUITS[a]

Consituent	Orange		Grapefruit		Lemon		Bitter orange (Seville)	
	Whole[b]	Juice[c]	Whole[b]	Juice[c]	Whole[b]	Juice[c]	Whole[b]	Peel[d]
Total ash	-	1800-4600	-	2300-4200	-	2400-2500	-	6700-8000
Potassium	920-2780	520-2840	1060-2350	670-2030	1290-3060	940-1930	1200-2290	2730-3063
Sodium	-	2-30	-	3-23	-	10-29	-	-
Calcium	-	49-206	-	-	-	31-74	-	-
Magnesium	-	50-147	-	-	-	10-60	-	-
Iron	-	0.5-6	-	-	-	1.6-10	-	-
Phosphorus	130-280	42-240	110-210	57-190	92-330	32-144	210-250	110-233
Nitrogen	1200-2460	570-1800	1000-2030	320-940	970-3050	350-840	-	-
Chlorine	-	20-55	-	24-136	-	-	-	-

[a]In mg/kg.

[b]Extremes of ranges from Money (1964, 1966) – fruit from Israel, Spain, South Africa, U.S.A., West Indies, and South America.

[c]Extremes of ranges from the following sources, recalculated where necessary:
Anonymous (1967b) – oranges and grapefruit from Israel.
Beisel and Kitchel (1967) – oranges from California and Arizona.

Benk (1965) – oranges from Cyprus, Greece, Israel, Italy, Morocco, Spain, Turkey, Brazil, and South Africa.
Birdsall *et al.* (1961) – oranges and lemons from California.
I. Calvarano (1961) – oranges from Italy.
I. Calvarano and M. Calvarano (1962) – oranges and lemons from Italy.
Dawes (1969) – lemons from New Zealand.
Floyd and Rogers (1969) – oranges from Florida.
S. H. Hart (1961) – oranges from Texas and Mexico.
Hopkins and Walkley (1967) – oranges from Israel and Greece.
Hulme *et al.* (1965) – oranges from Mediterranean Area, Brazil, and South Africa.
Primo and Royo (1965a,b) – oranges from Spain.
Royo and Aranda (1967) – oranges from Mediterranean Area, America, West Indies, and South Africa.
Sawyer (1963) – oranges from Mediterranean Area.
Sherratt and Sinar (1963) – grapefruit from Cyprus, Israel, South Africa, and Trinidad.

[d]From Osborn (1964).

where *k, p,* and *n* are the calculated fruit contents based on potassium, phosphorus, and nitrogen analyses, respectively.

In a similar regional survey using larger numbers of samples, Money (1966) found no significant differences between three regions–South Africa, Israel, and Spain–although he reported that the Shamouti orange from Israel was abnormally low in potassium. Moreover the potassium content of the whole Shamouti fruit (1230 mg/kg) was lower than that of the juice (1420 mg/kg), whereas generally this order is reversed. Other recent analytical data on Israel citrus fruits (Anonymous, 1967b; Hopkins and Walkley, 1967) confirm the low potassium content of Shamouti oranges.

Benk (1965) has drawn attention to the very wide range–5-550 ppm–in published values for the sodium content of orange juices. When he analyzed 46 juices extracted by hand from oranges from Cyprus, Greece, Israel, Italy, Morocco, Spain, Turkey, Brazil, and South Africa, Benk found sodium contents of only 4-20 mg/liter (0.1-0.6% of the ash) while the potassium contents (1140-1930 mg/liter; 34.9-48.8% of the ash) were in line with other data. In six samples of California oranges examined by neutron activation analysis, Castro and Schmitt (1962) also found very low sodium contents: in the juice 1.4-2.2, and in the peel 5.8-7.9 mg/kg. Commercial lemon juices extracted under higher pressures may have higher sodium and ash contents than juices extracted in the laboratory or kitchen (F. Guenther *et al.*, 1968). Citrus juices are commonly prescribed in diets designed to correct electrolyte balance because of their favorable sodium-to-potassium ratio, but it appears that this ratio is even more favorable than previously suspected and on Benk's values ranges from 1:70 to 1:475 in orange juice and from 1:49 to 1:310 in lemon juice (Benk, 1965, 1968b).

For the detection of adulteration in citrus juices, Primo and Royo (1967) advocated mineral analyses on the clear serum separated from the juice by centrifuging and filtering, because the variable proportions of pulp widen the variability of determinations on whole juice. In general, the standard deviations and coefficients of variability were lower for analyses of ash content and mineral constituents in serums than in whole juices (Primo and Royo, 1965a,b). Comparison of serum analyses of orange juices from Spain, California, and Florida revealed some surprising differences between these regions (Primo and Royo, 1967). Ash contents increased in the following order: Spain (3500), California (4500), and Florida (5700 mg/liter), with the American juices also markedly higher in potassium (about 2000 mg/liter) than the Spanish (1500 mg/liter). In phosphorus content there was a large difference between the California (163) and Florida (83) juices with the Spanish juices (123 mg/liter) in between. Sodium contents were all under 30 mg/liter but the American values (about 20 mg/liter) were clearly higher than the Spanish values (7.4 mg/liter). The alkaline earth metal contents (as calcium) were similar in all three regions.

A fractionation of the phosphorus compounds in orange, grapefruit, and lemon juices was undertaken by Vandercook and Guerrero (1969) in a search for useful indexes of juice content. In addition to total phosphorus, 3 fractions were determined–inorganic phosphorus, lipid phosphorus, and an ethanol-insoluble phosphorus fraction that contained nucleic acids and phosphoproteins. The coefficients of variation for all these fractions were, however, of the same order (approximately 20%) as for other citrus constituents. Nevertheless, a relationship that might be useful for characterizing citrus juices was demonstrated between the inorganic phosphorus (y) and the ethanol-insoluble phosphorus (x) when both were calculated as percentages of the total phosphorus; thus for orange juices, $y = 81.06 - 1.06x$ $(r = -0.826)$.

In a semiquantitative spectrographic survey of the mineral content of California Valencia and Navel oranges and Eureka lemons, Birdsall *et al.* (1961) reported that the following elements were generally present in amounts less than 1% of the ash: Na, Si, Fe, Mn, B, Sr, and Al. The following elements were present in amounts less than 0.01% of the ash: Cu, Li, Ti, Ni, Cr, V, Bi, Zr, Pb, Sn, Co, As, Ba, Mo, Ag, and Zn.

II. Horticultural Aspects

The expected relationship between the potassium content of citrus fruits and the potassium status of the tree was confirmed for grapefruit by Smith and Rasmussen (1961b), who also found that potassium continued to enter mature fruit on the tree, so that there was a 25% increase in potassium content from October to April in Florida. There was little change in the nitrogen, phosphorus, and magnesium content of the fruit during this period but the calcium content increased slightly. When potassium was withheld from orange trees, representing several levels of potassium status following a previous experiment, the potassium contents in the fruit declined in parallel fashion, but in a more gradual and uniform way than in the leaves which were more subject to seasonal variation (Koo, 1962).

Foliar sprays containing manganese and zinc increased the concentrations of those elements in the peel and juice of Valencia oranges (Labanauskas *et al.*, 1963).

The boron content of oranges is influenced by the level of boron in the soil and is increased by treatment of the fruit with borax; in untreated fruit Gounelle *et al.* (1965) found boron contents of 5 and 1.2-1.6 mg/kg in the flavedo and juice, respectively.

Current interest in atmospheric pollution has directed attention to the effects of fluorine on citrus trees. Even when the leaves have taken up substantial amounts of fluorine and the trees show characteristic symptoms of fluorine toxicity, only a little fluorine is found in the fruit. In polluted air in Florida, oranges had about 6 ppm of fluoride in the peel on a fresh weight basis and about 0.7 ppm in the juice (Leonard and Graves, 1966). Under experimental conditions in California, fruit from treatments involving sodium fluoride sprays and hydrogen fluoride fumigation had around 1 ppm of fluorine in the peel and around 0.3 ppm in the pulp and juice, while in the control fruit the fluorine contents were 0.27-0.55 ppm in the peel and 0.09-0.17 ppm in the pulp and juice (Brewer *et al.*, 1967).

Chapter 7 Nitrogen Compounds

It is well known that the nitrogen content of citrus fruits (Table VIII) is influenced by the nitrogen status of the tree. For instance, Smith (1963b) has calculated that the fruit contains about one-quarter of the total nitrogen in an orange tree, while about one-half of the nitrogen absorbed by the tree each season finds its way into the crop of fruit. W. W. Jones and Embleton (1967) confirmed that the nitrogen content of Washington Navel orange juice increased with increasing levels of nitrogen fertilization.

In a study of Shamouti oranges from fruit set to maturity, Goren and Monselise (1964a) found that the total and protein nitrogen contents in the fruit on a dry basis declined throughout growth just as previous workers had found for other varieties. In the young fruit, protein nitrogen amounted to 71% of the total nitrogen, but this proportion decreased to 58% in the mature fruit as soluble nitrogen compounds accumulated in the juice. Thus the total nitrogen content of the juice increases with advancing maturity and is higher in late-season oranges than in early or mid-season varieties which do not hang so long on the tree (Ting, 1967).

In a survey of citrus fruits from four regions, Hulme *et al.* (1965) found whole oranges from the eastern Mediterranean region (Israel and Cyprus) were significantly higher in total nitrogen content (2250 mg/kg) than oranges from South Africa (1900), Spain (1840), and Brazil (1780 mg/kg), but Money (1966) failed to confirm this difference among oranges from Israel, South Africa, and Spain.

TABLE IX AMINO ACIDS IN CITRUS JUICES[a]

Compound	Orange					Lemon			Tangerine	Grapefruit
	Valencia[b]	Navel[b]	Biondo di Sicilia[c]	Blood varieties[c]	Shamouti[d]	Eureka[b]	Lisbon[b]	Primo Fiore[c]	Dancy[b]	Marsh[b]
Alanine	13	12	7	7-10	36	9	10	13	7	9
γ-Aminobutyric acid	32	24	-	-	-	7	7	-	18	19
Arginine	57	54	37	47-50	45	3	3	2	84	47
Asparagine	50	67	-	-	-	16	17	-	85	42
Aspartic acid	33	27	28	32-41	115	36	32	51	36	81
(Cysteic acid)[e]	-	-	-	0.5	-	-	-	0.1	-	-
Cystine	-		-	0.7	-	-	-	0.4	-	-
Glutamic acid	18	12	14	14-21	28	19	18	26	16	22
Glycine	2	2	0.7	0.7-0.9	-	1	1	0.8	2	2
Histidine	-	-	0.9	1-2	-	-	-	-	-	-
Isoleucine	-	-	0.6	0.7-0.8	-	-	-	0.8	-	-
(Allo-isoleucine)[e]	-	-	0.4	0.4	-	-	-	0.1	-	-

Leucine	-	-	0.7	0.5-0.6	6	-	-	0.7	-	-
Lysine	4	3	4	3-5	10	1	1	1	4	3
Methionine	-	-	0.6	0.4-0.5	-	-	-	0.3	-	-
(Methionine sulfoxide)[e]	-	-	1	1	-	-	-	0.6	-	-
Phenylalanine	5[f]	3[f]	0.9	0.5-1	-	2[f]	3[f]	1	5[f]	3[f]
Proline	239	107	56	74-78	-	41	47	29	100	59
Serine	22	18	45	40-62	70	17	19	38	19	15
Threonine	-	-	1	2	-	-	-	1	-	-
Tyrosine	-		0.6	0.4-0.6	-	-	-	0.2	-	-
Valine	2	2	1	1	12	1	1	2	2	2

[a] In mg/100 ml.
[b] From Clements and Leland (1962a).
[c] From Casoli (1963).
[d] From Coussin and Samish (1968).
[e] Probably analytical artifacts.
[f] Phenylalanine plus tyrosine.

I. Amino Acids

More than 70% of the total soluble nitrogen in citrus juices is present in the form of free amino acids. Early in the period under review, Clements and Leland (1962a) applied the Moore and Stein ion exchange procedures to provide more definitive information than had previously been available about the amino acid composition of California oranges, grapefruit, lemons, and tangerines (Table IX). Proline was the dominant amino acid in all the fruits except grapefruit where it was second to aspartic acid.

The proline concentration was particularly high in Valencia orange juice. In another study, Clements and Leland (1962b) showed that this amino acid accumulated rapidly from about 9 months after blossoming and reached very high levels (440 mg/100 ml, 2.7% of solids) in oranges allowed to remain on the tree for about 21 months after blossoming. Arginine and γ-aminobutyric acid also increased in concentration with advancing maturity, while the asparagine, aspartic acid, and serine contents showed little change.

Similar surveys of the distribution of amino acids in Italian oranges, lemons, and mandarins were made by Averna (1960), I. Calvarano (1963b), Casoli (1963), and Di Giacomo *et al.* (1968d). Again proline predominated in common and blood oranges, but in lemons aspartic acid and serine were present in greater concentrations, while the arginine content was notably low (Table IX).

The high proline content of oranges is carried over into citrus molasses prepared by concentrating the extract from citrus processing residues (S. K. Long *et al.*, 1967).

When Ismail and Wolford (1967) examined the effect of concentrating Valencia orange juice to 65° Brix on the distribution of the nitrogen compounds they found that some of the insoluble nitrogen fraction passed into the soluble fraction, but there was little qualitative or quantitative change in the amino acid pattern.

Qualitative chromatographic studies revealed the presence in Assam rough lemons of alanine, γ-aminobutyric acid, asparagine, glutamic acid, and glycine (Chaliha *et al.*, 1964b), and in the Indian Mosambi orange *(Citrus sinensis)* the same five amino acids were found in addition to arginine, aspartic acid, proline, serine, and threonine (Srivastava and Tandon, 1966).

Silber *et al.* (1960) surveyed the amino acids in a wide range of fruits including oranges, lemons, limes, Temple tangors, and kumquats and found most of the amino acids listed in Table IX. Compounds not reported by other workers were α-aminoadipic acid and djenkolic acid, both thought to be present in the Temple tangor. Proline was absent from lime juice, an observation confirmed by Alvarez (1967).

On the evidence of microbiological growth tests, Shioiri and Katayama (1955) listed 17 amino acids in the Satsuma mandarin and the Natsudaidai.

II. Analytical Applications

The free amino acids of citrus juices have been widely investigated as analytical indices of citrus fruit content in juices, concentrates, and beverages (Bellomo, 1969; I. Calvarano, 1963b).

As a quantitative index of fruit content in orange products, Morgan (1966) recommended serine. After absorption of the amino acids from a juice or extract on a cation exchange column and elution with ammonia, serine is conveniently estimated by periodate oxidation of the terminal secondary alcohol group to formaldehyde, which is determined spectrophotometrically after reaction with chromotropic acid. Moveover, as already mentioned, the serine content remains reasonably constant during the maturation of oranges, and it is fairly uniformly distributed between the juice, pulp, and peel components. The latter consideration is important in England where comminuted beverages are popular and regulations specify "potable fruit content" which includes all components of the fruit (Charley, 1963). Based on the analyses of oranges from five countries as well as commercial juices and comminuted products, Morgan calculated an overall mean serine content of 19.2 mg/100 g and used this as the basis for the estimation of the orange content in beverages.

A gross measure of the amino acid content of citrus products is provided by the formol titration, in which the amino acids are titrated with alkali after blocking the free amino groups by reaction with formaldehyde to form *n*-methylene amino acids. The chemistry of the formol titration is somewhat empirical since the end point is taken at an arbitrary pH and there is some doubt as to the groups titrated at this point (Taylor, 1957). Nevertheless the procedure (which involves neutralization of the sample to pH 7, addition of neutralized formaldehyde, and titration to pH 8.4) was sufficiently reproducible in collaborative studies to be officially accepted at the first action stage by the Association of Official Analytical Chemists (H. Yokoyama, 1965). For Italian lemon juices, Di Giacomo *et al.* (1968a) concluded that a formol value of 1.4 with a tolerance of ±10% should be accepted as a standard.

As mentioned in Chapter 4, the amino acid content of lemon juices by formol titration was one of the parameters recommended by Rolle and Vandercook (1963) for assessing the integrity of lemon juices, since a highly significant correlation ($r = 0.794, p < 0.01$) had been established between the amino acid content and the citric acid content of the juices of lemons from coastal, inland valley, and desert regions in California and Arizona (Vandercook *et al.*, 1963). By paper chromatography 17 amino acids were identified in lemon juice and the relative concentrations of the more important ones determined. The total amino acid content of lemons held at 58°F and 85% relative humidity increased during storage up to 13 weeks (Vandercook *et al.*, 1966).

In contrast to lemon juice, orange juice showed no significant correlation between titratable acidity and amino acid content, but the latter was closely correlated with the betaine, polyphenolic, ash, and phosphorus contents (Coffin, 1968).

If adulteration of lemon juice with protein hydrolyzate, added to raise the formol value, is suspected, the relative concentrations of the amino acids may be a useful diagnostic aid. For instance, the concentration of γ-aminobutyric acid is approximately four times that of leucine-isoleucine, whereas in a protein hydrolyzate this relation is likely to be reversed (Vandercook *et al.*, 1963). Similarly, Alvarez (1967) pointed out that adulteration with glycine is readily detected because the natural glycine content of citrus fruits is low (Table IX). When orange, lemon, grapefruit, and lime juices were examined by thin-layer chromatography on silica gel, added glycine appeared as a pink ninhydrin spot adjacent to the yellow proline spot.

Thin-layer chromatography on silica gel was also used by Wucherpfennig and Franke (1966) to examine the distribution of amino acids in the juice, albedo, and flavedo of Navel and blood oranges. These workers were looking for a way of detecting adulteration of citrus juices with press juices or water extracts of peel and rag. The patterns of ninhydrin spots from juice and peel components were not, however, sufficiently different to be useful. Neither were the formol values, but the chloramine values of the albedo and flavedo components were considerably higher than for the juices and it was suggested that this be used as a basis for the detection of adulteration. For a similar purpose, Di Giacomo and Rispoli (1966) recommended calculation of the ratio of the chloramine and formol values.

The chloramine value is an empirical index based on the oxidation by chloramine T (sodium *N*-chloro-*p*-toluene-sulfonamide) of fruit constituents other than sugars and acids (Benk and Stein, 1959; I. Calvarano, 1966).

Gierschner and Baumann (1966) advocated performing the formol titration on both cloudy and clarified samples as a method for detecting whether peel components were present in a juice. The formol value was greater in the cloudy sample and the difference increased with increasing peel content.

In a peel extract from the bitter orange, Kunkar (1964a) identified β-alanine, γ-aminobutyric acid, asparagine, aspartic acid, glutamic acid, and proline.

III. Nitrogen Bases

In addition to amino acids, the soluble nitrogen compounds in citrus fruits include a number of bases. Betaine, putrescine, stachydrine, choline, ethanolamine (Silber *et al.*, 1960), and ammonia (Clements and Leland, 1962a,b) have been reported.

A survey of the betaine content of commercial juices and concentrates from six countries led W. M. Lewis (1966) to recommend it as a suitable index compound for the estimation of juice content. Following the usual procedure of absorption on a cation exchange column and elution with ammonia, betaine is separated from the other soluble nitrogen compounds by passage through a second column in which a strongly basic anion exchange resin mixed with a weakly acidic cation exchange resin retains the other ionic species present. Betaine is then determined as the reineckate salt.

In orange juices and reconstituted juices on the Canadian market, Coffin (1968) reported greater variability in betaine content than W. M. Lewis (1966) had found; the betaine content was, however, closely correlated with the amino acid, polyphenolic, ash, and phosphorus contents.

Betaine is unique to citrus fruits among the common fruits and is therefore useful for determining the citrus juice content of blended beverages.

A surprising recent discovery in the chemistry of citrus fruits was the identification of organic bases which are related to ephedrine and have sympathomimetic, vasopressor, and antihistamine activity. First detected in paper chromatograms from citrus leaves and later found also in the juices of a wide range of citrus fruits, two of these bases, synephrine and octopamine, had not previously been found in plants (Stewart, 1963; Stewart *et al.*, 1963; Stewart and Wheaton, 1964a; Gjessing and Armstrong, 1963).

Synephrine (3) was found in the juice of several varieties of orange (15-30 mg/liter), tangerine (58-152 mg/liter), and tangelo, and in the Temple and Murcott tangors, Troyer citrange, Meyer lemon, sweet lemon, rough lemon, and Calamondin, but not in the grapefruit, lemon, lime, or kumquat.

Octopamine (2) occurs in the Meyer lemon at a concentration of 4 mg/liter and in lower but detectable concentrations in tangerines and the Temple, Murcott, and Troyer hybrids. The principal base (25 mg/liter) in the Meyer lemon is tyramine (1) which also occurs in detectable quantities in tangerines and the Temple, Murcott, and Troyer hybrids (Stewart and Wheaton, 1964b).

Cleopatra mandarin juice had especially high concentrations of synephrine (280 mg/liter) and *N*-methyltyramine (58 mg/liter) and was the only citrus juice found to contain hordenine (*N*-dimethyltyramine, 7 mg/liter) (Wheaton and Stewart, 1965a).

Labeling experiments (Wheaton and Stewart, 1969) indicate that the most probable pathway for the accumulation of phenolic amines in citrus is tyramine → *N*-methyltyramine → synephrine rather than the tyramine → octopamine → synephrine route suggested for the biosynthesis of synephrine in animal metabolism by Pisano *et al.* (1961).

The general absence of the phenolic amines from lemons and limes strengthens the view that the Meyer lemon, sweet lemon, and rough lemon are hybrids.

In grapefruit, the only basic nitrogen compound detected was the aromatic amide, feruloylputrescine (4), which was present in concentrations of 15-41 mg/liter. It was also found in oranges (4-8 mg/liter) but not in tangerines,

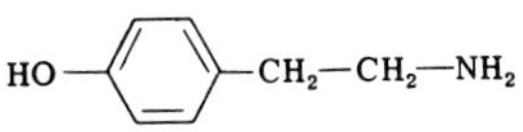

$HO-C_6H_4-CH_2-CH_2-NH_2$

Tyramine
(1)

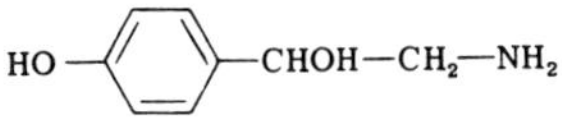

$HO-C_6H_4-CHOH-CH_2-NH_2$

Octopamine
(2)

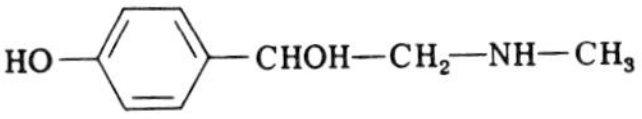

$HO-C_6H_4-CHOH-CH_2-NH-CH_3$

Synephrine
(3)

$HO(CH_3O)C_6H_3-CH{=}CH-CO-NH-(CH_2)_4-NH_2$

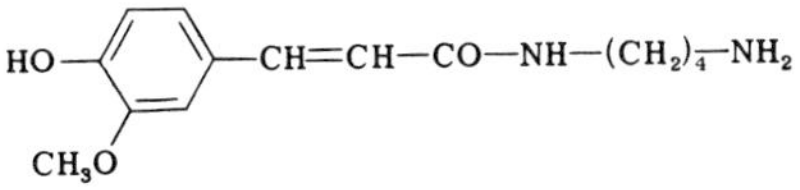

Feruloylputrescine
(4)

lemons, or limes (Stewart and Wheaton, 1964b; Wheaton and Stewart, 1965b). As a 4-hydroxy-3-methoxy aromatic amide present in foods, it should not be overlooked in metabolic investigations where the excretion of such compounds is measured.

The phenolic bases may be determined in the centrifuged serum of citrus juices by ion exchange chromatography and ultraviolet spectrophotometry or by gas chromatography of trifluoroacetyl derivatives (Coffin, 1969). They too have been proposed as index compounds for the estimation of citrus juice content in beverages and for the detection of adulteration.

IV. Proteins

Knowledge of the nature of the proteins in citrus fruits has been advanced by Clements (1966), who succeeded in extracting protein fractions (in a form close to the native state) from the vesicles, segment membranes, albedo, and flavedo of Washington Navel oranges, and the vesicles and peel of lemons and grapefruit. By freezing and homogenizing the tissues in acetone at low temperature, and subsequent extraction, washing, and drying (Clements, 1965), powders were obtained with nitrogen contents ranging from 0.9% in albedo powder to 3.2-3.7% in vesicle powders, the latter values representing calculated protein contents of about 20%. Most of this nitrogen was extracted in a medium consisting of a phosphate buffer pH 7.5 with the addition of sucrose and ethylenediaminetetraacetate (EDTA), and the extract was subjected to disk electrophoresis on polyacrylamide gels using both anionic and cationic systems.

In the anionic system each citrus tissue gave a characteristic and reproducible electrophoretic pattern and densitometric tracing containing about 20 bands, including 1-5 intense bands. Since there were broad similarities between the patterns for orange albedo and membrane, orange and grapefruit vesicles, and grapefruit and lemon peel, it is likely that the citrus tissues have a number of proteins in common. Cationic electrophoresis gave simpler profiles with fewer bands.

Scott *et al.* (1965) noted that orange juice and its cloud constituents were much higher in nitrogen and also in phosphorus, than the structural tissues in the albedo and the rag (Table V). The insoluble solids making up the fine cloud have now been shown by Baker and Bruemmer (1970) to contain 45% protein, while Vandercook and Guerrero (1969) demonstrated the presence of nucleotides and phosphoproteins.

In an examination of mixed orange and grapefruit seeds, a by-product of citrus processing operations, Ammerman *et al.* (1963) found 16.2% protein (dry basis) in the whole seeds, 19.5% in the kernels (which made up 74% of the weight of the seeds), and 6.1% in the hulls. When fed to lambs as 88% of the total protein in the ration, the protein of citrus seed meal was equal in digestibility and biological value to the protein of soybean meal. Kunjukutty *et al.* (1966) reported 9.8% protein in lime seeds.

V. Nucleic Acids

The nature and amounts of polymeric nucleic acids at different stages in the growth of citrus fruits were studied by Ismail (1966) using as raw material the peel of Calamondin fruits. Soluble ribonucleic acid (sRNA), deoxyribonucleic acid complex (DNA-RNA), and ribosomal ribonucleic acid (rRNA) were present in relative proportions that changed as the fruits developed. In young fruits, rRNA predominated but decreased in proportion with growth until at maximum size the levels of sRNA and rRNA were about equal. With advancing maturity the proportion of rRNA declined sharply. Treatment with ethylene increased the sRNA, while treatment with gibberellic acid retarded the loss of rRNA.

Chapter 8 Enzymes

I. Pectolyzing Enzymes

Citrus fruits contain a methylesterase which readily hydrolyzes polygalacturonic acid polymethylesters containing 10 or more units, but which is inactive towards, for instance, the mono- and diesters of digalacturonic acid, the diester of methyl digalacturonic acid, and the triester of methyltrigalacturonic acid (McCready and Seegmiller, 1954).

The pectinmethylesterase was known to be associated with the insoluble components of citrus fruits, and Jansen *et al.* (1960) established that it is bound to pectic substances in cell walls by ionic bonds in an enzyme-substrate relationship. The binding is not specific for pectinmethylesterase since α-chymotrypsin is bound to the same sites in even greater amounts.

Citrus pectinmethylesterase is dissociated from the cell-wall pectic substances by salts at slightly alkaline pH and also by soluble pectin in the presence of salts, even at acid pH. In the latter case, the soluble enzyme rapidly hydrolyzes the soluble pectin, leading to gel formation. The pectinmethylesterase bound to cell-wall pectin substances *in situ* is, however, inactive at pH 4.5 and below. Both soluble and bound enzymes showed the same optimum pH at 7.5, a value that was confirmed by Vas *et al.* (1967). These workers, who were primarily interested in the methodology of the assay of pectic enzymes, also reported an optimum temperature of 65°C and an optimum sodium chloride concentration of about 1% for the pectinmethylesterase of orange albedo.

Improved precision in the estimation of pectinmethylesterase in citrus juices is claimed by Primo *et al.* (1967) for a manometric method which measures the carboxylic acid groups, liberated by the enzyme from pectin, by evolution of carbon dioxide from sodium bicarbonate solution, rather than by continuous titration (Kertesz, 1955).

The relative pectinesterase activity of the component parts of citrus fruits has been reported by several workers. In Valencia and Pineapple oranges grown in Florida, Rouse *et al.* (1962a, 1964a) found the descending order of activity on a dry weight basis to be vesicle walls, segment walls, peel, seeds, and juice. In Silver Cluster grapefruit and in lemons the distribution of activity was slightly different since the peel was the most active component in about 70% of the samples examined (Rouse *et al.*, 1965b; Rouse and Knorr, 1969).

The changes with maturity were somewhat irregular but pectinesterase activity was generally greatest when the Brix:acid ratio was highest. In oranges from trees only 3 and 5 years old, the pectinesterase activity in all the component parts except the segment walls was less than in fruit from mature trees (Rouse *et al.*, 1965a; Rouse, 1967b). When oranges and grapefruit were examined shortly after the severe freeze in Florida in December, 1962, an increase in pectinesterase activity was observed (Rouse *et al.*, 1963).

In mandarin oranges from Coorg, India, Pruthi *et al.* (1960) found the descending order of pectinesterase activity on a wet basis to be peel, pulp, seeds, and juice.

The presence of polygalacturonase in oranges was considered to be unlikely by Sinclair and Jolliffe (1958b) when they were unable to find free galacturonic acid in oranges ripened in storage and in juice allowed to stand until the cloud precipitated. Primo *et al.* (1963c) have since, however, reported the presence of galacturonic acid in Spanish Valencia juice.

Further evidence for the absence of polygalacturonase in citrus fruits was provided by Hobson (1962) who failed to detect any activity in the flesh of the tangerine although he used a highly efficient method of extraction. Mannheim and Siv (1969) also found no polygalacturorase in Shamouti and Valencia oranges, mandarins, and lemons, but detected slight activity in grapefruit.

II. Esterases

The presence in citrus fruits of carboxylic ester hydrolase (esterase) activity was confirmed by Schwartz *et al.* (1964). In lemon, grapefruit, and orange peels they found enzymes capable of hydrolyzing a wide range of esters including acetylcholine bromide and 6-bromo-2-naphthylcarbonaphthoxycholine iodide. These same enzymes were, however, inactive against cucurbitacins, the acetylated bitter principles from Cucurbitaceae. Esterase activity in orange juice was very low. Application of starch gel electrophoresis to the citrus esterase extracts revealed several different enzymes capable of hydrolyzing α-naphthyl acetate.

There was only one cholinesterase, however, which appeared to be the same in the albedos and flavedos of all three fruits, but different from all of the α-naphthylacetylesterases. The probable identity of the cholinesterases in the different citrus tissues was supported by the observation that all were unaffected by diisopropyl fluorophosphate and all slightly stimulated by eserine.

A process using citrus acetylesterase in the preparation of deacetylcephalosporin C derivatives has been patented by Abraham *et al.* (1964).

III. Phosphatases

The presence of phosphatase activity in citrus fruits was first reported by Axelrod (1947) who examined the properties of a preparation from Navel orange juice. More recently, Schormueller *et al.* (1965a,b) studied the phosphatase system in orange peel and found it to be similar but not identical in nature with the orange juice enzyme.

Extraction of orange peel at pH 5 and fractional precipitation with acetone yielded a dry phosphatase preparation which was further separated by chromatography on Sephadex into 14 fractions. These fractions were sufficiently similar in specificity to be regarded as isoenzymes. About half of the total phosphatase activity in the eluates was contained in the final fraction. The enzymes hydrolyzed phosphoric acid mono- and diesters and pyro- and metaphosphates. Phosphoproteins (casein and phosvitin) and phosphopeptides were dephosphorylated to the extent of 75-100%. In tests with four substrates, there appeared to be for most of the enzymes two pH optima, a minor one in the range 3.5-4.3 and a major one at pH 5-6. The optimal temperature was 55°C. It is noteworthy that these phosphatase isoenzymes were strongly inhibited by EDTA. Phosphate esterification by mitochondrial preparations from orange juice vesicles was demonstrated by Vines and Oberbacher (1965) in their studies of the effect of arsenate on citrus metabolism.

IV. Oxidoreductases

The presence of ascorbic acid oxidase in oranges was confirmed by Vines and Oberbacher (1963) who extracted the enzyme from immature whole oranges and assayed the activity by manometric and spectrophotometric methods (Oberbacher and Vines, 1963). About 70% of the enzyme was soluble in 0.1 *M* potassium phosphate buffer at pH 5.6 and the rest remained with the cell-wall fraction. The enzyme was oxygen dependent, while the ratio of oxygen uptake to ascorbic acid oxidized was close to 1:2, which is the stoichiometric relation for oxidation to dehydroascorbic acid. Like ascorbic acid oxidases from other plants, the orange enzyme showed a pH optimum at 5-6 and was inhibited by cyanide and diethyldithiocarbamate. It was specific for ascorbic acid and inactive towards several phenolic substrates, including catechol.

The same workers (Vines and Oberbacher, 1962) surveyed ascorbic acid oxidase activity in several citrus fruits and found the order of decreasing activity to be Marsh grapefruit, Pineapple and Valencia oranges, Thompson grapefruit, Persian limes, and Villafranca lemons. On a fresh weight basis, the ascorbic acid oxidase content was highest in immature citrus fruit and decreased as the fruit matured, although the total ascorbic acid oxidase in the flavedo, reported only for lemons, increased up to maturity. By this time the enzyme had disappeared from the albedo and the juice. Peroxidase activity was also measured in the same fruits and was relatively greater in limes and lemons.

A comprehensive study of the metabolic enzymes in Shamouti oranges sampled from fruit set to maturity by Goren and Monselise (1965a,b) included ascorbic acid oxidase, peroxidase, and catalase activity. Also reported for the first time was indoleacetic acid oxidase activity in citrus fruits. All of these enzymes attained maximum activity during the first 3 months of growth in accord with higher metabolic activity during this period of cell division. In subsequent growth, enzyme activity on the basis of fresh or dry weight generally decreased owing to dilution as the fruit enlarged and to declining metabolic activity. However, the total activity of catalase and indoleacetic acid oxidase per whole fruit increased.

Peroxidase activity in citrus tissues was studied in greater detail by Goren and Goldschmidt (1966), who used four different hydrogen donors—pyrogallol, guaiacol, *p*-methylaminophenol sulfate, and *p*-phenylenediamine—to demonstrate that the concentrations of hydrogen donor and hydrogen peroxide for maximum enzyme activity were different for different hydrogen donors and different citrus fruits. The optimum pH range was 6.0-6.6. When measured under optimal conditions for each fruit, the descending order of activity in the flavedo on a fresh weight basis was lemon, Marsh grapefruit, sour orange, and Shamouti orange. Further illustration of the different behavior of the citrus peroxidase system towards different substrates was provided by a study of fluctuations in the enzymic activity in Shamouti oranges on the tree during 24 hours (Goren and Monselise, 1965b). Peroxidase activity tested with guaiacol showed a marked maximum during the day and a minimum during the night, while peroxidase activity tested with pyrogallol fluctuated in the opposite way.

Catalase and peroxidase activities in juice vesicle homogenates were measured by R. B. Clark and Wallace (1963) and the descending order of activity on a protein basis was Palestine sweet lime, Valencia orange, and Eureka lemon.

In the juices of Egyptian low-acid oranges (pH 6.85) and low-acid lemons (pH

6.10), Ragab (1962a,b) found that ascorbic acid was more stable to atmospheric oxidation than it was in lemon, lime, and Navel orange juices. Ragab attributed this stability to the presence of an L-dehydroascorbic acid reductase. The enzyme required –SH compounds and was inhibited by cyanide, EDTA, and aurintricarboxylic acid. The optimum conditions for activity were pH 7.55 and 42°-45°C.

From the juice vesicles of young oranges Bean *et al.* (1961) isolated an enzyme having the properties of a flavoprotein aerodehydrogenase and capable of catalyzing the oxidation of at least nine sugars (D-glucose, D-galactose, D-mannose, D-2-deoxyglucose, D-2-glucosamine, D-xylose, cellobiose, lactose, and maltose) to the corresponding aldonic acids, with simultaneous formation of hydrogen peroxide. It appeared likely that a single enzyme of broad specificity was involved rather than a family of enzymes. The enzyme showed maximal activity during the period of active cell division in the juice vesicles and disappeared entirely by the time the fruit reached a weight of about 15 g. A similar enzyme was present in the fruits of eight other citrus species and in the trifoliate orange.

Prompted by an interest in the mechanism of biosynthesis of oxygenated flavor components, Bruemmer (1969) examined citrus fruits for the oxidoreductase coenzymes NAD,[1] NADH, NADP, and NADPH. The juice vesicles of oranges and grapefruit respectively contained per 10 g of wet weight the following amounts of these coenzymes: NAD, 118 and 60 nmoles; NADH, 25 and 8 nmoles; NADP, 12 and 9 nmoles; and NADPH 16 and 12 nmoles. Bruemmer concluded that the enzymic capability exists in the juice vesicles to synthesize oxygenated flavor compounds. Moreover, since the NADH:NAD ratio increased with maturity, the coenzyme may exercise a controlling function in the synthesis of citric acid.

V. Nitrate Reductase

Bar Akiva and Sagiv (1967) demonstrated nitrate redutase activity in several citrus plant tissues including the flavedo and albedo of the sour orange. The enzyme resembled nitrate reductase systems in other plants since it was dependent on nitrate and molybdenum and had a specific requirement for NADPH as an electron donor.

[1]The following abbreviations will be used throughout this volume: ADP, adenosine 5′-phosphate; ATP, adenosine 5′-triphosphate; CoA, coenzyme A; EDTA, ethylenediaminetetracetate; NAD, nicotinamide adenine dinucleotide, oxidized; NADH, nicotinamide adenine dinucleotide, reduced; NADP, nicotinamide adenine dinucleotide phosphate, oxidized; NADPH, nicotinamide adenine dinucleotide phosphate, reduced; OA, oxalacetate; PEP, phosphoenolpyruvate; P_i, inorganic phosphate.

VI. Carboxypeptidase

From the flavedo of oranges, lemons, and grapefruit, Zuber (1964) isolated an exopeptidase which he designated carboxypeptidase C since it differed from the carboxypeptidases A and B from pancreas.

The enzyme was extracted from the flavedo with aqueous sodium chloride and purified by precipitation with ammonium sulfate followed by fractionation on Sephadex (Ciba Ltd, 1965; Zuber, 1968). The highest content of carboxypeptidase (3419 units per 100 g wet flavedo) was found in small immature lemons. The molecular weight of the enzyme was estimated to be about 148,500.

Carboxypeptidase C split off from peptides the terminal basic and acidic amino acids including proline, but did not hydrolyze peptides having a terminal amino acid with the D-configuration. The aromatic amino acids, phenylalanine and tyrosine, were released most readily, but glycine only very slowly. The enzyme was stable over the range pH 4-6, and showed optimal activity between 5.3 and 5.7. The optimal temperature range was 30°-40°C and the enzyme was rapidly inactivated at 60°C. Inhibition by diisopropyl fluorophosphate was observed but neither EDTA nor *o*-phenanthroline showed inhibitory activity.

Carboxypeptidase C showed some esterase activity but was different in pH optimum, stability, and inhibition reactions from the acetylesterase found in citrus tissues.

VIII. Enzymes of the Tricarboxylic Acid Cycle

A variety of enzymes catalyzing reactions in the tricarboxylic acid cycle have been found in citrus fruits, both as particulate (mitochondical) and soluble (cytoplasmic) fractions (cf. Chapter 4). Because of the highly acid nature of most citrus fruits the isolation of these enzymes requires experimental methods involving careful use of buffers and special grinding techniques at low temperatures, generally followed by acetone precipitation and ammonium sulfate fractionation.

In homogenates from the juice vesicles and flavedo of the Valencia orange, trifoliate orange, and rough lemon, Huffaker and Wallace (1959) demonstrated the presence of phosphoenolpyruvate (PEP) carboxylase and carboxykinase systems which catalyzed the fixation of carbon dioxide into organic acids according to the following reactions:

$$\text{PEP} + CO_2 \xrightarrow{\text{PEP carboxylase}} \text{OA} + P_i$$

$$\text{PEP} + \text{ADP} + CO_2 \xrightarrow{\text{PEP carboxykinase}} \text{OA} + \text{ATP}$$

As oranges matured, the activity of the two enzymes declined on a fresh weight basis but remained high on a protein basis. The activity of both enzymes from vesicles and flavedo was greater in the trifoliate orange than in the rough lemon on a fresh weight basis but similar on a protein basis, which indicates that there is a difference in the amounts of the enzymes in the two fruits. This observation may have a bearing on the fact that trifoliate orange as a rootstock consistently produces fruit of higher acidity than does rough lemon stock (see Chapter 2).

Later work, however, from the same school (R. B. Clark and Wallace, 1963) cast some doubt on the role of PEP carboxykinase in carbon dioxide fixation and acid synthesis in citrus fruits.

In the Eureka lemon and Tunisian low-acid lemon (limetta), Bogin and Wallace (1966a,b) confirmed the activity of PEP carboxylase and demonstrated carbon dioxide fixation by two other systems: NADP malic enzyme and isocitric dehydrogenase, according to the reactions:

$$\text{Pyruvate} + \text{NADPH} + CO_2 \xrightarrow{\text{Malic enzyme}} \text{malate} + \text{NADP}$$
$$\alpha\text{-Ketoglutarate} + \text{NADPH} + CO_2 \xrightarrow{\text{Isocitric dehydrogenase}} \text{isocitrate} + \text{NADP}$$

Greater activity of PEP carboxylase and the malic enzyme was observed in the soluble than in the mitochondrial fraction, but the latter fraction showed greater isocitric dehydrogenase activity.

Mitochondrial preparations catalyzing oxidative phosphorylation of intermediates in the tricarboxylic acid cycle have been obtained from sweet lemons (Bogin and Erickson, 1965), sour lemons (Bogin and Wallace, 1966b), and grapefruit (Vines and Metcalf, 1967).

Another enzyme isolated from grapefruit by Vines (1968) was NAD malic dehydrogenase which catalyzes the reaction:

$$\text{L-Malate} + \text{NAD} \rightleftarrows \text{OA} + \text{NADH}$$

Active preparations were obtained from both mitochondria and cytoplasm and each was specific for L-malate over D-malate and for NAD and NADH over NADP and NADPH. The grapefruit enzymes and malic dehydrogenases from animal sources were similar as to the Michaelis constant, inhibition by fumarate but not by malonate, and inhibition of the mitochondrial preparation by high concentrations of oxalacetic acid (the soluble preparation was not inhibited).

The citrate condensing enzyme which catalyzes the final step in citric acid synthesis

$$\text{Acetyl CoA} + \text{OA} \longrightarrow \text{citrate} + \text{CoA}$$

was demonstrated to be present in the peel and pulp of oranges and lemons by Srere and Senkin (1966).

When the distribution of radioactivity in lemons following fixation of $^{14}CO_2$ was examined by R. E. Young and Biale (1968), the first labeled products to appear were malic, citric, and aspartic acids, but these acids were presumed to have been formed from labeled oxalacetic acid by rapid single-step reactions catalyzed by malic dehydrogenase, citrate condensing enzyme, and oxalacetic-aspartic transaminase.

Chapter 9 Pigments

I. Changes during Maturation

Immature citrus fruits are characteristically green, because of the presence of chlorophyll in the flavedo. The green flavedo is therefore capable of photosynthetic fixation of carbon dioxide which yields products similar to those resulting from photosynthesis in leaves (Bean and Todd, 1960; Bean *et al.*, 1963). Photosynthesis in the fruit does not, however, make a significant contribution to its own nutrition (Todd *et al.*, 1961). At a certain stage in the development of the fruit, known as the color break, the flavedo begins to change in color from green to yellow or orange (Jahn and Sunday, 1965), because chlorophyll is lost and carotenoid pigments appear.

During about 2 months following the color break, chlorophyll a in the Navel orange peel was found by Lewis *et al.* (1964) to fall in concentration from 4.1 to 1.0 $\mu g/cm^2$ and chlorophyll b from 1.2 to 0.3 $\mu g/cm^2$. Although there is a net loss of chlorophyll, synthesis of chlorophyll evidently continues for a period after the color break and is stimulated by treatment with gibberellate (Coggins and Hield, 1962; Ismail, 1966), but retarded by low light intensity.

In the Valencia orange, which is subject to regreening, gibberellate applied to fully colored fruit stimulates the synthesis of protochlorophyll and chlorophyll a, and to a lesser extend chlorophyll b (Coggins and Lewis, 1962). When applied to green oranges, gibberellate retards the destruction of chlorophyll (Ismail *et al.*, 1967).

Loss of chlorophyll as citrus fruits ripen is accompanied by the accumulation of carotenoid pigments. For example, in the peel of Navel oranges the concentration of total carotenoids more than doubled from the color break (October in California) to March after which it leveled off (L. N. Lewis and

Coggins, 1964). Treatment with gibberellate just prior to the color break retarded carotenoid accumulation and lowered the level ultimately reached by about one-third, so that the peels of treated fruit were obviously pale in color. Carotenoid synthesis appeared to cease at the same point in the season for both treated and untreated fruit, rather than at a particular carotenoid level. Examination of the carotenoid pigments by thin-layer chromatography on silica gel showed that the effect of gibberellate was to inhibit the synthesis of all carotenoids rather than to change the relative concentration of individual pigments.

The case was different, however, for the influence of temperature on carotenoid synthesis. Accumulation of xanthophyll pigments in Valencia orange peels-leading to a bright orange color-was encouraged by low temperatures in the daytime air (20°C), the nighttime air (7°C), and the soil (12°C) whereas the carotene pigments were not influenced in this way (L. B. Young and Erickson, 1961).

In the juice as well as in the peel of oranges the concentration of total carotenoids increases with maturity, as I. Calvarano (1961) reported for Sicilian oranges, and Bowden (1968) for Valencia and Joppa oranges in Australia. Bowden also found for both these varieties that oranges on trifoliate orange and sweet orange stocks had significantly higher concentrations of total carotenoids in the juice than oranges on rough lemon stocks.

In Italian tangerine juice the total carotenoid and carotenoid ester contents increased from 11.7 and 3.5 mg/liter in November to maxima of 18.1 and 11.2 mg/liter respectively, in January, and then declined, while the β-carotene content increased steadily from 0.08 to 1.4 mg/liter in February (Di Giacomo *et al.*, 1968c).

Chemical spray treatments of citrus trees intended to encourage abscission of the fruit as an aid to mechanical harvesting may also influence pigment changes in the peel. Preliminary studies by Cooper *et al.* (1968) indicated that sprays of ascorbic acid and abscisic acid applied to Robinson tangerines and Temple and Murcott tangors accelerated the loss of chlorophyll from the peel and increased the carotenoid content of both the flavedo and the juice as compared with control treatments. The authors believe that the primary effect of the sprays was to increase ethylene production by the fruit and that ethylene in turn accelerated the pigment changes.

II. Oranges

The comprehensive studies of the carotenoids of citrus fruits by A. L. Curl and G. F. Bailey, based on separation of the pigments by countercurrent distribution and column chromatography (generally after saponification of the xanthophyll esters) were continued with an examination of some Florida Valencia orange juices and other processed products (Curl and Bailey, 1959).

As might be expected from the generally paler color of Florida juices, the total carotenoid content (about 16 mg/liter) was lower than in California Valencia orange juice (about 24 mg/liter), but the mixture of carotenoids present was similar in composition (Table X). Two minor constituents, lutein-5,6-epoxide and flavoxanthin, the isomeric 5,8-epoxide, had not been detected in California juice.

The same workers went on to survey the pigments in the pulp and peel of California Navel oranges (Curl and Bailey, 1961). Again the composition of the carotenoid mixture (Table X) was generally similar to that of Valencia oranges with some quantitative differences. Violaxanthin and its isomerization artifacts, the luteoxanthins, were the principal pigments in both the pulp and the peel. Phytoene, phytofluene, and ζ-carotene were the main hydrocarbons, the amounts of α- and β-carotenes being very low. *trans*-Neoxanthin (5) (Curl, 1965a) was found for the first time in a citrus fruit and the interesting pigment, reticulataxanthin (see Section VI), was also present.

III. Grapefruit

In the "white" Marsh variety of grapefruit, the flavedo again contained a highly complex mixture of carotenoid pigments (Table X), although the color was due mainly, as in other citrus fruits, to violaxanthin, cryptoxanthin, and reticulataxanthin (H. Yokoyama and White, 1967).

Carotenoid synthesis occurs in grapefruit peel, and the yellow color appears before all the chlorophyll has disappeared. As the fruit continues to mature, however, carotenoid synthesis evidently ceases, since the total content of colored carotenoids declined from 2.3 to 1.5 mg/kg (as β-carotene) between early and midseason. During the same period the colorless hydrocarbon, phytoene, accumulated from a trace to 50% of the total carotenoids and, the two colorless polyenes, phytoene and phytofluene, together accounted for 75% of the total carotenoids. Since these compounds are recognized intermediates in the biosynthesis of the colored carotenoids, it appears that in the white grapefruit the subsequent dehydrogenation steps in the biosynthetic sequence are blocked.

There is a further metabolic anomaly in the mutant "red" varieties of grapefruit, which owe their distinctive color mainly to the presence of lycopene, although β-carotene is also present. Both white and red varieties are able to synthesize these pigments, but the red varieties accumulate 200-400 times as much as the white varieties. The biochemical difference leading to this accumulation of pigments is evidently located in the fruit itself since fruits from

TABLE X CAROTENOID CONSTITUENTS OF CITRUS FRUITS[a]

	Valencia orange[b]	Navel orange[c]		Marsh grapefruit[d]		Eureka lemon[e]				Meyer lemon[f]		Sinton citrangequat[g]
						Mature-green		Yellow				
	Juice	Pulp	Peel	Early flavedo	Midseason flavedo	Pulp	Peel	Pulp	Peel	Pulp	Peel	Flavedo
Hydrocarbons	-	-	-	-	-	-	-	-	-	9	7	-
Phytoene	1	4	12	tr	51	-	-	-	-	31[h]	57[h]	-
Phytofluene	1	4	7	31	24	19	17	22	18	20[h]	22[h]	-
ζ-Carotene	1	9	8	-	2	-	-	4	17	32[h]	20[h]	1
Neurosporene	-	-	-	0.5	0.3	-	-	-	-	-	-	1
η-Carotene	-	-	-	5	-	-	-	2	3	-	-	6
β-Zeacarotene	-	-	-	4	-	-	-	-	-	-	-	5
α-Carotene	1	0.1	tr	1	-	6	9	-	-	0.5	-	0.2
β-Carotene	1	0.5	0.2	3	1	4	13	7	7	16	2	4
Neo-β-carotene-U	-	-	-	-	-	-	-	2	-	-	-	-
γ-Carotene	-	-	-	-	-	-	-	-	-	-	-	1
δ-Carotene	-	-	-	1	-	-	-	-	-	-	-	-
Epoxides	-	-	-	-	-	-	-	-	-	-	-	-
5, 6-Epoxy-β-carotene	-	-	-	-	-	-	-	14	-	-	-	-
Mutatochrome	-	-	-	3	-	-	-	2	-	-	-	2
Monols	-	-	-	-	-	-	-	-	-	51	18	-
Phytoenol	-	-	-	-	-	-	-	-	-	-	14[i]	-
Phytofluenol	-	-	-	-	-	-	-	-	-	-	5[i]	-
Hydroxy-ζ-carotene	-	-	-	-	-	-	-	-	1	-	3[i]	-
3-Hydroxy-α-carotene	3	0.5	0.2	-	-	3	3	2	3	-	1[i]	-
Cryptoxanthin	5	10	3	-	2	15	12	30	10	-	23[i]	0.1
Rubixanthin	-	-	-	-	-	-	-	-	-	-	1[i]	-

Monol epoxides	-	-	-	-	-	-	-	-	-	7	36	-
3-Hydroxy-5, 6-epoxy-α-carotene	-	-	0.1	-	-	-	-	-	-	-	1[i]	-
3-Hydroxy-5, 8-epoxy-α-carotene	-	-	0.1	-	-	-	-	-	1	-	-	-
5, 6-Epoxycryptoxanthin	-	0.4	1	-	0.5	1	3	-	-	-	27[i]	-
Cryptoflavinlike	-	0.4	5	-	1	6	1	3	13	-	4[i]	-
5,6,5′,6′-Diepoxycryptoxanthin	-	0.2	0.5	-	-	-	-	2	0.2	-	8[i]	-
5,6,5′,8′-Diepoxycryptoxanthin	-	-	tr	-	-	-	-	-	1	-	2[i]	-
5,8,5′,8′-Diepoxycryptoxanthin	-	-	-	-	-	-	-	1	2	-	0.3[i]	-
Diols	-	-	-	-	-	-	-	-	-	6	6	-
Zeaxanthin	9	2	1	-	1	-	2	-	-	-	-	1
Lutein	7	1	1	-	2	7	10	-	-	-	-	-
Diol monoepoxides	-	-	-	-	-	-	-	-	-	5	3	-
5,6-Epoxylutein	0.3	-	-	-	-	-	-	-	-	-	-	-
Flavoxanthin	0.1	-	0.2	-	-	-	-	-	-	-	-	-
Antheraxanthin	12	12	3	-	1	-	-	-	-	-	-	-
Mutatoxanthin	7	0.5	1	-	1	-	-	-	-	-	-	-
Diol diepoxides	-	-	-	-	-	-	-	-	-	18	26	-
Violaxanthin	15	46	26	-	6	12	4	1	10	-	-	-
Luteoxanthin	13	3	22	-	5	6	1	-	-	-	-	-
Auroxanthin	4	-	0.3	-	-	10	9	9	1	-	-	-
Polyols	-	-	-	-	-	-	-	-	-	3	4	-
Neoxanthin	-	0.5	-	-	-	1	4	-	-	-	-	-
Trollixanthin	10	1	1	-	-	-	-	-	-	-	-	-
Trollichrome	3	0.2	-	-	-	-	-	-	-	-	-	-
Trolliflor	-	0.4	-	-	-	-	-	-	-	-	-	-
Trollein	-	-	-	-	-	8	4	-	-	-	-	-
Sinensiaxanthin	2	2	3	-	-	-	-	-	-	-	-	-
Sinensiachrome	1	-	-	-	-	-	-	-	-	-	-	-
Valenciaxanthin	2	2	1	-	-	-	-	-	-	-	-	-
Valenciachrome	0.5	0.3	1	-	-	-	-	-	-	-	-	-

TABLE X *continued*

	Valencia orange[b]	Navel orange[c]		Marsh grapefruit[d]		Eureka lemon[e]				Meyer lemon[f]		Sinton citrangequat[g]
						Mature-green		Yellow				
	Juice	Pulp	Peel	Early flavedo	Midseason flavedo	Pulp	Peel	Pulp	Peel	Pulp	Peel	Flavedo
Aldehydes												
10-Apo-β-carotenal	-	-	-	-	-	-	-	-	-	-	-	2
8-Apo-β-carotenal	-	-	-	-	-	-	-	-	-	-	-	1
Citraurin	-	-	1.3[j]	-	-	-	-	-	-	-	-	3
Ketones												
Citranaxanthin (C)	-	-	-	-	-	-	-	-	-	-	-	11
Reticulataxanthin (R)	-	-	0.3(4.9)[j]	-	2	-	-	-	-	-	-	50
8-Hydroxy-7′,8′-dihydro-(C)	-	-	-	-	-	-	-	-	-	-	-	1
8-Hydroxy-7′-8′-dihydro-(R)	-	-	-	-	-	-	-	-	-	-	-	1
Sintaxanthin	-	-	-	-	-	-	-	-	-	-	-	0.1
3-Hydroxysintaxanthin	-	-	-	-	-	-	-	-	-	-	-	2
Total carotenoids (as β-carotene, mg/kg wet wt)	16[k]	23	67	2.3	1.5	1.1	2.1	0.6	1.4	2.4	5.6	-

[a]Except where otherwise indicated the values are approximate percentages of the total carotenoids; tr = trace amount.
[b]From Curl and Bailey (1959), Florida fruit.
[c]From Curl and Bailey (1961), California fruit.
[d]From H. Yokoyama and White (1967), California fruit.
[e]From H. Yokoyama and Vandercook (1967), California fruit.
[f]From Curl (1962a), California fruit.
[g]From H. Yokoyama and White (1966b), California fruit.
[h]Percentage of hydrocarbon fraction.
[i]Percentage of monol and monol epoxide fraction.
[j]From Curl (1965b).
[k]In milligrams per liter.

a red variety grafted on to "white" trees continued to synthesize lycopene and β-carotene, while fruits from a white variety grafted on to "red" trees accumulated no more pigment than normal white fruits (Purcell and Stephens, 1959; cf. Cameron *et al.*, 1964).

There are several red bud sports of grapefruit which increase in intensity of visual color and in lycopene content from the pink Thompson to the red Ruby Red and Burgundy varieties (Ting and Deszyck, 1958; Oberbacher *et al.*, 1960). The lycopene content declines with advancing maturity, whereas the β-carotene content increases to a maximum then falls in late maturity.

To explain the changes in lycopene content in red grapefruit, Purcell *et al.* (1968) postulated that there are two metabolic systems of lycopene synthesis and destruction operating simultaneously during maturation. Lycopene destruction appears to be initiated when the fruit reaches a certain age (Purcell and Schultz, 1964) and gradually increases in activity. High temperatures favor the accumulation and retention of lycopene in grapefruit, by encouraging synthesis or inhibiting destruction; low temperatures favor the loss of lycopene. The metabolism of β-carotene, however, is not influenced by temperature, as L. B. Young and Erickson (1961) previously observed in studies on the Valencia orange.

Changes in cell morphology are associated with coloration in red grapefruit (Purcell *et al.*, 1963). When lycopene is the major pigment, it is present in chromoplasts that are needlelike in form; but as the β-carotene content increases the chromoplasts become similar to platelets. The chromoplasts may be formed by deposition of carotenoids into bodies like the chromoplasts derived from fragmented chloroplasts. No structures resembling chromoplasts are found in white grapefruit.

There is also a pink bud sport of the Shamouti orange, the Sarah variety, which contains not anthocyanins like the blood oranges (see Chapter 13), but lycopene like red grapefruit (Monselise and Halevy, 1961). Although other carotenoids are present, β-carotene seems to be absent. The pink color is evident mainly in the inner albedo, the carpel walls, and the vesicle stalks, and dark-yellow to pink plastids are visible especially around the main conducting bundles. These observations together with those on red grapefruit suggest that a carotenoid precursor enters the fruit through the vascular system and diffuses into the surrounding parenchyma to produce the red coloration.

Earlier, Winterstein *et al.* (1960) had found the pigment bisdehydrolycopene in very small amounts (50 μg in 100 kg) in Spanish blood oranges. This pigment, which contains 15 conjugated double bonds, is the most unsaturated of all the naturally occurring carotenoid hydrocarbons and may be presumed to be derived from lycopene by successive dehydrogenations.

IV. Lemons

Although the lighter color of the peel and pulp of lemons as compared with oranges is mainly due to the much lower concentrations of total carotenoid pigments, there are also some interesting differences in the nature of the pigments as found by H. Yokoyama and Vandercook (1967) in a comprehensive study of mature green and yellow Eureka lemons (Table X). The carotenoid content decreased on ripening whether the lemons were ripened on or off the tree.

Cryptoxanthin was the most abundant pigment in the pulp and peel, as in other citrus fruits, while the colorless phytofluene (probably the *cis* form) was the principal hydrocarbon. Several carotenes were present in smaller amounts; in the yellow peel ζ-carotene became most abundant, and two unusual compounds appeared—neo-β-carotene-U and η-carotene (octahydro-β-carotene). Other unusual pigments were β-carotene-5,6-monoepoxide and neoxanthins a and b (5) (Curl, 1965a). The latter pigments and lutein, which are usually found in photosynthetic tissues, disappeared when the chlorophyll disappeared.

Pigments present in the peel of lemons which showed the skin condition known as "bronzing" are discussed in Chapter 9, Section VI.

V. Mandarins and Tangerines

The surveys of the carotenoid pigments of tangerines by Curl and Bailey were discussed in our previous review (Kefford, 1959). Curl (1962b, 1965b) subsequently returned to the study of these fruits because of interest in certain pigments not identified in their earlier work.

One pigment, tentatively designated as capsanthin-like or hydroxycanthaxanthin-like, was later named reticulataxanthin (6, R = OH) (Curl and Bailey, 1961) and proved to be a methyl ketone carotenoid (see Section VI). Another much less abundant pigment was named tangeraxanthin (Curl, 1962b) and tentatively assigned a structure (7) which included a terminal methylketone group in the conjugated side chain as well as an allylic 3-hydroxyl group.

A major pigment in the peels of Dancy tangerines and Clementine mandarins (amounting to 28% and 10% of the total carotenoids, respectively), was the aldehyde, β-citraurin, which is probably a natural degradation product of zeaxanthin. Present in small amounts was 8′-apo-β-carotenal (8) which may be a degradation product of β-carotene; this compound was also found in the juice and peel of Satsuma mandarins by Koch and Haase-Sajak (1965b).

The above aldehydes as well as three others—2′-apo-β-carotenal, 10′-apo-β-carotenal, and retinal—had been found earlier in Sicilian oranges and Clementines by Winterstein *et al.* (1960). Subsequently Thommen (1962) estimated the

8′-apo-β-carotenal contents of Mediterranean orange peels and juices to be 90-340 and 5-20 μg/100 g, respectively; there appeared to be no correlation between the concentrations in the peel and the juice. Curl (1965b), however, failed to detect 8′-apo-β-carotenal in the peels of California Navel and Valencia oranges.

Synthetic 8′-apo-β-carotenal (8) is now available as an orange-red food color (Bauernfeind and Bunnell, 1962); it has approximately three times the pigmenting intensity of β-carotene and about 70% of its vitamin A activity.

Tangerine peels also contained a new pigment tentatively identified as 12′-apo-violaxanthal (9), the first naturally-occurring carotenoid aldehyde with a 5,6 epoxide group (Curl, 1965b). Later, when examining pigment extracts from Valencia orange peel by countercurrent distribution and chromatography without prior saponification, Curl (1967) isolated minor amounts of 10′-apo-violaxanthal, which was present as a fatty acid ester of the 3-hydroxyl group.

In the Nagpur mandarin, Subbarayan and Cama (1965) found 15 carotenoid pigments; cryptoxanthin was the principal pigment in the pulp followed by cryptoflavin and prolycopene. The ratio of oxygenated carotenoids to hydrocarbons was 7:3; in the peel this ratio was 40:3 and violaxanthin, cryptoxanthin, and neoxanthin were the most abundant pigments.

VI. Hybrids

The Meyer lemon contains a mixture of carotenoid pigments that is distinctive in composition in several respects (Table X) and contains some unusual constituents (Curl, 1962a). The total carotenoid contents in the pulp and peel, 2.4 and 5.6 mg/kg, respectively, are only about one-tenth of those of the Valencia orange. Among the hydrocarbons, the colorless polyenes–phytoene and phytofluene–predominated with ζ-carotene as the most abundant carotene. There was an unusually high proportion of monols and monolepoxides, amounting to about 50% of the pigments in both pulp and peel, which is higher than previously reported for any other citrus fruit. Cryptoxanthin and its epoxides were the principal pigments. Compounds not previously found in citrus fruits were phytoenol (hydroxyphytoene), phytofluenol (hydroxyphytofluene), hydroxy-ζ-carotene, and an unusual polyene. The latter contained a conjugated system of about five double bonds, and two allylic hydroxyl groups, one of which was allylic to the conjugated system.

A Sicilian variety of low-acid lime called "dolce Romana," which is probably a hybrid of lime and low-acid lemon, was found by Benk and Bergmann (1963a) to have 16-37% of carotenes in the total carotenoids of the peel.

The Sinton citrangequat is a trigeneric hybrid with an orange-to-red fruit much more deeply colored than the fruits of its parents, the Rusk citrange

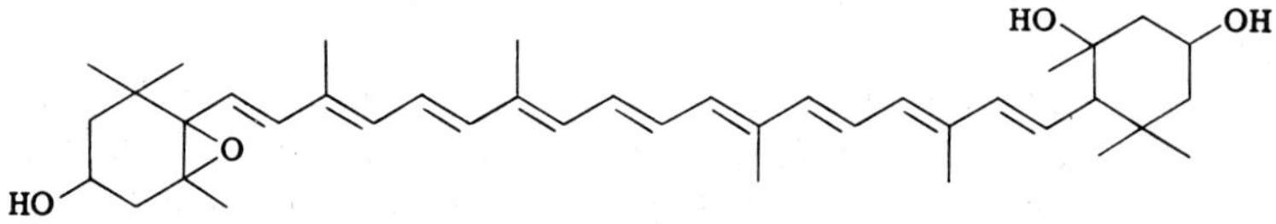

Neoxanthin
(5)

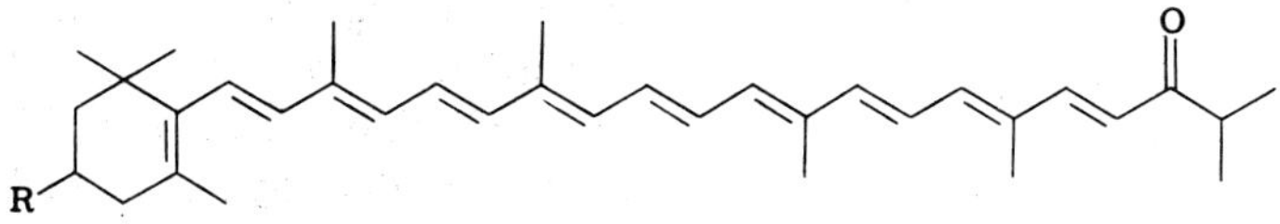

Citranaxanthin, R = H; Reticulataxanthin, R = OH
(6)

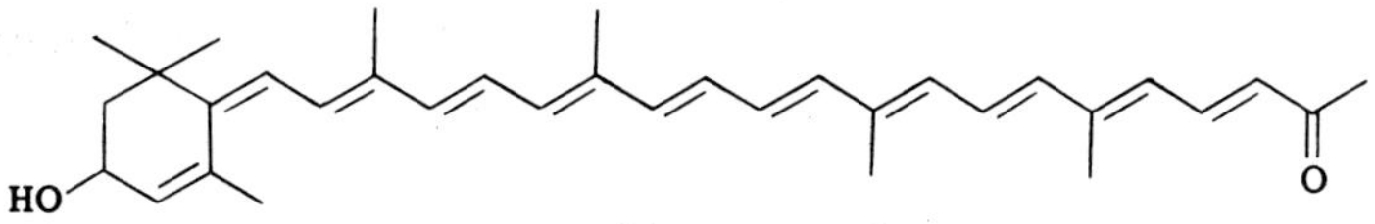

Tangeraxanthin
(7)

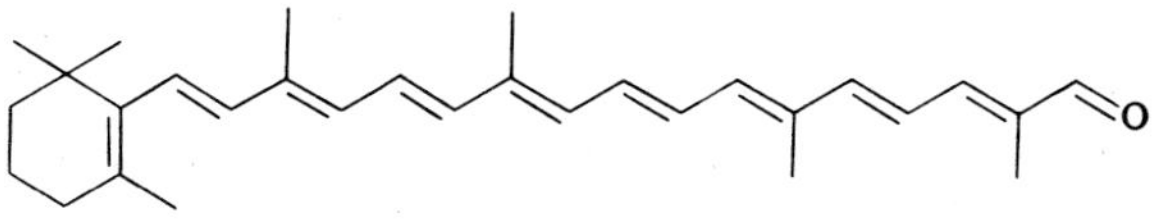

8′-Apo-β-carotenal
(8)

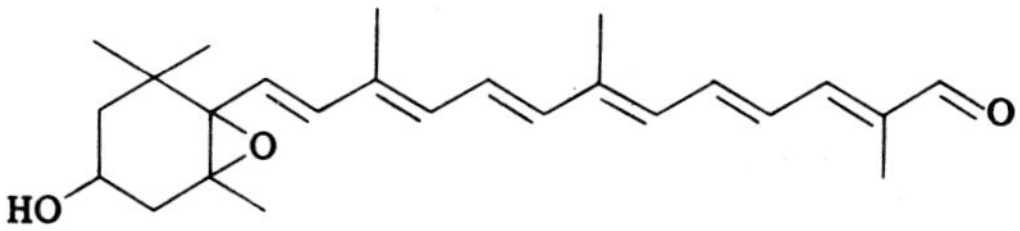

12′-Apo-β-violaxanthal
(9)

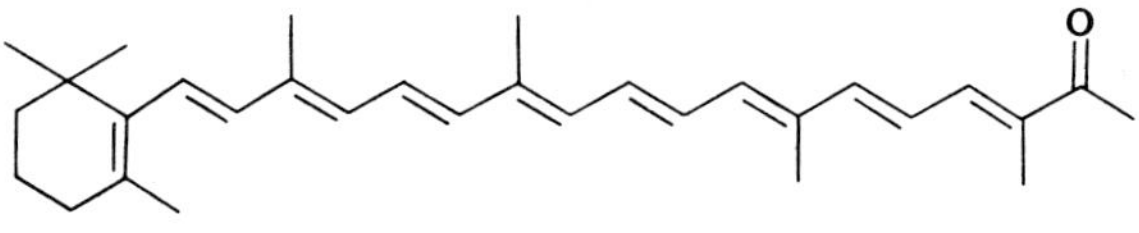

Sintaxanthin
(10)

(Poncirus trifoliata × *Citrus sinensis)* and the oval kumquat *(Fortunella margarita).* The deep color is due largely to the presence of several unusual pigments which are apo-carotenones containing a side chain terminating in a methylketone group (H. Yokoyama and White, 1966b). Most of these compounds have been found in no other natural product (Table X). Several carotenoid aldehydes are also present and among the carotene hydrocarbons is β-zeacarotene, which has not been reported in any other citrus fruit.

The principal pigment, making up almost 50% of the total carotenoids in the peel of the citrangequat, was reticulataxanthin (6, R = OH), which was proved by H. Yokoyama *et al.* (1965) to contain a decaeneone chromophore terminating in a methylketone group. The next most abundant pigment was citranaxanthin (6, R = H) of which reticulataxanthin is the 3-hydroxy derivative (H. Yokoyama and White, 1965a). Other methylketone carotenoids present in smaller amounts were sintaxanthin (10), which has a nonaeneone chromophore (H. Yokoyama and White, 1965b), 3-hydroxysintaxanthin, 8-hydroxy-7′,8′-dihydrocitranaxanthin (H. Yokoyama and White, 1966a), and the corresponding reticulataxanthin derivative.

Reticulataxanthin was also a major pigment in the peel of another hybrid, the Minneola tangelo (H. Yokoyama *et al.*, 1965). Hybridizing among some citrus species appears to promote biosynthetic mechanisms leading to the formation of methylketone carotenoid pigments.

It may be mentioned here that the red fruits of two species, *Murraya exotica* and *Triphasia trifolia,* which are rather remotely related to *Citrus,* have been found to contain other carotenoid ketones never encountered previously in nature. These compounds are the monocyclic diketone, semi-β-carotenone, and the acyclic tetraketone, β-carotenone, both partial oxidation products of β-carotene (H. Yokoyama and White, 1968b).

The chemical nature of these carotenoid ketones found in rather uncommon citrus hybrids and relatives might be considered to have only academic interest but for the fact that they have very recently been identified in the pigments responsible for the skin discoloration in lemons known as "bronzing." In lemons showing pink bronzing, H. Yokoyama and White (1968c) found reticulataxanthin and semi-β-carotenone, and in lemons with a more golden color 3-hydroxysintaxanthin was also present.

VII. Analytical Aspects

Analytical interest in citrus pigments arises partly from attempts to use these constituents as index compounds for the determination of juice content in citrus products, and partly from the desire to detect adulteration with synthetic carotenoids (Bauemfeind *et al.*, 1962) or adulteration of one kind of citrus juice with another (I. Calvarano, 1963c).

The proportion of carotene hydrocarbons in the total carotenoid pigments is one analytical relation that has been studied. Doubt was cast by W. K. Higby (1962) upon many of the published values for the concentration of carotenes in orange juices because the analytical methods made inadequate provision for separation of the carotenes from esterified oxygenated xanthophylls. Higby devised a simple quantitative procedure in which two separate samples of juice are extracted for the estimation of total carotenoids and carotenes, respectively, using a foaming solvent technique with different mixtures of isopropyl alcohol and hexane. The absorbance of the hexane layer is a measure of total carotenoid content, and the carotenes are similarly estimated after saponification and column chromatography on magnesia. Both fractions are expressed in terms of β-carotene standards although ζ-carotene is commonly the dominant carotene present.

Using this method, W. K. Higby (1962, 1963) found in the juices of a number of California orange and mandarin varieties total carotenoid contents ranging from 3.2 to 35 mg/liter and carotene contents from nil to 3.3 mg/liter as β-carotene. Only in Kara mandarin juice did the ratio (9.6) of total carotenoids to carotenes fall below 10; Higby therefore concluded that the finding of more than 10% of carotenes in the total carotenoids of an orange juice product indicated addition of β-carotene.

Higby's results generally agree with those of workers in other regions. For instance, his ranges encompass most of the values reported for orange juices from Calabria and Sicily (I. Calvarano, 1961; M. Calvarano, 1962; Di Giacomo *et al.*, 1968b), Florida (Ting, 1961), Greece (Catsouras, 1965), Spain and South Africa (Benk, 1961c), and Australia (Bowden, 1968).

The total carotenoid content of lemon juices was examined by Vandercook and Yokoyama (1965) as a possible index of authenticity. While the ranges of values in American (0.29-0.74 mg/liter) and Italian (0.24-0.75 mg/liter) juices were very similar, with average values about 0.45 mg/liter, the correlation (r = 0.290) between total carotenoid content and citric acid content was not close enough to provide a useful index.

Another analytical relation that has been studied in citrus juices is the proportion of carotenoid esters in the total carotenoids. In the juices of mandarins, tangerines, and Clementines from Mediterranean countries and Florida, the carotenoid esters made up 30.5 to 51.9% of the total carotenoids (Benk and Seibold, 1966; Koch and Haase-Sajak, 1965b; Di Giacomo *et al.*, 1968b) whereas in Mediterranean oranges the proportion was only 9.8 to 15.9%. On the basis of this comparison, Benk and Seibold (1966) considered that an ester content exceeding 20% of the total carotenoids in an orange juice product pointed to adulteration with mandarin or tangerine juice. However, Di Giacomo *et al.* (1968b) demonstrated that heat processing of orange juice, and specifically concentration by evaporation, increased the percentage of esters in the total carotenoids to levels above 30%, and accordingly the proportion of esterified carotenoids was not a reliable indication of adulteration in processed orange products. Another possibility for the detection of the addition of mandarin juice to orange juice, suggested by Koch and Haase-Sajak (1965b), is estimation of the cryptoxanthin content, which is much higher in mandarin juices (Table X).

For the separation of added carotenoid colors from natural pigments in citrus products, thin-layer chromatography is a useful technique (Benk *et al.*, 1963), although it appears to be necessary to go to two-dimensional chromatography for good results. For instance, Primo and Mallent (1966) demonstrated clean separations of added 8′-apo-β-carotenal, methyl 8′-apo-β-carotenoate, and canthaxanthin from the natural pigments in orange juice on two-dimensional chromatograms on silica gel.

Chapter 10 Lipids

The fine cloud of suspended material in commercially extracted orange juice was found by Scott *et al.* (1965) to consist of 25% lipids and 75% alcohol-insoluble solids (Table V), which led them to conclude that this cloud was derived mainly from the contents of the juice cells and not from the lipid-deficient structural tissues of the orange. A detailed study of the juice lipids from oranges, grapefruit, lemons, and limes by Nordby and Nagy (1969) revealed a large number of glyceride fractions and from 80 to 100 different fatty acids including iso and anteiso branched acids.

The principal accumulation of lipids in citrus fruits occurs in the seeds, however, which contain oils in which the glycerides of unsaturated fatty acids predominate. But in spit e of the current dietary interest in unsaturated fats, citrus seed oils are not widely utilized; it has been stated that at least 10 million lb are available as by-products of citrus processing in Florida alone. Citrus seed oils are often bitter owing to the presence of limonoid bitter principles (see Chapter 14), but they may be refined to bland products by treatment with alkali.

During the maturation of citrus fruits, the moisture content of the seeds decreases to a level of 50-55% of the fresh weight at maturity while the oil content increases to 35-45% of the dry weight of the seeds (Hendrickson and Kesterson, 1961, 1963a).

Information on the composition of citrus seed oils reported in the last 10 years is given in Table XI. The most active workers in the field have been R. Hendrickson and J. W. Kesterson of the Citrus Experiment Station, Lake Alfred, Florida. Their values for the proportions of component fatty acids in citrus seed oils, determined by gas chromatographic separation of the methyl esters, may be regarded as superseding earlier data obtained by less selective analytical procedures. Among the more than 60 different fatty acids in the lipids of citrus

TABLE XI CHARACTERISTICS OF CITRUS SEED OILS

Fruit	Growing region	Refractive index n^{25}_D	Iodine number	Component fatty acids (% by wt. of total fatty acids)				
				Palmitic	Stearic	Oleic	Linoleic	Linolenic
Lemon[a]	Italy	1.4742	109	27.3	5.7	11.6	38.5	10.3
Lemon (39 varieties)[b]	Florida	1.4707	109.9	23.2	3.7	29.8	32.5	10.9
Lime (Key)[b]	Florida	1.4701	106.2	29.4	4.5	20.6	35.1	10.4
Lime (Persian)[b]	Florida	1.4699	101.6	28.8	4.9	22.1	39.1	5.1
Tangerine (Dancy)[c]	Florida	1.4700	103.7	31.6	3.3	20.5	40.0	4.6
Mandarin (King)[c]	Florida	1.4695	101.1	31.7	3.0	18.2	43.9	3.0
Mandarin (Satsuma)[c]	Florida	1.4695	99.7	28.8	5.0	22.9	40.0	3.5
Clementine[c]	Florida	1.4697	100.6	31.5	3.9	21.0	40.8	2.9
Mandarin[d]	Assam	1.4693	94.1	36.1		24.5	36.0	3.5
Grapefruit (Marsh)[e]	Florida	1.4696	101.0	36.4	3.1	19.5	36.4	4.6
Grapefruit (Duncan)[e]	Florida	1.4696	99.5	35.9	3.4	20.6	35.4	4.7
Grapefruit (Walters)[e]	Florida	1.4695	99.6	34.8	3.8	20.6	36.3	4.5
Grapefruit (Excelsior)[e]	Florida	1.4695	97.8	34.4	4.8	21.2	34.8	4.8
Pummelo (Siamese)[e]	Florida	1.4692	99.4	26.0	4.9	27.2	37.0	4.9
Orange (Hamlin)[f]	Florida	1.4688	95.1	31.7	4.1	24.4	36.6	3.1
Orange (Parson Brown)[f]	Florida	1.4688	93.6	30.9	4.1	24.9	36.6	3.4
Orange (Pineapple)[f]	Florida	1.4687	93.0	30.0	4.6	25.0	36.8	3.7
Orange (Valencia)[f]	Florida	1.4686	92.9	31.8	3.2	26.4	35.9	2.6
Orange[g]	Ceylon	-	91.6	21.8	6.4	27.4	34.2	2.1

Hybrids:								
Calamondin[c]	Florida	1.4701	104	24.4	4.6	31.0[h]	29.8	5.2
Tangelo[c]	Florida	1.4699	103.2	27.2	4.3	21.6	42.7	4.4
Temple tangor[c]	Florida	1.4695	100.6	25.9	3.8	27.2	39.3	3.8
Murcott tangor[c]	Florida	1.4695	99.9	31.3	2.0	20.8	41.3	4.5
Ponderosa[b]	Florida	1.4682	91.7	23.8	3.7	50.1	18.7	3.7

[a]From Franguelli and Mariani (1959) who also reported 0.5% of arachidic plus lignoceric acids.
[b]From Hendrickson and Kesterson (1963b).
[c]From Hendrickson and Kesterson (1964a).
[d]From Chaliha *et al.* (1963c).
[e]From Hendrickson and Kesterson (1961).
[f]From Hendrickson and Kesterson (1963a).
[g]From Weerakoon (1960) who also reported 6.8% of myristic acid and 0.6% each of lauric and arachidic acids.
[h]Also contained 5.0% of palmitoleic acid.

seeds, about half are iso and anteiso branched chain acids present in proportions from 0.001 to 0.2% (Nordby and Nagy, 1969).

Within each kind of citrus fruit the seed oils of different varieties are generally very similar in fatty acid composition. Lemon and lime seed oils are high in linolenic acid and show the highest refractive indexes and iodine numbers; mandarin seed oils tend to be high in linoleic acid and are next in order in refractive index and iodine number; grapefruit seed oils follow; and orange seed oils are lowest in refractive index and iodine number (Hendrickson and Kesterson, 1963a). The Ponderosa fruit is very different from the true lemons in the composition of the seed oil, which is low in iodine number and has a notably high content of oleic acid (Hendrickson and Kesterson, 1963b).

During maturation, citrus seed oils show little change in composition except for a tendency toward increasing unsaturation in grapefruit and mandarin oils with advancing fruit maturity (Hendrickson and Kesterson, 1961, 1964a).

Shigeyama and Murakami (1962) recovered 37% oil from the seeds of *Fortunella;* the seed meal contained 45% protein.

Chapter 11 Volatile Flavoring Constituents

In this section, perhaps more than in any other in the present survey, spectacular progress has been made by the imaginative application of a wide range of separation and identification techniques. As a result, an astonishing array of volatile compounds has been identified in citrus fruits (Table XII).

Apart from the intrinsic interest of the compounds identified, these investigations tell a fascinating story of the evolution of techniques towards greater subtlety of separation and certainty of identification. In the early stages investigators merely counted peaks and relied on gas chromatographic retention values for tentative identifications. But very soon they were satisfied with nothing less than unequivocal identification, based on evidence from chemical tests, and physical methods which included ultraviolet and infrared spectroscopy, nuclear and proton magnetic resonance, and mass spectrometry (McFadden *et al.*, 1963). At the same time, sharper and sharper resolution of constituents was achieved by refinement of separation methods involving spinning band and molecular distillation; paper, column, and thin-layer chromatography; and notably gas chromatography, where techniques have been continually perfected by developments in column technology, stationary phases, detectors, and the programming of column temperature and carrier gas pressure.

In his pioneering applications of gas chromatography to citrus oils, Bernhard (1958b) found only five peaks in California lemon oil; within 2 years, however, he had separated 32 peaks and tentatively identified 25 of them (Bernhard, 1960). By first absorbing the oxygenated compounds on silicic acid to separate them from the terpenes, and then subjecting the separate fractions to gas chromatography, J. R. Clark and Bernhard (1960a,b) extended the list of tentative identifications to 40. Parallel British work by Slater (1961a) revealed at least seven terpenes, nine sesquiterpenes, and 24 oxygenated compounds in lemon oil.

TABLE XII VOLATILE FLAVORING CONSTITUENTS OF CITRUS FRUITS[a]

Hydrocarbons	
Aliphatic	*Sesquiterpenes*
Nonane	Farnesene
Undecane	Sesquicitronellene
Tridecane	
Tetradecane	β-Bisabolene
Pentadecane	α-Bergamotene
	δ-Cadinene
Monoterpenes	γ-Cadinene
Myrcene (2-methyl-6-methylene-2,7-octadiene)	ζ-Cadinene
Ocimene (2,6-dimethyl-1,5,7-octatriene)	Calamenene
	Calacorene
p-Cymene (*p*-isopropyltoluene)	α-Cubebene[b]
p,α-Dimethylstyrene (*p*-isopropenyltoluene)	β-Cubebene[b]
α-Terpinene (*p*-mentha-1,3-diene)	α-Copaene[b]
β-Terpinene [*p*-mentha-1(7),3-diene]	β-Copaene[b]
γ-Terpinene (*p*-mentha-1,4-diene)	
Terpinolene [*p*-mentha-1,4(8)-diene]	α-Elemene
Isoterpinolene (*p*-mentha-2,4-diene)	β-Elemene[c]
α-Phellandrene (*p*-mentha-1,5-diene)	δ-Elemene
β-Phellandrene [*p*-mentha-1(7),2-diene]	
Dipentene (*dl*-*p*-mentha-1, 8-diene, or *dl*-limonene)	α-Selinene
d-Limonene (*d*-*p*-mentha-1,8-diene)	γ-Selinene
α-Thujene (3-thujene)	Valencene
Sabinene [4(10)-thujene]	
3-Carene	α-Humulene
α-Pinene (2-pinene)	β-Humulene
β-Pinene [2(10)-pinene]	β-Caryophyllene
Camphene	Aromadendrene
	Longifolene
	β-Sesquiphellandrene
Hydroxy Compounds	
Aliphatic	*Terpenoid*
Methanol	Geraniol (*trans*-3,7-dimethyl-2,6-octadien-1-ol)
Ethanol	Nerol (*cis*-3,7-dimethyl-2,6-octadien-1-ol)
Propanol	Tetrahydrogeraniol
Butanol	Linalool (3,7-dimethyl-1,6-octadien-3-ol)
2-Butanol	Citronellol (3,7-dimethyl-6-octen-1-ol)
Pentanol	Thymol (cymen-3-ol)
2-Pentanol	α,α,*p*-Trimethylbenzyl alcohol (cymen-8-ol)
3-Methylbutanol	α-Terpineol (*p*-mentha-1-en-8-ol)
1-Penten-3-ol	β-Terpineol (*p*-mentha-8-en-1-ol)
3-Methyl-3-buten-2-ol	1-Terpinenol (*p*-mentha-1-en-4-ol)
Hexanol	4-Terpinenol (*p*-mentha-4-en-1-ol)
4-Methylpentanol	Isopulegol [*p*-mentha-4(8)-en-3-ol][d]

TABLE XII *continued*

2-Hexen-1-ol	Neoisopulegol (*p*-mentha-8-en-3-ol)
3-Hexen-1-ol	*p*-Mentha-1-en-9-ol
Heptanol	*p*-Mentha-1,8-dien-7-ol (as acetate)
3-Hepten-1-ol	*p*-Mentha-1,8-dien-9-ol
Octanol	*p*-Mentha-2,8-dien-1-ol
3-Octanol	Carveol (*p*-mentha-6,8-dien-2-ol)
Methylheptanol	*p*-Mentha-8-ene-1,2-diol
Nonanol	Sabinene hydrate [4(10)-thujen-3-ol]
2-Nonanol	Borneol (2-bornanol)
Decanol	Fenchol (1,3,3-trimethyl-2-norbornanol)[e]
Undecanol	Nerolidol (3,7,11-trimethyl-1,6,10-dodecatrien-3-ol)
Dodecanol	Elemol
Benzyl alcohol	
Carbonyl Compounds	
Aliphatic	*Terpenoid*
Acetaldehyde	Geranial (α-citral)
Butyraldehyde	Neral (β-citral)
Hexanal	Citronellal
2-Ethylbutyraldehyde	α-Sinensal
2-Hexenal	β-Sinensal
Heptanal	Cuminaldehyde (*p*-isopropylbenzaldehyde)
Octanal	Perillaldehyde (*p*-mentha-1,8-dien-7-al)
2-Octenal	Menthone (*p*-menthan-3-one)
Nonanal	Piperitenone (*p*-mentha-1,4(8)-dien-3-one)
Decanal	Carvone (*p*-mentha-6,8-diene-2-one)
Undecanal	Camphor (norbornone)
Dodecanal	Nootkatone (selinenone)
2-Dodecenal	
Tridecanal	**Ethers and Epoxides**
Tetradecanal	Acetal (1,1-diethoxyethane)
Pentadecanal	Thymyl methyl ether
Hexadecanal	α,α,*p*-Trimethylbenzyl methyl ether[e]
Heptadecanal	
1,2-Dialkylacroleins	
Benzaldehyde	1,4-Cineole (1,4-epoxy-*p*-menthane)
Furfuraldehyde	1,8-Cineole (1,8-epoxy-*p*-menthane)
Jasmone	Pinol (6,8-epoxy-*p*-mentha-1-ene)[e]
Acetone	Limonene oxides
2-Butanone	
2-Hydroxy-2-butanone	2,2-Dimethyl-5-(1-methyl-1-propenyl)-tetrahydrofuran[e]
Diacetyl	2,6,6-Trimethyl-2-vinyl-tetrahydropyran[e]
4-Methyl-2-pentanone	Linalool oxides
2-Octanone	
5-Octen-2-one	
Methylheptanone	Caryophyllene oxide
Methylheptenone	
6-Methyl-5-hepten-2-one	
2-Decanone	

TABLE XII *continued*

Acids and Esters	
Aliphatic	*Terpenoid*
Ethyl formate[f]	Geranyl formate
Ethyl acetate[f]	Geranyl acetate
2-Butyl acetate	Geranyl propionate
Octyl acetate	Geranyl butyrate
Nonyl acetate	Neryl formate
Decyl acetate	Neryl acetate
Undecyl acetate	Linalyl acetate
Benzyl acetate	Linalyl propionate
Ethyl propionate[f]	Citronellyl formate
Methyl butyrate[f]	Citronellyl acetate
Ethyl butyrate	Citronellyl butyrate
Octyl butyrate	Terpinyl formate
Ethyl valerate[f]	Terpinyl acetate
Methyl isovalerate[f]	*p*-Mentha-1-en-9-ol acetate
Ethyl isovalerate	*p*-Mentha-1,8-dien-7-yl (perillyl) acetate
Octyl isovalerate	*p*-Mentha-1,8-dien-9-yl acetate
Methyl hexanoate[f]	Bornyl acetate
Ethyl hexanoate	
Methyl 3-hydroxyhexanoate	*Other Esters*
Ethyl 3-hydroxyhexanoate	Diethyl citrate (sym. and asym.)[g]
Isohexanoic acid	γ-Decanolactone
Ethyl heptanoate[f]	Methyl *N*-methylanthranilate
Ethyl octanoate[f]	
Octyl octanoate	
Methyl 2-ethylhexanoate	
Nonanoic acid	
Decanoic acid	
Undecanoic acid	
Dodecanoic acid	

[a]A catalog of substances identified, though some only on gas chromatographic evidence, from the following principal sources: Attaway *et al.* (1962, 1964a,b), Bernhard (1961), D'Amore and Calabro (1966), Fernandez *et al.* (1963), Hunter and Brogden (1964a-d, 1965a,b), Hunter and Moshonas (1965, 1966), Ikeda *et al.* (1962a), Kugler and Kovats (1963), MacLeod *et al.* (1966), Moshonas and Lund (1969a,b), Nursten and Williams (1967), Ohta and Hirose (1966), Schultz *et al.* (1964, 1967), Scora *et al.* (1966, 1968), Sundt *et al.* (1964), Teranishi *et al.* (1963), Wolford and Attaway (1967), and Wolford *et al.* (1962, 1963).

[b]Veldhuis and Hunter (1967).

[c]Hunter and Parks (1964).

[d]May be artifacts by cyclization of citronellol during silica chromatography (Teranishi *et al.*, 1966).

[e]Found only in distilled lime oil (Kovats, 1963).

[f]Free acid also found.

[g]Neurath and Luettich (1968).

When Bernhard (1961) applied these techniques to California Valencia orange oil, he was able to separate more than 50 components and to identify seven in the terpene fraction and 36 in the oxygenated fraction, including 14 which had not previously been reported. Then Fernandez *et al.* (1963) indicated the presence of 72 components in Florida orange oil.

And so the refinement of techniques has added more and more compounds to the amazing variety present in citrus oils. For example, in lime and mandarin oils, Kovats (1963) and Kugler and Kovats (1963) found more than 100 constituents and identified about half of them. Most recently, Goretti *et al.* (1967b) separated approximately 180, 170, and 150 compounds from lemon, orange, and tangerine oils, respectively.

Notable centers of active research on citrus volatiles during this period were the Florida Citrus Experiment Station at Lake Alfred, under R. W. Wolford and J. A. Attaway, and two laboratories of the U.S. Department of Agriculture: the Fruit and Vegetable Products Laboratory at Winterhaven, Florida, under M. K. Veldhuis and G. K. Hunter, and the Western Regional Research Laboratory at Albany, California, under W. L. Stanley and R. Teranishi.

The principal outcome of all this work is that it is now possible to put together, as in Table XII, an impressive catalog of the volatile flavoring constituents of citrus fruits. What the food scientist is most interested in, however, is the role of these volatile compounds in the aroma and flavor of citrus products. This may be a positive contribution to characteristic flavor, or a negative contribution in the sense that some volatiles of little intrinsic flavor are degraded chemically into compounds that contribute off-flavors.

In the text that follows an attempt will be made to indicate the extent of present knowledge about the significance of particular volatiles in the flavors characteristic of different citrus fruits. All too little is known, however, about this aspect (Stanley, 1965), and the use of citrus oils for flavoring purposes remains largely an empirical art based on direct organoleptic testing.

In commercial practice the essential oils of the peels of citrus fruits are extracted either by cold-pressing or by distillation. Cold-pressing refers to extraction in various mechanical devices which rupture the oil sacs in the flavedo and express the oil as an aqueous emulsion from which it is separated by centrifuging. Distilled oils are recovered by steam distillation of citrus fruits or residues, followed by decantation or centrifuging of the condensate. In both processes the oil is in contact with large volumes of water which may cause leaching of water-soluble constituents. For instance, in a comparative test on Sicilian lemons, Slater (1961a) found that the peel oil extracted with light petroleum contained 18.9% of oxygenated compounds while the oil extracted with a process using water contained only 12.5%.

Volatile flavoring compounds are also present in the juice of citrus fruits; "juice oils" may therefore be separated by direct centrifuging. These oils are derived mainly from oil glands in the peel that are broken during extraction of the juice (Huet, 1969). Some interesting observations on the state of the essential oil in orange juice were made by Scott *et al.* (1965). If the juice is centrifuged immediately after extraction, the juice oil is recovered in the lightest phase of the effluents. On the other hand, if the juice is held for a few hours, and then centrifuged, the oil is found mainly in the heavy sediment. The authors regarded this occurrence as evidence that the essential oil enters the lipid fraction in the cloud components of the juice, and conjectured that the subsequent decrease in volatility might explain the loss of fresh aroma from orange juice which takes place soon after extraction. Huet (1969) confirmed that most of the essential oil in citrus juices is attached to the solid particles in suspension.

When citrus juices are evaporated, the volatile flavoring compounds are removed to a greater or less extent. They may be recovered from the evaporator condensate, either as an oil phase (condensate oil) or as an aqueous phase, commonly called "essence" in the American literature (Coleman *et al.*, 1969).

Present knowledge concerning the composition of citrus volatiles is summarized in Tables XII and XIII. Table XII lists by chemical categories all the volatile compounds that have so far been identified, either tentatively or conclusively, in citrus peel oils, juices, or essences. Table XIII contains miscellaneous quantitative data about major fractions and components of the peel oils from several kinds of citrus fruits. This latter table is not as useful as it might have been for comparative purposes because of the diverse ways in which results have been reported.

The volatile flavoring constituents of citrus fruits were also the subject of a recent review by Huet (1968).

The nonvolatile constituents of citrus oils, mainly coumarin compounds, are discussed in Chapter 12. Citrus oils are commonly subjected to a "dewaxing" or "wintering" procedure which involves storage at low temperatures, to permit the nonvolatile compounds to separate out.

I. Oranges

A. Whole Oranges Citrus fruits have distinctive aromas because they send into the atmosphere small amounts of volatiles. The nature of the volatiles from whole Valencia oranges, for example, was studied by Norman *et al.* (1967). As might be expected, the amounts of volatiles emanated increased with increasing temperature, and increased greatly when the peel was injured in such a way as to rupture oil sacs, although the number of constituent compounds did not change. The components identified were *d*-limonene, β-myrcene, α-pinene, acetaldehyde, octanal, ethanol, and ethyl acetate.

From intact Hamlin oranges, Attaway and Oberbacher (1968) recovered a much greater variety of esters. Thus, ethyl acetate, butyrate, hexanoate, and octanoate were positively identified, while ethyl formate, propionate, isovalerate, valerate, and heptanoate were probably present. In addition, these authors found in the cuticular "wax" extracted with methylene chloride the sesquiterpene valencene (more than 90%) as well as five other sesquiterpenes, which they believe contribute to the aroma of fresh oranges.

B. Orange Peel Oils The volatiles in orange oils and essences have been examined to a greater extent than those from any other citrus fruit and to date most of the compounds listed in Table XII have been found.

The yield of peel oil from Valencia oranges grown in Florida increased with maturity from 3.7 lb per ton of small immature fruits at 3 months from blossom to 5.8 lb per ton at 8 months and then declined to about 5 lb per ton at 12 months. It was not until the fruit reached commercial maturity (11-12 months) that the oils achieved true orange character (Kesterson and Hendrickson, 1962).

Orange peel oil is high in monoterpenes, and has as its major component *d*-limonene, which makes up about 90% of the oil (Table XIII). The components of the terpene fraction do not differ greatly among the different kinds of citrus fruits, but orange oils are distinguished by the presence of valencene as the principal sesquiterpene, a compound which is probably derived from the cuticular "wax" (Attaway and Oberbacher, 1968). California Valencia peel oil was higher in hydrocarbons than Florida Valencia oil, and especially high in valencene (Hunter and Brogden, 1965a,b). Florida Pineapple orange peel oil was also rich in valencene and distinctive because of the presence of the straight-chain hydrocarbons, tetradecane and pentadecane. Valencene has been detected in orange juice as well as in the peel oil (Wolford and Attaway, 1967).

It is the oxygenated compounds that provide the character of citrus oils. Linalool is the principal component of this fraction in orange oils (Table XIII). Its concentration declines with maturity (see Chapter 11, Section VII). As a commercial index of the quality of orange oils, the aldehyde content is commonly specified and the major aldehydes are the even-carbon saturated aliphatic members: octanal, decanal, and dodecanal (Table XIII). The concentrations of total aldehydes and of octanal increase with maturity but decline again with overmaturity (Kesterson and Hendrickson, 1962). The saturated aliphatic aldehydes have a sweet pungent fatty aroma.

Citral is a minor component of orange oils, present in greater amounts in Navel than in Valencia oils (Stanley, 1962). It is likely, however, that another α,β-unsaturated aldehyde, α-sinensal, contributes significantly to orange aroma since it has a sweet pungent penetrating aroma and a very low odor threshold and is present in cold-pressed orange oils in a concentration of about 0.1%

TABLE XIII THE AMOUNTS OF THE PRINCIPAL CONSTITUENTS IN CITRUS PEEL OILS[a]

Fractions	Orange		Mandarin			Grapefruit		Lemon		Lime	
	Valencia (% oil)		Medit. (% oil)[b]	Satsuma (% oil)[c]	Dancy (% oil)[d,e]	(% oil)		U.S.A. (% oil)	Sicily (% oil)[f,g]	Cold-pressed (% oil)	Distilled (% oil)[h]
Monoterpenes	89-91[i]		91	98	-	88[i]		81-85[i]	-	69[i]	77
Sesquiterpenes	-		0.07	-	-	-		0.04[j]	-	-	1.9
Oxygenated compounds	-		3.2	-	-	-		-	4-14	-	17
Aldehydes	1.8[k]		0.7	-	-	1.2-1.8[l]		-	3.6	-	-
Citral[m]	0.05-0.2		-	-	-	0.06		1.9-2.6	-	3.1-5.3	0.3
Free alcohols	0.9[k]		-	-	-	0.3-1.3[l]		1-2[n]	-	-	-
Esters	2.9[k]		-	-	-	3-4[l]			-	-	-
Constituents:	(% oil)[k,o]					(% oil)[d,e]		(% terpenes)[i]	(% oil)[f]	(% oil)[d]	(% oil)[h]
Limonene	83-93		65-68	94	87-93	93-95		72-80	70	64	60
α-Pinene	0.5		2.7	0.8	1	1.6		2	3	1.2	0.8
β-Pinene	1		1.2	-	0.4	-		7-13	14	8.3	0.8
Myrcene	2		1.3	2	1.2	1.9		2	2	-	0.8
γ-Terpinene	0.1		9-17	-	3.4	0.5		10	10	22	0.6
p-Cymene	-		1-8	2.8	0.4	0.4		-	0.6	1.9	12
	(% oil)[o]	(% aldehydes)[p]		(% terpeneless fraction)[c]		(% oil)[e]	(% aldehydes)[p]	(% aldehydes)[p]	(% oxygen compounds)[f]		
Heptanal	-	3	-	-	-	-	4	1	-	-	-
Octanal	2.8	39	0.04	-	4[q]	0.6	16-35	4	3	-	0.03
Nonanal	-	5	-	-	-	-	7	6	3	-	-
Decanal	-	42	0.04	5	-	-	43-54	3	1	-	0.09
Undecanal	-	2	-	0.4	-	-	4	2	-	1	-
Dodecanal	-	8	-	0.5	-	-	9	-	-	-	-
Citral	-	-	-	-	-	-	-	85	-	-	-
	(% oil)[o]	(% terpeneless fraction)[k]				(% oil)[e]	(% terpeneless fraction)[l]	(% alcohols and esters)[n]	(% oxygen compounds)[f]		
Octanol	-	2	0.09	-	-	0.8	1-2	-	-	-	0.01
Decanol	-	1	0.04	-	-	-	1-2	-	-	-	0.06
Geraniol	-	2	-	-	0.04	-	-	-	31-45	-	-

Linalool	5.3	26	0.25	2	4.2	0.4	0-3	7-20	2-3	-	0.1
Nerol	-	-	-	-	-	-	-	-	15-18	-	-
α-Terpineol	0.5	8	1.1	2	0.9	0.2	2-5	8-22	-	-	5.9
4-Terpinenol	0.2	-	0.1	-	0.3	0.1	-	2-19	-	-	1.6
Thymol	-	-	0.08	-	0.2	-	-	-	-	-	-
Octyl acetate	-	1	-	1	-	-	1-3	-	1	-	-
Nonyl acetate	-	2	-	-	-	-	3-4	-	4-13	-	-
Citronellyl acetate	-	-	-	4	-	-	-	0-2	4-10	-	-
Geranyl acetate	-	2	0.003	4	-	-	1	9-44	-	-	0.3
Linalyl acetate	-	2	-	-	-	-	2-3	0.2	-	-	-
Neryl acetate	-	-	-	3	-	-	-	29-54	-	Fenchol	1.2
Methyl N-methyl-anthranilate	-	-	0.85	-	-	-	-	-	-	1,4-Cineole	1.8
β-Elemene	-	-	-	25	-	-	-	-	-	1,8-Cineole	0.7
β-Sesquiphel-landrene	-	-	-	17	-	-	-	-	-	-	-

[a]The values given in this table are rounded off from published values; some are means and some refer to single samples only. The data in the same column do not necessarily refer to the same oil sample. The absence of a value does not mean that the constituent is absent but that no quantitative data are available.

Similar information for some other citrus fruits has been reported in the following sources: bergamot (Huet and Dupuis, 1968); Clementine and Rangpur (Scora *et al.*, 1968); Meyer lemon (Ikeda *et al.*, 1962b); Hamlin orange (Attaway *et al.*, 1967a); Temple tangor (Ashoor and Bernhard, 1967; Stanley *et al.*, 1961a); Natsudaidai and other Japanese varieties (Kadota and Nakamura, 1967; Maekawa *et al.*, 1967; R. Okada and Nakamura, 1967); and trifoliate orange (Scora *et al.*, 1966, 1968).

[b]From Kugler and Kovats (1963) and D'Amore and Calabro (1966).
[c]From Yamanishi *et al.* (1968) and Kita *et al.* (1969).
[d]From Ashoor and Bernhard (1967).
[e]From Attaway *et al.* (1967a).
[f]From Di Giacomo *et al.* (1962, 1965).
[g]From Slater (1963).
[h]From Kovats (1963).
[i]From Ikeda *et al.* (1962b).
[j]From Hunter and Brogden (1965a).
[k]From Kesterson and Hendrickson (1962).
[l]From Kesterson and Hendrickson (1963, 1964).
[m]By the vanillin-piperidine method, from F. Yokoyama *et al.* (1961).
[n]From Ikeda and Spitler (1964).
[o]From Attaway *et al.* (1968).
[p]From Stanley *et al.* (1961a).
[q]Octanal and γ-terpinene.

(Stanley, 1965). α-Sinensal has a farnesane skeleton with four double bonds (Stevens *et al.*, 1965; A. F. Thomas, 1967) and it is accompanied by a double bond isomer, β-sinensal (Flath *et al.*, 1966). Recently a number of other carbonyl compounds have been identified in orange oil–the ketones, piperitenone, and 6-methyl-5-hepten-2-one (Moshonas, 1967), and a group of 7 α,β-dialkylacroleins in which the alkyl groups are hexyl, heptyl, octyl, or nonyl (Moshonas and Lund, 1969a,b).

The concentration of aldehydes in the peel of Valencia oranges grown in Florida is influenced by the amount of rainfall during the growth of the fruit (Kesterson and Hendrickson, 1966). During 14 years of Florida production, the seasonal averages for the aldehyde content of Valencia orange oil ranged from 1.4 to 1.8% for which there was a positive correlation ($r = 0.603$, $p < 0.05$) between the average aldehyde content and the total rainfall for the preceding 15 months (March to May), the period of growth of the Valencia orange; i.e., the aldehyde content rose with increased rainfall and fell in the drier years. There were some exceptions that suggested that the distribution of rainfall within a season also had a significant influence. The Florida workers envisaged the possibility of predicting the average aldehyde content of orange oil from rainfall data, and of controlling the aldehyde content by irrigation.

However, when Scora and Newman (1967) made a survey of the composition of distilled Valencia peel oils from fruit grown in several regions of the United States–ranging in climate from humid, subtropical Florida through warm, semiarid California and Arizona to the cool California coast–they concluded that the composition of the oil was little influenced by climate provided that samples were examined at equivalent maturities.

C. Orange Juice Volatiles The volatile compounds in orange juice are generally similar to those in orange oil, with the exception that there is a greater representation of saturated and unsaturated alcohols, aldehydes, and esters up to six carbon atoms (Wolford *et al.*, 1963; Attaway *et al.*, 1964b; Schultz *et al.*, 1964, 1967). Ethyl butyrate is the major ester present and its concentration increases with advancing maturity of the fruit.

A gas chromatographic procedure for the estimation of linalool and α-terpineol in orange juices was devised by Swift (1961) as part of a method to detect the addition of peel juice. The concentrations of linalool (23 ppm) and α-terpineol (8.5 ppm) in peel juices were high enough to contribute to unpleasant flavor since a tasting panel reported objectionable flavor in orange juice when both linalool and α-terpineol were added at levels of 8 and 3 ppm, respectively.

The amount of α-terpineol in orange juice volatiles may be increased by bacterial action during processing (Murdock *et al.*, 1967).

Another constituent of citrus juice volatiles associated with microbial deterioration is diacetyl, which imparts a buttermilk off-flavor. Hill *et al.* (1960)

reported that fresh orange, grapefruit, and tangerine juices, showing no evidence of bacterial growth and no off-flavor gave values up to 4 ppm when assayed for diacetyl. This was probably due to the tasteless precursor, acetylmethylcarbinol. Murdock (1968), however, found only 0.15 ppm of diacetyl in juice extracted by hand from sound oranges.

In an attempt to correlate the chemical composition of orange juice volatiles with organoleptic characteristics, Wolford and Attaway (1967) made a gas chromatographic study in which they recorded parallel responses from electron capture and flame ionization detectors. The results led them to conclude that there were substances contributing significantly to the aroma of orange juices that still eluded isolation and identification.

When the volatile flavors from canned Florida Valencia oranges juices stored 27 months at 4° and 27°C were compared by Rymal *et al.* (1968), it appeared that the principal changes occurring at the higher temperature were (1) conversion of much of the *d*-limonene to α-terpineol by acid-catalyzed hydration, (2) disappearance of most of the linalool, and (3) accumulation of furfural, probably from nonvolatile precursors such as ascorbic acid.

II. Mandarins

In the peel oil of Mediterranean mandarins, Kugler and Kovats (1963) identified 48 compounds which accounted for 99.2% of the volatile fraction. However, they estimated that there were at least 93 constituents still unidentified. Again, the oil contained mainly terpene hydrocarbons (Table XIII), with α-terpineol (1.1%) the major oxygenated constituent. Uncommon compounds found were sabinene hydrate in two forms and α,α,*p*-trimethylbenzylalcohol. It is likely, however, that the distinctive notes in the aroma of Mediterranean mandarin oil are due to the presence of methyl *N*-methylanthranilate (0.85%) and thymol (0.08%).

Because of the presence of methyl *N*-methylanthranilate, Mediterranean mandarin oil exhibits an intense blue-violet fluorescence at 415 mμ which is useful for characterization of the oil and for detecting adulteration (D'Amore and Corigliano, 1966).

Neither methyl *N*-methylanthranilate nor thymol was found in the peel oil of the Japanese Satsuma mandarin, and the characteristic aroma of this fruit in comparison with other citrus fruits was attributed to larger contents of sesquiterpenes, notably β-elemene and β-sesquiphellandrene, and the acetate esters of geraniol, nerol, citronellol, p-mentha-1,8-dien-7-ol, and p-mentha-1,8-dien-9-ol (Yamanishi *et al.*, 1968; Kita *et al*, 1969).

Marked differences in the monoterpene composition of the peel oils of Mediterranean mandarins and Florida tangerines were pointed out by Ashoor and Bernhard (1967); the tangerine oils contained much more *d*-limonene than the mandarin oils and less γ-terpinene, α-pinene, and camphene.

Florida tangerine oils were distinguished by the presence of the sesquiterpenes α- and δ-elemene (Hunter and Brogden, 1965b); thymol was identified among the hydroxyl compounds (Hunter and Moshonas, 1966) but methyl *N*-methylanthranilate was evidently not present.

The peel oil of the Murcott tangor resembled tangerine oil more than orange oil in character and in gas chromatographic pattern (Kesterson and Hendrickson, 1960).

III. Grapefruit

The peel oil of grapefruit (Table XIII) resembles orange oil in having a high content of the monoterpene fraction, which is mainly limonene (Hunter and Brogden, 1964d, 1965b) and it is also broadly similar in the composition of the oxygenated fraction with some interesting differences (Hunter and Moshonas, 1966).

The yield of oil from Marsh grapefruit in Florida declined with advancing maturity from 1.9 lb per ton of small, immature fruit (May) to 1.4 lb per ton of mature fruit (October) (Kesterson and Hendrickson, 1963). Only the mature fruit yielded oil of satisfactory organoleptic quality. In the experience of the industry, optimal quality in grapefruit oil is attained only after "curing," i.e., holding the oil at 65°-70°F for at least 6 months (Kesterson and Hendrickson, 1967).

As grapefruit ripened the peel oil increased in total carbonyls, which were principally octanal, nonanal, and decanal. Oils from white grapefruit varieties in Florida had higher aldehyde contents than those from red varieties and the relative proportions of the constituent aldehydes were different. In the white varieties the decanal content was 1.1 to 1.4 times the octanal content, while in the red varieties this relation was reversed (Kesterson and Hendrickson, 1964). In Arizona grapefruit, however, decanal was the dominant aldehyde in both white and pink varieties (Stanley *et al.*, 1961a).

In contrast to the aldehydes, the alcohols in grapefruit oil decreased in concentration with maturity and the linalool content, in particular, fell to undetectable levels in mature grapefruit of the white varieties, although in the mature red varieties it made up 3% of the terpeneless fraction (Kesterson and Hendrickson, 1964). Some of the linalool is converted into *cis*- and *trans*-linalool oxides (Hunter and Moshonas, 1966). Disappearance of linalool is the principal chemical change during the "curing" of grapefruit oils (Kesterson and Hendrickson, 1967), and the difference in linalool content is the major difference in the composition of grapefruit oil and orange oil.

In 1953, Kirchner and Miller had reported an unknown sesquiterpene ketone in grapefruit juice, and subsequently C. G. Beisel and his colleagues drew the attention of MacLeod and Buigues (1964) to an unidentified compound of distinctive grapefruit aroma that followed the sesquiterpene hydrocarbons on gas chromatograms of California grapefruit oil. This compound was identified as nootkatone, a sesquiterpene ketone having the same skeleton as valencene, the major sesquiterpene of oranges (MacLeod, 1965). Valencene is readily converted into nootkatone by oxidation with *t*-butyl chromate (Hunter and Brogden, 1965c). Nootkatone is stable in air, in acid juice, and at high temperatures in the absence of air.

The nootkatone content of grapefruit oil increases with advancing maturity. In Duncan grapefruit oil in Florida the nootkatone content increased from 0.07% at legal maturity in November to 0.08% in June, but in the same period the oil yield fell from 2.6 to 0.5 lb per ton of fruit (Kesterson *et al.*, 1965a). In California grapefruit oil, nootkatone levels as high as 1.8% have been recorded (MacLeod, 1966). When grapefruit oils with different nootkatone and aldehyde contents were examined by three different panels of experts, there was some lack of agreement, but the oils most generally preferred for grapefruit character had the highest aldehyde contents, around 1.8% as decanal, and nootkatone contents in the middle range (0.5-0.7%) (Kesterson *et al.*, 1965b). Small amounts of nootkatone are present in grapefruit juice free from peel oil. Only traces of nootkatone have been found in other citrus oils: orange, tangerine, lemon, lime, and bergamot (MacLeod and Buigues, 1964).

Nootkatone has an astringent taste and an odor variously described as pungent, aromatic, and musty. Its flavor threshold in water is about 1 ppm and in grapefruit juice 4-6 ppm (Berry *et al.*, 1967). At 6-7 ppm nootkatone improved the flavor of grapefruit juice but at levels about 8 ppm it imparted an unpleasant bitter taste.

IV. Lemons

Production of volatiles from whole lemons was greater in quantity and variety from yellow fruit than from green fruit, and was much greater from both green and yellow lemons treated with ethylene (1:1000) (Norman and Craft, 1968). The principal components detected were γ-terpinene, terpinolene, *d*-limonene, and citral.

A sample of lemon oil taken directly from an oil sac by means of a glass capillary was compared by Fischer and Still (1967) with oils extracted commercially. Even in freshly extracted commercial oils there was some loss of aldehydes and esters and some oxidation and isomerization of terpenes; these changes were greatly accentuated on long storage.

In comparison with orange, mandarin, and grapefruit oils, lemon oil has lower proportions of monoterpenes and limonene, and higher proportions of β-pinene and γ-terpinene (Table XIII). The γ-terpinene in lemon oil contributes to flavor deterioration because it is degraded to *p*-cymene (Ikeda *et al.*, 1961). Platt (1967) has patented a procedure for removing these two terpenes by distillation.

The hydrocarbon fraction of lemon oil is distinguished by the presence of the straight-chain members, tetradecane and pentadecane, and the sesquiterpenes, α-bergamotene, β-bisabolene, and caryophyllene (Hunter and Brogden, 1965b).

Lemon oil also differs markedly from the oils of the orange, mandarin, and grapefruit group in the composition of the oxygenated fraction (Table XIII). Citral is the major aldehyde and the two geometrical isomers are present in the approximate proportions, geranial 75% and neral 25%. Normal saturated aliphatic aldehydes from C-6 to C-17 are present but only in small amounts (Ikeda *et al.*, 1962a). Contrary to the position in orange and grapefruit oils, nonanal is present in higher proportion than octanal or decanal.

The alcohol and ester fractions in lemon oils are very variable in composition (Table XIII) with marked differences between American and Sicilian oils. Only small proportions of aliphatic alcohols are present but large proportions of α-terpineol, 4-terpinenol, and nerol (Hunter and Moshonas, 1966). It is likely that the tertiary terpene alcohols are formed to a greater or less extent by hydration of terpenes during contact with acidic juice and water during extraction (Ikeda and Spitler, 1964). The primary alcohols are present mainly as acetate esters.

According to H. Guenther (1968), lemon oils of high organoleptic quality tend to have low citral contents (around 2%) but tend to contain a wide variety of components, especially the high-boiling members, geraniol, geranyl acetate, neryl acetate, and bergamotene. Oils of moderate quality have higher citral contents, and low-quality oils lack high-boiling components.

Regional effects on the composition of lemon oils were observed by Ikeda *et al.* (1962b), who found oils from lemons grown on the California coast to have a total monoterpene content of about 80% while oils from desert-grown lemons had about 85% monoterpenes and only about half the β-pinene content of the coastal lemon oils. The higher citral content of the coastal lemons, already well-known, was confirmed in analyses reported by F. Yokoyama *et al.* (1961). Somewhat similar climatic influences on the composition of bergamot oil were reported by Huet and Dupuis (1968). The linalool and linalyl acetate contents were low in Moroccan samples and high in Corsican samples although the trees were similar in horticultural origin. The authors conclude that a high content of oxygenated compounds in bergamot oil is favored by atmospheric humidities above 70% and low temperatures during the later development of the fruit.

V. Limes

Cold-pressed oils from the peel of limes resemble lemon oil in the distribution and constituents of terpene, sesquiterpene, and oxygenated fractions (Table XIII) (Hunter and Brogden, 1965b; Hunter and Moshonas, 1966).

More common in commercial use than the cold-pressed oil is distilled lime oil, the product of a traditional West Indian procedure in which the oil is in contact with the highly acid juice for some weeks prior to distillation. Under these conditions the oil develops a distinctive "terpeney" character that the consuming public enjoys. Although Kovats (1963) identified 44 compounds in West Indian distilled lime oil (accounting for 97.6% of the volatile fraction), he stated that there were at least 100 unidentified trace components. Among the most abundant oxygenated constituents (Table XIII) are α-terpineol, 1,4-cineole, and 1,8-cineole, which Slater (1961b,c) and Slater and Watkins (1964) have shown to be derived from limonene, the major hydrocarbon, by hydration and dehydration reactions under the acid conditions during processing. Citral, present in the natural oil, is also degraded during processing and largely disappears by transformation to *p*-cymene and *p*-α-dimethylstyrene (Loori and Cover, 1964).

The latter workers doubted whether *p*-α-dimethylstyrene was a naturally occurring compound in any citrus oils, but a biogenetic origin is made more likely by its presence in the leaf and blossom oils of tangerine and tangelo (Attaway *et al.*, 1966a,b).

Indian limes yielded 0.5-0.6% oil when fresh, but the oil was almost entirely lost during storage at room temperature or at 8°-10°C (Srivas *et al.*, 1963).

VI. Other Citrus and Related Fruits

A. Natsudaidai In a qualitative study of the constituents of Natsudaidai peel oil, Ohta and Hirose (1966) showed that the sesquiterpene fraction was distinguished by the presence of α-selinene and aromadendrene, and possibly also calamenene and α-calacorene. The authors consider that perillaldehyde, carvone, and decyl acetate make a particular contribution to the character of the Natsudaidai.

B. Trifoliate Orange The essential oils of the trifoliate orange were studied by Scora *et al.* (1966, 1968) as a potential aid in taxonomic identification. In spite of some variability among strains, they concluded that the gas chromatographic patterns were sufficiently characteristic to permit recognition of this fruit. The effects of maturity on the composition of the oils were, however, considerable and it was important to compare fruit of the same physiological age. There were no consistent effects of position on the tree.

The distilled peel oil of the trifoliate orange (Scora *et al.*, 1966) is lower in limonene content than orange, lemon, and grapefruit oils and the terpene fraction is made up of larger proportions of β-myrcene and γ-terpinene. Linalool, nonanol, and citronellal are the principal oxygenated constituents.

Hybrids of trifoliate orange with Ruby orange, Clementine mandarin, Sukega orangelo, and Rangpur were also examined, but there were no consistent relations in composition between the peel oils of the hybrids and those of the parents (Scora *et al.*, 1968).

VII. Biogenetic Aspects

The biogenetic origins of the volatile flavoring components of citrus fruits were explored by Attaway *et al.* (1966a,b, 1967a, 1968) in a series of studies of the oils from the leaves, blossoms, and fruit of several kinds of citrus trees. They found that the alcohol linalool was the single constituent present in greatest amount in almost all of the leaf oils studied. Linalool was also a major constituent of the peel oils of immature Hamlin oranges and Dancy tangerines, but decreased in relative concentration with maturity from 62% to 4% of the oil, in the period 1 to 11 months from blossom. During the same period the hydrocarbon limonene showed a closely corresponding increase in relative concentration from 21 to 87% of the oil. In immature Valencia oranges the linalool content was lower (20-30%) but the same loss of linalool and rapid increase in limonene content occurred. The Marsh grapefruit showed similar but less marked trends with maturity, the concentration of linalool and linalool oxides decreasing from about 5% to 0.4% of the oil, as Kesterson and Hendrickson (1963) had also observed.

Accordingly, Attaway *et al.* (1967a) propose a biogenetic route to linalyl pyrophosphate from isopropenylpyrophosphate, the active isoprene unit, and then with linalool as a precursor for terpenes, as has been suggested in other plants, they outline a scheme by which the major terpene hydrocarbons in citrus fruits are derived (Attaway *et al.*, 1966a).

While limonene is the major terpene in citrus fruit oils, it is only a minor component in the leaf oils where, by contrast, the dominant terpene is ocimene, which has been found in peel oils only as a very minor component in tangerine and grapefruit oils (Attaway *et al.*, 1967a). Sabinene is also a major terpene in orange leaves and it is the dominant compound in immature Valencia oranges, amounting to more than 30% of the oil in fruit about 6 weeks from blossom.

The oils from the pistils of orange blossoms, the part from which the fruit is formed, had large amounts of the sesquiterpenes valencene and β-caryophyllene which are major components of the sesquiterpene fraction in orange peel oils (Attaway *et al.*, 1966b).

VIII. Analytical Aspects

For many years commercial standards and specifications for citrus essential oils have included a number of physical and chemical parameters: specific gravity, refractive index, optical rotation, evaporation residue, aldehyde content, acid content, and ester content. It is often difficult, however, to relate these parameters to the organoleptic quality of the oils. The optical rotation of California lemon oil was inversely related to the concentration of β-pinene (α_D^{20} −23.4°), and directly related to the concentration of *d*-limonene (α_D^{20} +126.6°), and to the percentage of citral in the total aldehyde fraction (Stanley *et al.*, 1961b; Stanley, 1962).

More recently, ultraviolet and infrared (Vernin, 1967) spectrometric measurements have been added to the classical parameters. Slater (1963) reported that the ratio of certain peak heights in infrared spectra provided a useful indication of adulteration and storage deterioration in citrus oils.

Now, however, gas chromatography has become the major tool for the chemist involved in quality control as well as for those engaged in research on essential oils. The ultimate development in methodology in this field is perhaps represented by the achievement of Macleod (1968a,b), following Macleod *et al.* (1966), in obtaining chromatograms of whole citrus oils, with good resolution of minor components, in about 15 minutes, by programming both the temperature and the flow of carrier gas in a capillary column. A technique for the qualitative identification of essential oils (Rasquinho, 1965), by means of gas chromatographic profiles known as "standard retention masters," was applied by Huet (1967) to the oils from 35 varieties of orange, bitter orange, mandarin, lemon, lime, citron, and bergamot. Linear temperature programmed chromatograms of the oils were calibrated in relation to the positions of the peaks, under identical conditions, of a standard series of homologous compounds, in this case the methyl esters of the normal aliphatic acids from C-1 to C-18.

Sources of information about the characteristics of citrus oils from many parts of the world are listed in Table XIV.

A method devised by Scott and Veldhuis (1966) that permits the estimation of the recoverable oil content of a citrus juice in less than 7 minutes has recently been adopted by the Association of Official Analytical Chemists as official (Scott, 1968; Gerritz, 1969). Propanol added to the sample assists rapid distillation of the oil which is determined in the acidified distillate by titration with potassium bromate-bromide. The bromine liberated reacts with *d*-limonene in

TABLE XIV SOURCES OF INFORMATION ON PHYSICAL AND CHEMICAL PROPERTIES OF CITRUS PEEL OILS

	Orange Oil
Assam	Chaliha *et al.* (1963a).
Florida	Kesterson and Hendrickson, (1962); Kesterson *et al.* (1959).
Israel	Basker (1967).
Italy	Di Giacomo *et al.* (1963); Rispoli *et al.* (1963).
Pakistan	Ahmad *et al.* (1962a,b).
	Bitter Orange Oil
Greece	Mehlitz and Minas (1964, 1965a,b).
Italy	Di Giacomo *et al.* (1963, 1964); Kunkar (1964b).
	Bergamot Oil
Review	Muller (1966).
Corsica	Huet and Dupuis (1968).
Italy	Benk and Wildfeuer (1961); M. Calvarano (1963, 1965); M. Calvarano and I. Calvarano (1964); D'Amore and Corigliano (1966); Di Giacomo *et al.* (1963); La Face (1961).
Morocco	Huet and Dupuis (1968).
	Mandarin Oil
India	S. K. Mukherjee and Bose (1962a,b); Parekh *et al.* (1961); Pruthi *et al.* (1961).
Italy	M. Calvarano (1958); Di Giacomo *et al.* (1963); D'Amore and Calabro (1966); D'Amore and Corigliano (1966).
	Grapefruit Oil
Florida	Kesterson and Hendrickson (1963, 1964, 1967); Kesterson *et al.* (1965a,b).
Israel	Basker (1967).
Italy	Di Giacomo *et al.* (1963).
	Lemon Oil
Argentine	Di Giacomo and Rispoli (1964); Rispoli *et al.* (1965).
Borneo lemon	Peyron (1966).
Israel	Lifshitz *et al.* (1969).
Italy	Cavoli (1966); D'Amore and Corigliano (1966); Di Giacomo and Rispoli (1962a); Di Giacomo *et al.* (1963, 1969); Pennisi and Di Giacomo (1965); Rispoli *et al.* (1963); Rispoli and Di Giacomo (1965); Slater (1961a, 1963).
Pakistan	Ahmad *et al.* (1962a,b).
Spain	Carpena and Laencina (1967); Goretti *et al.* (1967a).
	Lime Oil
Dominica	Slater (1961c).
Mexico	Slater (1961c).
Ghana	Talalaj (1966).
	Tangelo Oil
Florida	Kesterson and Hendrickson (1969).
	Trifoliate Orange Oil
Italy	Nobile and Pozzo-Balbi (1968).

the oil until at the end point excess bromine bleaches the color of methyl orange indicator.

The standards for g rades for citrus products in the United States have recently been revised to provide for the use of the bromate method, and the maximum limits for recoverable oil increased by 0.005% to allow for higher recoveries by this method (Anonymous, 1968).

One of the most widely used chemical procedures in the examination of citrus oils is the estimation of aldehyde content, usually calculated as decanal in orange and grapefruit oils, and as citral in lemon and lime oils. A critical comparison of several assay procedures was made by F. Yokoyama *et al.* (1961). Satisfactory agreement in the determination of citral in lemon, lime, orange, and grapefruit oils was observed between the methods of Stanley *et al.* (1958) and Levi and Laughton (1959). These methods depend on the formation of specific colored condensation products by reaction of citral with vanillin-piperidine and barbituric acid reagents, respectively. The nonspecific phenylhydrazine and hydroxylamine titrations and the *m*-phenylenediamine condensation method gave distinctly higher values.

Another procedure, advocated by Calapaj and Sergi (1967) as more rapid than the hydroxylamine titration, was based on the spectrophotometric measurement of the yellow color produced by reaction of citral with isonicotinic acid hydrazide.

Following the observations of Leach and Lloyd (1956), it appeared that the citral content of foods might have toxicological significance, since citral was alleged to promote glaucoma and cardiovascular disease. Subsequent tests in Canada, however, failed to confirm these toxic effects (Laughton *et al.*, 1962).

For the rapid estimation of flavoring constituents in citrus essences, Attaway *et al.* (1967b) recommended four procedures based on the determination of particular functional groups: (1) oxygenated terpenes as $C_{10}H_{18}O$, by colorimetric reaction with vanillin in sulfuric acid; (2) saturated aliphatic aldehydes, as octanal, by catalytic oxidation of *p*-phenylenediamine with hydrogen peroxide; (3) α,β-unsaturated aldehydes, as citral, by formation of a colored Schiff base with *o*-dianisidine; and (4) esters, as ethyl butyrate, by reaction with hydroxylamine and ferric ion. The ranges of values in essences from mature oranges in Florida were $C_{10}H_{18}O$, 70-180; octanal, 120-520; citral, 10-45; and ethyl butyrate, 10-30 ppm.

A gross measure of the water-soluble volatile constituents in citrus products is given by the chemical oxygen demand (COD), determined by oxidizing the organic matter in an oil-free distillate with potassium dichromate-sulfuric acid. For example, a typical range of COD values for fresh Florida Valencia orange juice was 435-655 ppm (Dougherty, 1968).

Chapter 12 Simple Polyphenolic Compounds

I. Phloroglucinol and Derivatives

Phloroglucinol occurs only rarely in nature, and the isolation of its β-D-glucoside, phlorin, from oranges and grapefruit (Horowitz and Gentili, 1961a) is the first report of the natural occurrence of this compound in plant material. The subsequent identification of phloroglucinol as the second most abundant aglycone in lemon juice after eriodictyol (Vandercook and Stephenson, 1966) suggests that it may be a constituent of all citrus fruits. Its presence is of interest because it would provide a more active substrate for enzymic browning reactions than any other polyphenolic citrus constituent, and in addition it has been reported to produce glucosuria. The co-occurrence of phloroglucinol with considerable amounts of inositol (Table VII) suggests a biosynthetic link between these two cyclic compounds (Horowitz and Gentili, 1961a), but it is more likely that phloroglucinol is synthesized by condensation of acetate groups, as is the phloroglucinol ring of flavonoids.

The peel of the grapefruit contains seven times more phloroglucinol than the endocarp (Maier and Metzler, 1967b).

II. Phenolic Acids

The simplest phenolic acid identified as a citrus constituent is gentisic acid which occurs as the ester of β-D-glucose in the bark of sweet orange trees affected by the greening virus and in the albedo of the fruit (A. W. Feldman, 1968). Orchard studies in South Africa by Schwarz (1968) showed that the presence of this compound was a reliable diagnostic test for greening virus infection in oranges and mandarins, but that further observations were necessary before the test could be applied with confidence to lemons, grapefruit, and tangelos.

In their exhaustive investigation of lemon phenolics, Horowitz and Gentili (1960b) reported the presence of *p*-coumaric and sinapic acids, while Chopin *et al.* (1963) identified *p*-coumaric, ferulic, and caffeic acids as constituents of this fruit. Studies on the biosynthetic relationships between the aglycones of the grapefruit by Maier and Metzler (1967b) revealed the presence of *p*-coumaric, caffeic, and ferulic acids, in order of increasing amount; the concentrations of each acid were comparable in both peel and endocarp and significant amounts of caffeic and ferulic acids were present as glucosides. These three acids are important components of the "metabolic grid" linking grapefruit phenolics, all of which can be derived by accepted biogenetic processes from *p*-coumaric acid (see Chapter 13). In addition, Maier and Metzler suggested that the orthohydroxycinnamic acids corresponding to the coumarins bergaptol, umbelliferone, esculetin, and scopoletin were present as glycosides or esters, but under the conditions of their extraction procedure were converted by spontaneous lactonization to the coumarins.

Chlorogenic acid, the 3-caffeic acid ester of quinic acid, was identified by Freedman *et al.* (1962) as the constituent of oranges responsible for intradermal reactions in persons allergic to oranges. This allergen was present in such low concentrations that special techniques were required for its extraction and identification (Siddiqi and Freedman, 1963). Subsequently it was shown (Freedman *et al.*, 1964) that chlorogenic acid was transformed in the gastrointestinal tract to inactive substances, thus explaining why oranges give positive results in tests for skin and inhalation allergies but not in those for food allergies.

The presence of the substituted cinnamic acid amide, feruloylputrescine, in grapefruit and oranges has been mentioned in Chapter 7.

III. Coumarins

Compared with the biogenetically related flavonoids, the coumarin compounds in citrus fruits have received little attention from food chemists, despite the important physiological properties possessed by coumarins in general (Soine, 1964). These compounds are present in the peel, juice, and oil sacs of citrus fruits, and can be isolated from cold-pressed citrus oils, but the amounts are small and better sources for most of them are known. An elegant technique for the examination of citrus coumarins was reported by Peyron (1963) who made direct chromatographic studies on the fluid removed from the oil sacs of the fruit with a microsyringe.

Present knowledge of the identity and distribution of coumarins in citrus fruits is summarized in Table XV, and the subject has been briefly reviewed by Kefford (1959), Sarin and Seshadri (1961), Stanley (1964), Kunkar (1965) and

Stanley and Vannier (1967). As their names indicate, several of the compounds listed in Table XV, such as citropten, bergapten, and aurapten, were first isolated from citrus fruits, and some, e.g., auraptenol, have not yet been reported as constituents of other plants. Four such coumarins have been isolated from citrus fruits in recent years, one from Seville orange oil (Stanley *et al.*, 1965) and three from grapefruit oil (Fisher and Nordby, 1966; Fisher *et al.*, 1967); two of the grapefruit coumarins are unusual in carrying a formyl side-chain.

It is perhaps unfortunate that the new coumarin from the Seville orange should have been named auraptenol as there is already some confusion in the literature respecting three different citrus coumarins given the similar names, aurapten (or auraptene) and auraptin. The term "auraptene" was first used by Komatsu *et al.* (1930) for a coumarin isolated from a fruit designated "grapefruit, *Citrus aurantium* L.," and was given the structure 4-heptoxycoumarin. Several years later, Böhme and Pietsch (1938) used the same name for a coumarin from orange peel oil which was shown to be 7-methoxy-8-(2,3-epoxyisopentyl)-coumarin (Böhme and Schneider, 1939). In a review of the lactones of citrus oils, Dodge (1938) recognized the duplication of terms and renamed the later product "meranzin." From the Natsudaidai, Nomura (1950) isolated a coumarin similar to the aurapten of Komatsu *et al.* (1930) and showed it to be different from meranzin; Kariyone and Matsuno (1953) later proved this compound to be 7-geranoxycoumarin. Another coumarin isolated from the Natsudaidai by Nomura (1950) was called "auraptin" but this compound was shown to be identical with isoimperatorin (5-isopentenoxycoumarin) by Matsuno (1956).

The citrus fruit richest in coumarins is the lime, in the oil of which Cieri (1969) found 5-geranoxy-7-methoxycoumarin (2.2-5.2%), bergamottin (2.2-2.5%), citropten (0.89-1.70%) and bergapten (0.17-0.33%) with traces of bergaptol and isopimpinellin, while other workers reported seven additional coumarins (Table XV). Lime oil from Guinea contained 50-150% more 5-geranoxy-7-methoxycoumarin and 100% more citropten than lime oils from Florida and Mexico although the bergamottin contents of the oils from all three regions were similar.

The bergamot approaches the lime in yield of coumarins though not in the number or complexity of those present. Although M. Calvarano (1961b) and Labruto and Di Giacomo (1963) had reported that bergapten and citropten were the principal coumarins in the bergamot, Sundt *et al.* (1964) established and Cieri (1969) confirmed that bergamottin in fact is the dominant coumarin, the concentrations in the oil being 1.1-1.9% bergamottin, 0.30-0.36% bergapten, and 0.24-0.32% citropten. The bergamottin content varied little in oils from different sources.

The similarity in the ultraviolet spectra of bergamottin and bergapten pointed out by Cieri (1969) had probably misled earlier workers in their analyses of bergamot oil, and the usefulness of spectrophotometric measurements in the

TABLE XV COUMARIN COMPOUNDS IN CITRUS FRUITS[a]

Compound		Sweet Orange	Sour Orange	Mandarin	Bergamot	Natsudaidai	Grapefruit	Lemon	Lime
Common to several citrus species:									
Umbelliferone	7-Hydroxycoumarin		*[b]			+[a]	+	*	
Scopoletin	6-Methoxy-7-hydroxycoumarin						*	*	
Citropten (limettin)	5,7-Dimethoxycoumarin	*	*	*	+		*	+	+
	5-(2′-Isopentenoxy)-7-methoxycoumarin				*			*	
	5-Geranoxy-7-methoxycoumarin				+			+	+
Bergaptol	5-Hydroxypsoralen	*			+	+	*	*	+
Isoimperatorin (aurapten)	5-(2′-Isopentenoxy)-psoralen	*	*			+		*	*
Bergamottin	5-Geranoxypsoralen				+		*	+	+
Byakangelicin	5-Methoxy-8-(2′,3′-dihydroxyisopentoxy)-psoralen						*	+	
Apparently not present in lemons and limes:									
Aurapten	7-Geranoxycoumarin	*	*						
Meranzin (auraptene)	7-Methoxy-8-(2′,3′-epoxy-isopentyl)-coumarin	+	*			+			
Osthol	7-Methoxy-8-(2′-isopentenyl)-coumarin		*				*		
Auraptenol	7-Methoxy-8-(2′-hydroxy-3′-methyl-3′-butenyl)-coumarin		*						
Bergaptene	5-Methoxypsoralen		*		+		*		*
Apparently present only in lemons and limes									
Imperatorin	8-(2′-Isopentenoxy)-psoralen							*	*
	8-Geranoxypsoralen							+	+
Isopimpinellin	5,8-Dimethoxypsoralen							*	+
Oxypeucedanin hydrate	5-(2′,3′-Dihydroxyisopentoxy)-psoralen							*	*
Phellopterin	5-Methoxy-8-(2′-isopentenoxy)-psoralen							*	*
	5-Geranoxy-8-methoxypsoralen							*	*
	5-Isopentoxy-8-methoxypsoralen								*

Reported only in grapefruit		
Herniarin	7-Methoxycoumarin	*
Esculetin	6,7-Dihydroxycoumarin	*
	7-(6′,7′-Dihydroxy-3′,7′-dimethyl-2′-octenoxy)-coumarin	*
	7-Methoxy-8-(2′-formyl-2′-methylpropyl)-coumarin	*
	5-(3′,6′-Dimethyl-6′-formyl-2′-heptenoxy)-psoralen	*

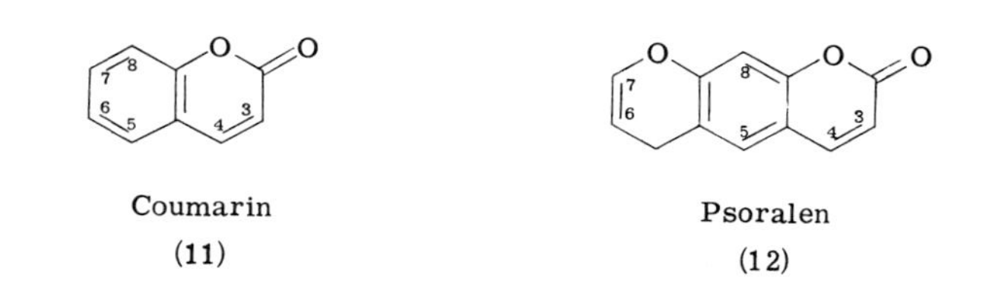

Coumarin
(11)

Psoralen
(12)

[a]+, Occurrence reported in previous review (Kefford, 1959).

[b]*, Occurrence reported since previous review by the following authors: Caporale and Cingolani (1958) – bergamots. Cieri (1969) – bergamots, limes, lemons. D'Amore and Calapaj (1965) – lemons, mandarins, bergamots, sweet and sour orange. Farid (1968) – bergamots. Fisher and Nordby (1965, 1966) – grapefruit. Fisher *et al.* (1967) – grapefruit. Horowitz and Gentili (1960b) – lemons. La Face (1961) – bergamots. Latz and Madsen (1969) – limes, lemons. Maier and Dreyer (1965) – grapefruit. Maier and Metzler (1967b) – grapefruit. Matsuno (1959a) – limes. Peyron (1963) – lemons, bergamots, sweet and sour oranges, grapefruit. Rotondaro (1964) – bergamots. Slater (1961b) – limes. Stanley (1964) – lemons, limes, bergamots, sour oranges, grapefruit. Stanley and Vannier (1959) – lemons. Stanley and Vannier (1967) – limes. Stanley *et al.* (1965) – sour oranges. Vandercook and Rolle (1963) – lemons. Vandercook and Stephenson (1966) – lemons. Venturella *et al.* (1964) – sour oranges.

assessment of the quality of bergamot oil (Rudol'fi and Sharapova, 1965) must be regarded as doubtful.

Bergapten alone among the citrus coumarins has the property of photosensitizing the skin, and both bergamot and lime oils are known to cause photodermatitis (Cieri, 1969). The strong mutagenic activity of bergamot oil (Gori, 1958) is also believed to be due to bergapten.

Among the other citrus oils, quantitative coumarin analyses are available only for lemon oil which contains 0.1% bergamottin, 0.5% citropten, and 0.03% 5-geranoxy-7-methoxycoumarin (Cieri, 1969) together with 12 other coumarins in smaller amounts (Table XV). The identification of five coumarins, including limettin (citropten), in lemon juice specially prepared from carefully separated juice sacs indicates that these compounds may be present in intact juice vesicles (Bernhard, 1958a). More recently, however, Vandercook and Rolle (1963) were unable to detect limettin in hand-reamed lemon juice, although it was present in all commercial samples and peel extracts examined.

The use of the characteristic ultraviolet absorption and fluorescent spectra of the coumarin components to identify cold-pressed citrus oils and to detect adulteration has been greatly extended in recent years by advances in thin layer chromatography and instrumentation. Lemon and lime oils give markedly different chromatograms from other citrus oils and can be differentiated from each other by two-dimensional chromatography (Macleod and Buigues, 1966), even though they have nine coumarin components in common (Stanley and Vannier, 1967). Still greater sensitivity and specificity in both qualitative and quantitative studies of citrus oils and their coumarin constituents are possible by observations of fluorescence and phosphorescence at ambient and low temperatures (Latz and Madsen, 1969).

The application of new chromatographic and spectrophotometric techniques increases the possibility of solving problems in citrus taxonomy by the examination of the cold-pressed oils. For instance, the qualitative and quantitative similarity in the coumarins, and also in the aldehydes and monoterpenoids, of lemons and limes support their classification by Linnaeus in a group apart from oranges and grapefruit (Stanley and Vannier, 1967). However, although examination of Table XV indicates that there is one group of coumarins common to all citrus fruits, a second group found so far only in lemons and limes, and a third group not found in either of these two fruits, it is difficult to find any single structural feature that can be used to distinguish one group chemically from another, despite the varied patterns shown by citrus coumarins in hydroxylation, substitution, and modification of isopentyl and geranyl side-chains. More positive conclusions may be drawn when all citrus varieties have been examined with the same degree of thoroughness and by the same techniques. At this stage, such differences as exist cannot be used as guides in citrus taxonomy. For instance, the Borneo lemon, which has some of the characteristics of the grapefruit, lemon, and citron, cannot be classified into any group on the basis of the presence of citropten and 5-geranoxy-7-methoxycoumarin in its oil and the absence of aurapten and bergamottin (Peyron, 1966). So far, the only coumarin reported (Nigam *et al.*, 1958) in the oil of the Calamondin, also of uncertain taxonomy, is citropten (0.3%), the most widely occurring citrus coumarin.

In their study on the biosynthetic relationships among the grapefruit aglycones, Maier and Metzler (1967b) detected esculetin and scopoletin in the hydrolysis products of peel and endocarp extractives, along with the previously reported bergaptol and umbelliferone. Scopoletin and umbelliferone were dominant coumarins in the endocarp and peel, respectively, in approximately the same concentration as apigenin. Except for bergaptol, which existed in the free form in the peel and endocarp and in the bound state only in the endocarp, the coumarins were initially present as glycosides and esters, and to a considerable extent, if not entirely, as derivatives of the corresponding orthohydroxycinnamic acids. Modifications in the basic structure of the coumarins probably occur at this cinnamic acid stage, but the ease with which the free acids undergo lactonization makes difficult their detection as natural plant constituents.

Citrus seeds were not recognized as a source of coumarins until Nikonov and Molodozhnikov (1964) reported the occurrence of imperatorin in the seeds of the Meyer lemon (0.08%) and the trifoliate orange (0.02%). However in an extensive survey of sources of limonoids, Dreyer (1965b, 1966a) found few coumarins in citrus seeds, except in the trifoliate orange which contained both imperatorin and bergapten, as well as smaller amounts of xanthotoxol (8-hydroxypsoralen) and alloimperatorin (5-isopentyl-8-hydroxypsoralen). Also studied was *Aeglopsis chevalieri,* a close citrus relative, which contained isopimpinellin. Alloimperatorin is a rare coumarin, previously isolated with imperatorin from the fruits of another citrus relative, *Aegle marmelos* (Saha and Chatterjee, 1957).

Chapter 13 Flavonoids

The flavonoids of citrus continue to be a very important subject for study for a number of reasons: (1) they play a significant role in the technology of processed citrus products; (2) as by-products of the citrus industry they offer material for exploitation by the chemical industry, particularly since claims have been made for their therapeutic and pharmacological activity; (3) they present a challenge to the natural product chemist through the complexity of the structural variations on the standard C-15 flavonoid molecule; and (4) no satisfactory definition has yet been made of their function in the developing fruit, nor of the factors that govern their occurrence and their structural variations within the maturing fruit.

The occurrence of citrus bioflavonoids may be summarized as follows: The flavanones (13) (Table XVI) usually predominate among the citrus flavonoids. The flavones (14) and the related flavon-3-ols (15) are present in considerably smaller amounts. The anthocyanins [flavylium ions (16)] are restricted to one variety of one group. Only one chalcone (17) has been detected but not isolated. The only aurone (18) has since been found to be an artifact. Leucoanthocyanins, catechins, isoflavones, and dihydrochalcones have yet to be detected in citrus.

The flavonoids usually occur as the glycosides and are distributed throughout all the tissues of citrus fruits, but the permethoxylated flavones and flavanones are exceptional in that they occur only as the free compounds, and only in the oil sacs of the flavedo of certain citrus fruits.

Since the appearance of the previous review various aspects of the subject of citrus flavonoids have been summarized in English by Sarin and Seshadri (1961), the Agricultural Research Service (1962), and Horowitz (1961a,b, 1964); in German by Herrman (1962), and Hörhammer and Wagner (1962); in French by Huet (1962a); in Italian by Di Giacomo and Rispoli (1962b); and in Japanese by Nakabayashi (1962b). Swain (1962) and Davidek (1963) have made reference to

TABLE XVI
NOMENCLATURE AND STRUCTURE OF CITRUS FLAVONOIDS

Common Name	Chemical Structure
Acacetin	5,7-Dihydroxy-4′-methoxyflavone
Apigenin	4′,5,7-Trihydroxyflavone
Auranetin	3,4′,6,7,8-Pentamethoxyflavone
Cirantin	Identical with hesperidin
Citromitin	3′,4′,5,6,7,8-Hexamethoxyflavanone
Cyanidin	3,3′,4′,5,7-Pentahydroxyflavylium ion
Chrysoeriol	4′,5,7-Trihydroxy-3′-methoxyflavone
Citrofolioside	Identical with poncirin
Delphinidin	3,3′,4′,5,5′,7-Hexahydroxyflavylium ion
Didymin	Isosakuranetin-7-rutinoside
Dihydrokaempferol	3,4′,5,7-Tetrahydroxyflavanone
Diosmetin	3′,5,7-Trihydroxy-4′-methoxyflavone
Diosmin	Diosmetin-7-rutinoside
Eriocitrin	Eriodictyol-7-rutinoside
Eriodictyol	3′,4′,5,7-Tetrahydroxyflavanone
Fortunellin	Acacetin-7-neohesperidoside
Hesperetin	3′,5,7-Trihydroxy-4′-methoxyflavanone
Hesperidin	Hesperetin-7-rutinoside
Isolimocitrol	3,3′,5,7-Tetrahydroxy-4′,6,8-trimethoxyflavone
Isorhamnetin	3,4′,5,7-Tetrahydroxy-3′-methoxyflavone
Isosakuranetin	5,7-Dihydroxy-4′-methoxyflavanone
Kaempferol	3,4′,5,7-Tetrahydroxyflavone
Limocitrin	3,4′,5,7-Tetrahydroxy-3′,8-dimethoxyflavone
Limocitrol	3,4′,5,7-Tetrahydroxy-3′,6,8-trimethoxyflavone
Luteolin	3′,4′,5,7-Tetrahydroxyflavone
Naringenin	4′,5,7-Trihydroxyflavanone
Naringin	Naringenin-7-neohesperidoside
Narirutin	Naringenin-7-rutinoside
Neohesperidin	Hesperetin-7-neohesperidoside
Neoponcirin	Isosakuranetin-7-rutinoside
Nobiletin	3′,4′,5,6,7,8-Hexamethoxyflavone
Poncirin	Isosakuranetin-7-neohesperidoside
Ponkanetin	Identical with tangeretin
Quercetin	3,3′,4′,5,7-Pentahydroxyflavone
Rhoifolin	Apigenin-7-neohesperidoside
Rutin	Quercetin-3-rutinoside
Sinensetin	3′,4′,5,6,7-Pentamethoxyflavone
Sudachitin	4′,5,7-Trihydroxy-3′,6,8-trimethoxyflavone
Tangeretin	4′,5,6,7,8-Pentamethoxyflavone
Tetramethylscutellarein	4′,5,6,7-Tetramethoxyflavone
Vitexin	Apigenin-8-C-glucoside

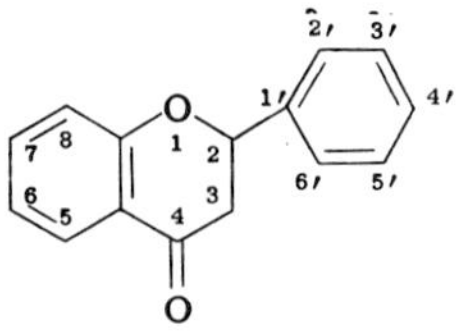

Flavanone
(13)

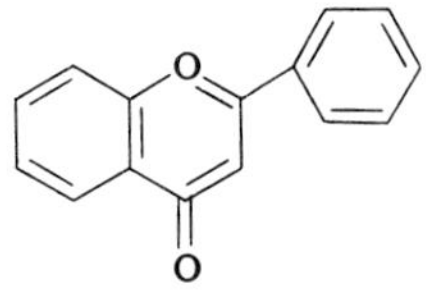

Flavone
(14)

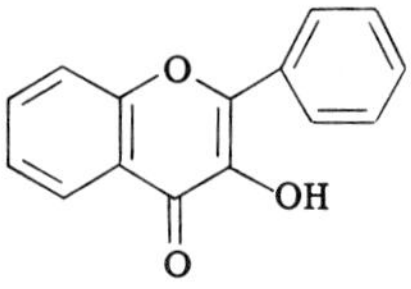

Flavon-3-ol
(15)

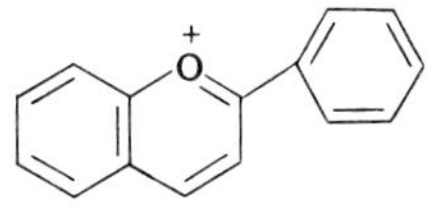

Flavylium ion
(16)

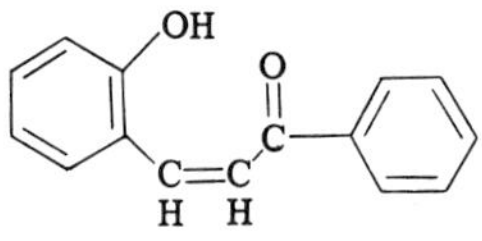

Chalcone
(17)

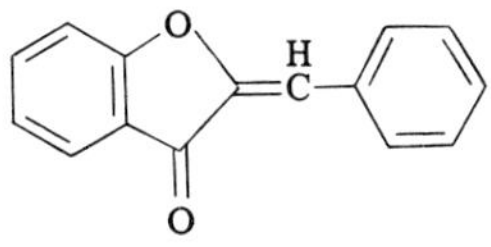

Aurone
(18)

citrus flavonoids in their general reviews on the importance, isolation, analysis, and utilization of flavonoids in foodstuffs.

Throughout this section the flavonoids will be referred to by common names where such exist. The corresponding chemical structures are given in Table XVI.

I. The Isomeric Rhamnoglucosides

One intriguing problem in the chemistry of citrus flavonoids needs to be discussed early in this review. Briefly, this problem may be stated as involving the structural differences between the tasteless hesperidin and the bitter neohesperidin, both of which occur in a number of citrus varieties. These compounds are both 7-glycosides of hesperetin, yielding hesperetin, rhamnose, and glucose on hydrolysis, and hesperetin-7-glucoside on partial hydrolysis. There is some evidence that the two flavonoids are mutually exclusive, and often where one occurs the other does not, although more frequently the one predominates considerably over the other. Thus sweet and sour oranges contain both hesperidin and neohesperidin in ratios as high as 40:1 (Row and Sastry, 1962a).

As early as 1938, Zemplen *et al.* proposed that the difference between hesperidin and neohesperidin lay in the structure of the saccharides attached at the 7-position. They identified rutinose (6-rhamnosylglucose) as the moiety in hesperidin and suggested that the disaccharide of neohesperidin, which they named neohesperidose, was a 4-rhamnosylglucose, with the 3- and 2-isomers as the other possibilities. The presence of a rhamnosylglucose differing from rutinose in other citrus flavonoids was also reported by Nakabayashi (1961a) in his study of fortunellin, the 7-rhamnoglucoside of acacetin isolated by Matsuno (1958) from *Fortunella* fruits. Nakabayashi noted that fortunellin differed from linarin (acacetin-7-rutinoside) and suggested that the rhamnose and glucose in fortunellin were linked in a way that differed from the 1,6 combination in linarin. He subsequently exhaustively methylated the three rhamnoglucosides—neohesperidin from sour oranges, naringin from grapefruit, and poncirin from the trifoliate orange—and obtained on hydrolysis of the three products the same trimethyl ethers of rhamnose and glucose, thus confirming that neohesperidose was the rhamnosylglucose in each case (Nakabayashi 1961b). From the structure of the methylated sugars he described neohesperidose as 4-L-rhamnopyranosyl-D-glucose. Rutinose had previously been identified as 6-*O*-α-L-rhamnopyranosyl-D-glucose by Gorin and Perlin (1959).

At about the same time, Horowitz and Gentili (1961a) isolated poncirin from grapefruit and showed it to be a neohesperidoside like naringin and neohesperidin. They noted also that poncirin was, like naringin and neohesperidin, intensely bitter and postulated that the bitterness was associated with the neohesperidose moiety of these compounds. They subsequently reported details of their studies on the structure of this disaccharide (Horowitz and Gentili, 1963a) after following an experimental approach similar to that of Nakabayashi. However, their results were different in that they proved that neohesperidose was 2-*O*-α-L-rhamnopyranosyl-D-glucopyranose, a configuration established by

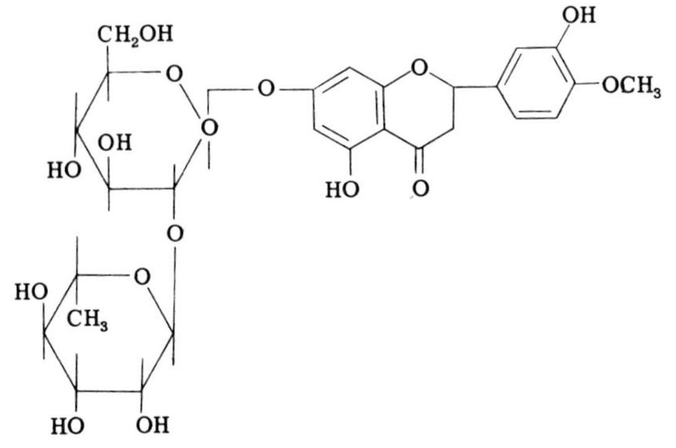

Neohesperidin

(19)

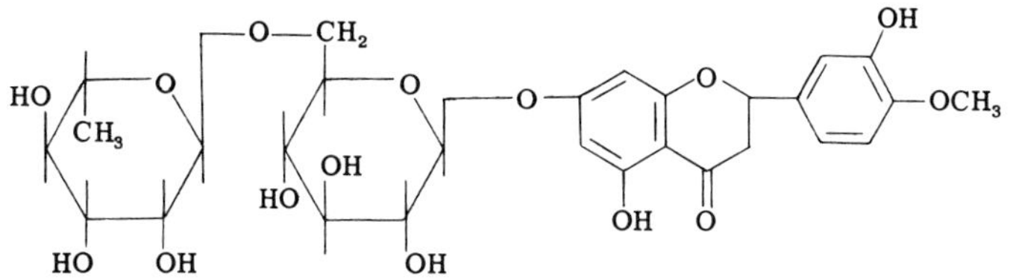

Hesperidin

(20)

optical rotation measurements. Previously Hardegger and Braunshweker (1961) had proved the identity of the absolute configurations of the aglycones of hesperidin and neohesperidin and thus eliminated another possible source of isomerism between the two compounds. Since the configuration of hesperetin had already been determined (Arakawa and Nakazaki, 1960), and since enzymic studies proved the aglycone-glycoside linkage to have the β-configuration, the structure of neohesperidin (19) was completely fixed as the 7-β-neohesperidoside of hesperetin-2S, hesperidin (20) being the corresponding rutinoside. This identification was later confirmed, with proof of each glycosidic configuration, by Christiansen and Boll (1964).

Horowitz and Gentili (1963a) further showed that neohesperidosides could be easily differentiated from rutinosides by their smooth cleavage in alkali to form phloracetophenone-4′-neohesperidoside in high yields, while, on the other hand, rutinosides yielded no identifiable carbohydrate fraction under these conditions. This reaction confirmed that the neohesperidosides are 2-rhamnoglucosides since the glycosidic linkage remains intact only if the C-2 hydroxyl group of the glucose is blocked, thus preventing the ionization at this group which precedes disruption of the sugar molecule. The bitterness of the phoroacetophenone neohesperidoside was further evidence of the importance of the point of attachment of the rhamnose to the glucose in determining the taste response to citrus flavonoids.

The presence of neohesperidose in a compound, however, does not necessarily endow it with bitterness, although all rutinosides are tasteless. In fact, Horowitz (1964) has reported that rhoifolin, the tasteless flavone analog of naringin that is also present in grapefruit, depresses the bitterness of naringin solutions. This observation further complicates the problem of measuring the bitterness of grapefruit juice by determining its naringin content. Other types of flavonoid glycosides besides the neohesperidosides are bitter, and a full discussion of the relation between the taste and structure of such glycosides has been given by Horowitz (1964) and Horowitz and Gentili (1969).

Some of the transformation products of the flavonoid neohesperidosides are tasteless while others are intensely sweet. The dihydrochalcones related to neohesperidin and naringin are comparable in sweetness to saccharin, and patents have been granted to Horowitz and Gentili (1963b, 1968) for the production of such sweetening agents from the bitter flavonoids–more recently for the conversion of the readily available naringin into neohesperidin dihydrochalcone which is 20 times sweeter than saccharin. In addition, Feldman *et al.* (1968) have patented a variant on the original procedure which was claimed to give higher yields. It is interesting that pure crystalline neohesperidose has been synthesized and found to be completely tasteless (Koeppen, 1968).

II. Analytical Methods

The most important flavonoids from the commercial point of view are the predominating flavanones, which not only affect the quality of processed citrus products but have also received considerable attention as potential pharmaceuticals. Consequently methods for flavanone determination have formed the subject of numerous papers, and analytical values reported for "citrus flavonoids" almost always refer to unspecific colorimetric methods for the flavanones as a group, without consideration of other flavonoid constituents. Analyses of flavonoid content have been considered as a means of characterizing the authenticity of lemon juice (Rolle and Vandercook, 1963), of measuring the adulteration of citrus juices (Wagner and Monselise, 1963), and of identifying the presence of orange juice in fruit drinks (Koch and Haase-Sajak, 1965a).

A. The Davis Method The most widely used method for flavonoid analysis, developed by W. B. Davis in 1947, is based on the yellow color given by flavanone glycosides in alkali. Unless otherwise stated all analyses discussed in this review were made by this method. This procedure is used for the analysis of the intensely bitter naringin and hence for assessing the bitterness of grapefruit products, and also for analysis of hesperidin which has been the subject of many pharmacological studies.

In 1959, Griffiths and Lime reported that the Davis method failed to provide a measure of bitterness in grapefruit juice treated with enzymes to convert the bitter naringin (naringenin-7-rhamnoglucoside) into the comparatively nonbitter prunin (naringenin-7-glucoside), since both compounds responded similarly to the Davis test. Horowitz and Gentili (1959) pointed out that the method depends on the ready conversion to colored chalcones of flavanones having both a protected 7-hydroxyl group and a free 4′-hydroxyl group; thus naringin and prunin react in the Davis test while naringenin, hesperidin, and hesperetin do not give color responses that can be used for accurate analyses.

When Ting first reported the enzymic hydrolysis of naringin by naringinase in 1958, he suggested that the Davis procedure for naringin determination in grapefruit products could be improved by analyzing samples before and after enzyme treatment, in the belief that the product of hydrolysis was naringenin, which does not interfere in the analysis. However, Griffiths and Lime (1959) found that the naringin contents obtained by this technique did not correlate with juice bitterness because the product of hydrolysis was prunin (naringenin-7-glucoside), which interfered with naringin estimation. D. W. Thomas *et al.* (1958) used emulsin to hydrolyze the prunin to naringenin, and this method gave satisfactory results in studies by Olsen and Hill (1964) on debittering grapefruit juice concentrate using naringinase. Okada *et al.* (1964b) removed the interfering prunin by selective extraction with ethyl acetate or butyl alcohol, and according to Shimoda *et al.* (1966a) this method was more satisfactory than emulsin treatment.

Although more selective methods have been developed for naringin and hesperidin determination, the Davis procedure is still employed for those analyses where its simplicity offers sufficient compensation for its inaccuracy. For instance, Hendrickson and Kesterson (1962) used this method in their study of oranges as a commercial crop for hesperidin production, and Hagen *et al.* (1966)

in their work on the effect of maturity on grapefruit composition. The latter authors reported that, although the Davis method usually gave high values compared with other methods and was clearly unsuitable for measuring bitterness in grapefruit products, it accurately reflected the decrease (which attends maturity) in the flavanone concentration in the juice sacs of Ruby Red grapefruit. Recent modifications of the Davis procedure have been suggested by Tomas *et al.* (1966) and Fishman and Gumanitskaya (1966).

B. Other Colorimetric Methods General dissatisfaction with the Davis test led Babin (1957) and Kwietny and Braverman (1959) to make a critical evaluation of the "cyanidin reaction" for flavanones, which produce a pink color when treated with magnesium and hydrochloric acid. Although the reaction is governed by a number of factors, including the purity of the magnesium, accurate results were obtained for pure hesperidin solutions provided control tests were run simultaneously. This reaction forms the basis for a method for the determination of hesperidin in "vitamin P" preparations (Romanenko, 1966). A similar color reaction, borohydride reduction followed by acidification, was recommended by Rowell and Winter (1959) as a specific reaction offering a rapid and simple means of assaying flavanones in amounts as low as 0.02 mg per milliliter.

A more involved process, specific for hesperidin in mandarin orange syrup (Nakabayashi and Kamiya, 1959), required selective extraction to overcome interference by other flavonoids and depended on the formation of a strong color (λ max. 630 mμ) on reaction of hesperidin with dimethyl-*p*-phenylenediamine in the presence of hypochlorite.

Gerngross and Renda (1966) proposed a method for naringin estimation in citrus fruits whereby albedo extracts were treated with 1-nitroso-2-naphthol in the presence of nitric acid to give a blood red color. Since neither the extraction nor the reaction are selective, the results should be regarded as applying to total flavanones rather than specifically to naringin. The method is simple, however, and may find application in control laboratories lacking photometric instrumentation since the analysis is carried out by visual reference to standard solutions.

C. Direct Spectrophotometric Methods In discussing the shortcomings of the Davis method, Horowitz and Gentili (1959) suggested that more reliable hesperidin analyses might be made by direct ultraviolet spectrophotometric measurements similar to the procedure developed by Hendrickson *et al.* (1958b) for naringin in grapefruit juice. Such a method was later applied to the estimation of hesperidin in orange juice (Hendrickson *et al.*, 1959) and orange tissues (Goren and Monselise, 1964b). A similar measurement, absorption by the clarified juice at 326-332 nm was used by Vandercook and Rolle (1963) for total polyphenols in lemon juices, and was subsequently adopted as official, first action by the Association of Official Analytical Chemists (H. Yokoyama, 1965).

D. Chromatographic Methods In a study of the composition of commercial naringinase, Dunlap *et al.* (1962a) developed a quantitative paper chromatographic procedure for the analysis of naringin and prunin in small amounts of mixtures of naringin, prunin, and naringenin. Oashi (1964) followed changes in the flavonoid composition of naringinase-treated canned Natsudaidai by paper chromatography, measuring the fluorescence of the naringin, prunin, and naringenin spots after spraying with aluminium chloride. Both procedures had the disadvantages inherent in methods involving paper chromatography of flavonoids.

The development of thin-layer chromatography offered a possible solution to the problem of the multiplicity of similar compounds encountered in the analysis of citrus flavonoids. Hörhammer and Wagner (1962) were the first to apply the technique to citrus flavonoids, in procedures for estimating naringin, naringenin, hesperidin, and eriodictyol. A more detailed procedure, evolved by Hagen *et al.* (1965), was specific for the determination of naringin and related flavanone glycosides in grapefruit juice and juice sacs which contain the three bitter neohesperidosides, naringin, neohesperidin, and poncirin, and the three nonbitter isomeric rutinosides. Since six similar constituents are to be estimated, the method is naturally lengthy, and involves preliminary column chromatography, thin-layer chromatography, and finally fluorimetry. This method was simplified by Fisher *et al.* (1966) into a relatively rapid and reasonably accurate procedure for comparing the amounts of naringin and its isomer in grapefruit juice, without consideration of the other flavanones. Similar rapid procedures were developed by Drawert *et al.* (1966) for several flavonoids of oranges, and by Fontanelli and Silvestri (1968) for hesperidin alone.

Where interest centers on the flavonoid aglycones, and not in the individual glycosides, as is the case with lemon flavonoids, the procedure of Vandercook and Stephenson (1966) presents a satisfactory approach. These workers enzymically hydrolyzed the glycosides in lemon juice, chromatographed the extracted aglycones, and completed the analysis by colorimetry or densitometry.

E. Extraction Procedures Except in the analysis of juices and syrups, extraction of the flavonoids is an essential preliminary to analytical procedures. Optimum conditions have been set for the extraction of hesperidin from orange tissues (Hendrickson and Kesterson, 1962) and of naringin from grapefruit tissues (Hendrickson and Kesterson, 1956). Soxhlet procedures are also satisfactory for naringin (Davis, 1966), but extraction with alkali is required for the more insoluble hesperidin. The problems encountered in procedures involving caustic alkali were fully discussed by Kwietny and Zimmermann (1960) and by Hendrickson and Kesterson (1964b) whose results should be considered before such methods are applied to flavonoid extractions.

III. Lemons

A. *O*-Glucosides and Aglycones The active team working on citrus flavonoids at the U.S. Department of Agriculture Fruit and Vegetable Chemistry Laboratory at Pasadena began their studies of lemon flavonoids (Horowitz, 1956) with the isolation of diosmin, the least soluble glycoside in lemon peel, and the flavone analogue of the ubiquitous flavanone, hesperidin. It is noteworthy that although diosmin may occur in concentrations up to 0.5% of the dry weight of lemon peel, it is present, if at all, only in trace amounts in orange peel, whereas hesperidin is present in large amounts in both fruits.

The studies of lemon flavonoids were prompted by an earlier report that the peel contained glucosides of eriodictyol which had been demonstrated to cause involution of the thymus gland in rats (Masri and De Eds, 1958). A commercial preparation, "Calcium Flavonate Glycoside, Lemon," later shown to have the same effect as eriodictyol (Masri *et al.*, 1959), was used as starting material by the Pasadena workers. After hydrolyzing the flavonoid glycosides to the aglycones with a hemicellulase, Horowitz (1957) was able to identify eriodictyol as the major constituent of the ether-insoluble aglycones and a new flavonol, limocitrin, as the major constituent of the ether-soluble aglycones.

Subsequently the eriodictyol glycoside, named eriocitrin, was established as a 7-β-rutinoside and found to be identical with 4′-demethylhesperidin (Horowitz and Gentili, 1960a). Despite the earlier report, neither eriodictyol itself nor its glucoside could be detected in lemon peel, and no eriodictyol compounds could be detected in orange peel. Eriodictyol was prepared in good yield by the enzymic hydrolysis of commercial lemon flavonoid products, in a process which was covered by a patent (Horowitz, 1958).

The treatment of commercial flavonoid preparations with hemicellulase to obtain aglycones (Horowitz and Gentili, 1960b) was the basis for further isolation procedures by the Pasadena team. Acid hydrolyses resulted in tarry products only from which no crystalline aglycones could be isolated, while β-glucosidase was apparently totally inactive. In the Pasadena procedure, the commercial flavonoid preparation was adjusted to pH 4.5 and treated with fungal hemicellulase (3.5 g/100 g) to give complete hydrolysis within 3 days. Extraction of the aglycones with various solvents was followed by chromatography on silicic acid columns using mixtures of chloroform, methanol, and acetone as eluants.

In this way commercial lemon flavonoid preparations yielded apigenin, luteolin, chrysoeriol, quercetin, isorhamnetin, limocitrin, limocitrol, and simpler polyphenolic compounds. Limocitrin had been previously isolated from lemon aglycones by selective extraction, but the other compounds were obtained from lemons for the first time, although glycosides of apigenin (rhoifolin) and quercetin (rutin) had previously been reported as constituents of the sour orange and the Satsumelo tangelo, respectively. Chopin *et al.* (1964) later reported the presence of a quercetin derivative in lemon peel responding to color tests for the 3,5-diglucoside, a luteolin derivative provisionally identified as the 7-rutinoside, and the 3-arabinosides of limocitrin, limocitrol, and isorhamnetin. In light of the results from Pasadena the identifications of the sugar moieties are suspect, although each aglycone may exist in several glycosidic combinations, usually with rutinosides predominating. Thus hesperidin, eriocitrin, and diosmin, in that order, are the most abundant flavonoids in lemon peel.

Limocitrin and limocitrol were new flavonoids, not isolated from any other natural source. Limocitrin was later shown to be 3,4′,5,7-tetrahydroxy-3′,8-dimethoxyflavone (Horowitz and Gentili, 1961b), the first representative of the 5,7,8-hydroxylation (gossypetin) pattern found in citrus, while limocitrol was shown to be 3,4′,5,7-tetrahydroxy-3′,6,8-trimethoxyflavone (Gentili and Horowitz, 1964). A third new flavonoid, isolimocitrol, isolated as its 3-glucoside from commercial lemon flavonoid products, was identified as 3,3′,5,7-tetra-hydroxy-4′,6,8-trimethoxyflavone, isomeric with limocitrol. This isolation was achieved by application of silicic acid column chromatography without prior enzymic hydrolysis. At the same time the 3-glucosides of limocitrin (from lemon and orange flavonoids) and limocitrol (from lemon flavonoids only) were isolated. Except for the anthocyanins of the blood oranges these three compounds were the first glucosides reported as citrus constituents, all other glycosides isolated being rhamnoglucosides.

Limocitrol and isolimocitrol are the lemon constituents most closely approaching the permethoxylated flavonoids found in oranges and mandarins but not in the lemon (Walther *et al.* 1966; D'Amore and Calapaj, 1965; Chaliha *et al.* 1965b). These compounds resemble the flavones, sudachitin and 3-demethoxysudachitin, isolated from *Citrus sudachi* by Horie *et al.* (1961, 1962). The pattern of 6,8 dimethoxylation shown by these four compounds and the permethoxylated flavonoids is not otherwise recorded in nature and may present an instance of flavonoid elaboration peculiar to citrus species. Limocitrol and limocitrin were later synthesized by Dreyer *et al.* (1964), and sudachitin and 3′-demethoxysudachitin by Farkas *et al.* (1967).

The flavonoid and other polyphenolic constituents of lemon juice were studied by Vandercook and Stephenson (1966), who developed a method for the analysis of the aglycones obtained after enzyme hydrolysis. These workers reported that eriodictyol was by far the predominant aglycone, followed by quercetin, phloroglucinol, hesperidin, and the coumarin, umbelliferone. While these results were contrary to previous reports that hesperidin was the major constituent of lemon juice, the discrepancy could be explained by the use of nonhomogeneous samples and nonspecific analytical methods in previous work. Hesperidin, being comparatively water-insoluble, was not removed from pulp particles during extraction or processing whereas the soluble eriodictyol glycoside, eriocitrin, passed completely into solution in the juice.

The isolation of the 6-rhamnoglucoside of aureusidin, the only aurone reported as a citrus constituent (Chopin *et al.*, 1963), has since been shown to be the result of spontaneous oxidation of eriocitrin under the alkaline extraction conditions (Chopin and Dellamonica, 1965). The identification of naringin as a constituent of lemons (Gerngross and Renda, 1966) is suspect since it was based on a nonspecific colorimetric analysis.

On the basis of chromatographic studies only, the oil sacs of the lemon have been reported to contain 2,4,6-trihydroxy-4′-methoxychalcone, the chalcone corresponding to hesperidin (Peyron, 1963). Confirmation of this report is required since chalcones are otherwise unknown as constituents of citrus fruits, the latter being regarded as too acid to prevent spontaneous cyclization of chalcones to flavanones (Maier and Metzler, 1967b).

B. *C*-Glycosides The *C*-glycosyl flavonoids are compounds of flavonoids and sugars in which the two moieties are linked not by means of carbon-to-oxygen bonds as in the normal glycosides, but by means of carbon-to-carbon bonds, usually C-6 or C-8 of the flavonoid and C-1 of the sugar. Such glycosides are resistant to acid and enzymic hydrolysis, although they present many similarities, e.g., in their chromatographic behavior, to the *O*-glycosides. Lemons provide interesting examples of the natural co-occurrence of several *O*- and *C*-glycosyl derivatives of the same flavones, apigenin and diosmetin.

The first detailed report of the presence of flavonoid *C*-glycosides in citrus was made by Chopin *et al.* (1964) who provisionally identified three lemon peel extractives as diosmetin-*C*-glycoside, apigenin-*C*-glycoside, and a luteolin-*C*-glycoside different from orientin (luteolin-8-*C*-glucoside). About the same time Horowitz and Gentili (1964) remarked briefly on the occurrence in citrus of vitexin, vitexin xyloside, and a number of closely related flavone *C*-glycosides; they subsequently identified the vitexin glycoside from oranges as 2′-*O*-β-D-xylopyranosylvitexin (Horowitz and Gentili, 1966).

In a recent paper by Gentili and Horowitz (1968), which presents a fuller account of their studies, they report that the *C*-glycosylflavones are present in very low concentration in citrus and can be isolated in milligram quantities only with considerable difficulty. Starting with commercial flavonoid preparations, they applied the technique of silicic acid chromatography developed for the *O*-glycosides and obtained 8-(*C*-β-D-glucopyranosyl)diosmetin from both orange and lemon flavonoids, and 6-(*C*-β-D-glucopyranosyl)diosmetin from lemon flavonoids only. Although partial structures for these compounds were derived by classical degradation methods, the small quantities available made necessary the application of nuclear magnetic resonance measurements to locate the position of the glycosyl units. Lemons also yielded, in even lower concentrations, two flavones which apparently contained more than one glycosyl residue; these compounds were provisionally identified as 6,8-di-*C*-glycosylapigenin and

6,8-di-*C*-glycosyldiosmetin. Subsequently, by direct comparison of natural and synthetic compounds, Chopin *et al.* (1968, 1969) confirmed the presence in lemons of 6-(C-β-D-glucopyranosyl)diosmetin and 6,8-di-C-(β-D-glucopyranosyl)apigenin.

IV. Grapefruit

The intensely bitter naringin was the only flavonoid isolated from grapefruit until 1961 when, in studies leading to the elucidation of the relationship between the bitter and nonbitter flavonoid rhamnoglucosides, Horowitz and Gentili (1961a) reported that a second bitter principle, poncirin, was present in concentrations about one-fifth those of naringin.

A more extensive examination of the grapefruit flavonoids, carried out by Dunlap and Wender (1962) at the Chemistry Department of the University of Oklahoma, involved the use of a technique of magnesol column chromatography previously developed for the separation of orange flavonoids. From commercial naringin they isolated naringin, neohesperidin, and rhoifolin, together with three unidentified compounds—a naringenin rhamnoglucoside, an isosakurenetin rhamnoglucoside, and a second apigenin glycoside. Kaempferol was also identified among the products of the hydrolysis of the grapefruit flavonoids, in addition to the aglycones of the glycosides mentioned.

The complex glycosidation pattern of the grapefruit flavonoids was finally elucidated as involving both the tasteless rutinosides (6-rhamnoglucosides) and the bitter neohesperidosides (2-rhamnoglucosides) of naringenin, hesperetin, and isosakuranetin (Mizelle *et al.*, 1965). Separation of these six closely related compounds was achieved only by the use of column and thin-layer chromatography in a complicated procedure that could also be employed for quantitative analysis (Hagen *et al.*, 1965). The neohesperidosides of hesperetin (neohesperidin) and isosakurenetin (poncirin) constitute about 8% of the total flavanone neohesperidosides and thus, compared with naringin, make only a small contribution to the bitterness of grapefruit products (Hagen *et al.*, 1966). Further fractionation of the column eluates led to the isolation of two new compounds, the 4′-β-*O*-glucosides of the two naringenin rhamnoglucosides, the first trisaccharides obtained from citrus (Mizelle *et al.*, 1967).

The flavonoids and related polyphenolics in grapefruit form one of the most complete "metabolic grids" yet found in a single plant tissue. First kaempferol (Dunlap and Wender, 1962) and then dihydrokaempferol (Maier and Metzler, 1967a) were identified among the hydrolysis products from grapefruit flavonoids. Further intensive search (Maier and Metzler, 1967b) yielded another flavanone, eriodictyol, the corresponding flavonol, quercetin, and its 3′-methyl ether, isorhamnetin, in addition to a number of related simple polyphenolics.

Thus the dominant aglycone in grapefruit, the flavanone naringenin, is accompanied by apigenin, kaempferol, and dihydrokaempferol, the related flavone, flavonol, and dihydroflavonol, respectively, and by three other flavanones, isosakuranetin, eriodictyol, and hesperetin, which represent increasing substitution of the naringenin molecule. In addition, kaempferol is accompanied by two other flavonols, quercetin and isorhamnetin, which represent similar increasing substitution of the kaempferol molecule. The importance of these relationships in the metabolism of citrus flavonoids and particularly in the disappearance of bitterness from the maturing grapefruit will be discussed later.

Rhoifolin, the apigenin rhamnoglucoside first isolated from commercial naringin by Dunlap and Wender (1962), attained pharmacological importance with the reported spasmolytic activity of the aglycone (Janku, 1957). A procedure that avoided the use of column chromatography was developed by Rowell and Beisel (1963) for its isolation in gram quantities from grapefruit peel in which it is present to the extent of 1 mg/100 g dry weight.

Grapefruit peel oil has been reported to contain 3,3′,4′,5,6,7,8-heptamethoxyflavone and two other highly methoxylated flavones (Stanley *et al.*, 1967). Such compounds, which are of restricted occurrence in citrus, are discussed in more detail in the next section.

V. Oranges

A. Simple Flavanones and Flavones Despite the commercial importance of the fruit, the flavonoids of the orange have not received the intensive study given to those of the lemon and grapefruit in recent years. Hesperidin predominates among the orange flavonoids, and most workers continue to report the results obtained by nonspecific colorimetric analyses as "hesperidin" content.

The pharmacological properties ascribed to hesperidin and the difficulties in obtaining pure samples for testing prompted Dunlap and Wender (1960) to develop a procedure for the separation of orange flavonoids by chromatography on a magnesol column using aqueous ethyl acetate as eluant. They were successful in obtaining chromatographically homogeneous hesperidin whereas even the purest samples of commercial hesperidin were contaminated with isosakuranetin-7-rhamnoglucoside. From orange peel they also isolated a naringenin rhamnoglucoside which they were convinced was the neohesperidoside, naringin. However, Gentili and Horowitz (1965) subsequently isolated from the peel of both Navel and Valencia oranges, in addition to hesperidin, glycosides of naringenin and isosakuranetin, which they identified as 7-β-rutinosides like hesperidin. These structures were subsequently confirmed by synthesis (H. Wagner *et al.*, 1969). The three compounds were thus the isomers of the bitter neohesperidosides, neohesperidin, naringin, and poncirin, none of which they were able to find in oranges. Gerngross and Renda (1966) reported the presence of naringin in orange peel but they used a nonspecific colorimetric method for its detection.

Dunlap and Wender (1960) found evidence for the presence of other flavonoid glycosides in oranges, but only two have so far been isolated: the 3-glucoside of limocitrin (Gentili and Horowitz, 1964) and the 8-*C*-glucoside of diosmetin (Gentili and Horowitz, 1968), both of which were obtained from crude flavonoids from both oranges and lemons. Mizelle *et al.* (1967) reported chromatographic evidence for the presence in orange peel of the two isomeric trisaccharides, the 4′-glucosyl-7-rhamnoglucosides of naringenin, previously isolated from grapefruit.

From orange peel Rahman and Khan (1962) obtained a compound which they believed identical with "citronin," previously reported as a constituent of the Ponderosa lemon by Yamamoto and Oshima (1931). "Citronin" was considered to be the 7-rhamnoglucoside of 5,7-dihydroxy-2′-methoxyflavanone, but Horowitz and Gentili (1960c) noted that some of its recorded chemical properties were incompatible with this structure. Although Rahman and Khan obtained similar melting points for both the aglycone from hydrolysis of "citronin" and the 5,7-dihydroxy-2′-methoxyflavanone synthesized by Shinoda and Sato (1931), the actual structure of their product and that of Yamamoto and Oshima remains in doubt, particularly since the assigned structure would make the compound unique among the citrus flavonoids by virtue of an oxygenated substituent at the 2′-position.

The compound "cirantin" isolated from orange peel and reported to possess antifertility properties (Ghosh *et al.*, 1955) was later shown to be identical with hesperidin by Manwaring and Rickards (1968).

B. Anthocyanins of Blood Oranges The pigments of blood oranges are anthocyanins, the only occurrence of this type of flavonoid in citrus, other red pigments being carotenoids. Chandler (1958a) identified cyanidin-3-glucoside as the predominant anthocyanin in the Moro variety of the blood orange; traces of a second pigment, provisionally identified as delphinidin-3-glucoside, were also found. Investigation of three other varieties, Tarocco, Florida, and Ruby, indicated no major varietal differences in anthocyanin pigmentation.

The results relating to the Moro variety were confirmed by Koch and Haase-Sajak (1965c), but a variation in the anthocyanin components was reported in three other varieties, Sanguinella, Paterno, and Spanish Half-blood. All contained cyanidin-3-glucoside, but delphinidin-3-glucoside occurred in the Sanguinella alone and there in only trace amounts. On the other hand, all three varieties contained two unidentified anthocyanins, one of which was present in the Sanguinella in a concentration similar to that of the cyanidin glucoside. Subsequently four unidentified pigments were reported to accompany cyanidin-3-

glucoside in the Moro variety but in amounts too small to permit identification (Poretta *et al.*, 1966). A publication has recently appeared on the anthocyanins in Calabrian orange juice (Kunkar, 1968), but no details of the work are available.

It is surprising that no anthocyanin rhamnoglucosides have so far been reported in the blood orange in view of the predominance of such glycosides among the citrus flavonoids. Probably the unidentified blood orange pigments will fall into this category, with the pigmentation pattern of aglycone, glucoside, and rhamnoglucoside of cyanidin and delphinidin common to all varieties.

C. Permethoxylated Flavonoids Citrus fruits are notable for the presence in their oil receptacles of a number of highly methoxylated flavonoids, many of which occur in no other plants; in many cases the 5-hydroxyl group is methylated despite the extreme resistance of this group to chemical methylation. Of all the citrus fruits, the orange contains the greatest number and the most highly methoxylated of such compounds, and in addition it contains only one representative of the moderately methoxylated flavones occurring in other citrus. Permethoxylated flavones account for 80% of the neutral fraction of the benzene-soluble components of orange peel juice and are responsible for 75% of the bitterness of this material (Swift, 1965b,c).

The hexamethoxyflavone, nobiletin, is the major permethoxylated flavonoid in oranges. It was present in orange peel juice at a level of 20 ppm (Swift, 1960), tentatively identified as a constituent of Navel oranges by Curl and Bailey (1961), and isolated from dried orange peel in 0.8% yield by Sastry *et al.* (1964). In addition, Böhme and Volcker (1959) isolated from orange oil tangeretin, the major permethoxylated flavonoid of mandarins 3,3′,4′,5,6,7,8-heptamethoxyflavone was also isolated.

A tangeretin isomer, 3′,4′,5,6,7-pentamethoxyflavone, was isolated from orange peel by Born (1960), and from orange peel juice by Swift (1964), who named it sinensetin and reported it to be, like nobiletin, very bitter. Swift (1965a) later isolated the tetramethoxyflavone, tetra-*O*-methylscutellarein, from peel juice, while D'Amore and Calapaj (1965) obtained chromatographic evidence for the occurrence of tangeretin and the moderately methoxylated compound, 5,8-dihydroxy-3,3′,4′,7-tetramethoxyflavone, with nobiletin in orange peel oil. Nobiletin, sinensetin, and the heptamethoxyflavone, together with β-sitosterol glycoside, were the only compounds positively identified by Tatum *et al.* (1965) in a chromatographic comparison of orange concentrate and foam-mat dried powder.

Seasonal variations in the permethoxylated flavones of orange peel juice were studied by a method involving thin-layer chromatography and spectrophotometry (Swift, 1967). Even though fruit of mixed varieties were used, the results obtained were reasonably consistent. Nobiletin and sinensetin were present in the greatest amount—about 20% of the total neutral fraction—followed by tetra-*O*-methylscutellarein (8%), heptamethoxyflavone (5%), and tangeretin (3%). There was only a slight upward trend in the concentration of the flavones with maturity; nobiletin showed the greatest increase.

VI. Other Citrus Species

It is regrettable that many workers on citrus constituents, particularly those of the less known varieties, have reported the presence and content of specific flavonoids, e.g., naringin or hesperidin, when their data were obtained by unspecific colorimetric methods. Thus, hesperidin has been reported as a constituent of the Clementine mandarin (Benk, 1961a) and of the bergamot (M. Calvarano, 1961a), without satisfactory evidence of identification.

The need for specific methods in flavonoid identifications has been emphasized, particularly procedures for the isomeric rhamnoglucosides, by Horowitz (1964) because of the taxonomic significance that may be attached to the results. Horowitz suggested that citrus fruits could be classified into two main groups on the basis of the predominance of one of the rhamnoglucoside types, neohesperidosides or rutinosides, over the other. Thus, lemons, oranges, and mandarins were placed in the rutinoside category, and grapefruit and sour oranges in the neohesperidoside category. In an examination of over 60 Japanese citrus fruits and hybrids, Nishiura *et al.* (1969) found the flavonoid components of about 30 recognized varieties to conform to this classification. Of 30 fruits of doubtful taxonomy, one half could be placed in the rutinoside (hesperidin) category and the other half in the neohesperidoside (naringin/neohesperidin) category, while the hybrids fell into one or other of these categories or else into a group which contained both types of flavonoids. For instance, the Natsudaidai, a hybrid with parents from both categories, contains both naringin and hesperidin (Nomura, 1953) while two hybrids of more doubtful parentage, the Meyer lemon and the Ichang lemon, contain hesperidin only (Nishiura *et al.*, 1969).

A. Sour Oranges Based on the evidence of their flavonoid components, there seem to be two groups of sour orange varieties. The early report of Kolle and Gloppe (1936) that neohesperidin was the dominant flavonoid in the immature fruit (up to 15% dry weight) was confirmed by Nakabayashi (1961c) who, however, found that the species he examined were distinguished by the virtual disappearance of neohesperidin with maturity, while the concentration of the second major flavonoid, naringin, remained unchanged. Nakabayashi also isolated a third neohesperidoside, rhoifolin, from this fruit, as well as a compound named lonicerin about which no further information is available.

On the other hand, according to Karrer (1949), immature fruit of European varieties contained both hesperidin and neohesperidin, with hesperidin dominant in Italian samples and neohesperidin in Spanish samples. This result was confirmed in part by Venturella *et al.* (1964, 1965) who isolated hesperidin but no neohesperidin from two Italian sour orange varieties, Foetifera and Alphonsii, the latter also containing some unidentified isorhamnetin and acacetin glycosides. Finally, hesperidin was the only flavonoid found in two Japanese sour orange varieties (*C. myrtifolia* and *C. canaliculata*) by Nishiura *et al.* (1969), while another variety contained neohesperidin and naringin in the immature fruit.

Thus there appears to be a chemical basis for dividing *C. aurantium* varieties into two groups: sour oranges in which hesperidin is the dominant flavonoid, and bitter oranges in which neohesperidin and naringin predominate. Further studies are necessary however to elucidate this interesting problem in chemotaxonomy.

The permethoxylated flavone, auranetin, was isolated from Indian varieties of sour orange, together with two related flavonoids in which the 5-hydroxyl group had escaped methylation–5-*O*-demethylnobiletin and its isomer, 5-hydroxyauranetin (Sarin and Seshadri, 1960). From chromatographic studies, D'Amore and Calapaj (1965) reported four other permethoxylated flavonoids, tangeretin and nobiletin positively and auranetin and sinensetin provisionally, as constituents of sour orange oil. More recent studies by Schneider *et al.* (1968) confirmed the presence of nobiletin (in largest amount), tangeretin, and sinensetin, together with 4′,5,7,8-tetramethoxyflavone (an auranetin isomer) and 3′,4′,5,7,8-pentamethoxyflavone, neither of which have previously been identified as plant constituents.

An isomer of 5-*O*-demethyltangeretin, 5-hydroxy-3′,4′,7,8-tetramethoxyflavone, was found by Farid (1968) in oil of bergamot, a sour orange hybrid.

B. Mandarins and Related Fruits The isolation of hesperidin from several groups of mandarins–common, Satsuma, and Cleopatra mandarins and tangerines by Nishiura *et al.* (1969), and Mediterranean mandarins by Di Giacomo and Lo Presti (1959)–confirms the placing of this fruit by Horowitz (1964) in the same category as oranges and lemons. Moreover several hybrids thought to have mandarin parentage have also been reported to contain hesperidin: the Calamondin (Sastry and Row, 1960), the Temple tangor, and the Kusaie and Otaheite Rangpurs (Chaliha *et al.*, 1964a; Nishiura *et al.*, 1969).

The issue is complicated however by the report (Chaliha *et al.*, 1965a,b, 1967) that neohesperidin and hesperidin may sometimes occur together in the mandarin depending upon the origin of the fruit; these variations may have had their origin in similarly appearing but physiologically different forms. Obviously, close examination should be made of otherwise similar fruit that differ markedly in their neohesperidoside:rutinoside ratios.

The mandarin resembles the orange in the complexity of its permethoxylated flavonoids. Tangeretin and nobiletin have long been recognized as constituents of tangerines and mandarins; a method for their identification in tangerine oil using thin-layer chromatography has been developed by D'Amore and Calapaj (1965). Nobiletin, which was first obtained from the Chinese drug "Chen pi" derived from dried mandarin peel (Tseng, 1938), has also been isolated together with tangeretin from the Mediterranean mandarin (Venturella *et al.*, 1961; Bellino *et al.*, 1962). Tangeretin was detected in the Murcott tangor (Kesterson and Hendrickson, 1960). Like the rough lemon, the mandarin contains 5-*O*-demethyl analogues of its principal permethoxylated flavonoids. Nobiletin is accompanied by 5-*O*-demethylnobiletin in the Calamondin (Sastry and Row, 1960), and nobiletin and tangeretin by 4′,5-*O*-demethylnobiletin and 4′,5-*O*-demethyltangeretin in the Cleopatra mandarin and Dancy tangerine (Pinkas *et al.*, 1968).

The compound, "ponkanetin," earlier reported as a constituent of the Ponkan mandarin (Ichikawa and Yamashita, 1941) and described as the pentamethoxyflavanone corresponding to tangeretin, has since been shown not to be a flavanone (Sehgal *et al.*, 1955) but tangeretin itself (S. Matsuura, 1957a,b). Nevertheless, permethoxylated flavanones have been isolated from the mandarin and related citrus fruits, the first authenticated report of their occurrence in nature. Sastry and Row (1961a,b) isolated two new flavonoids from the Calamondin, namely, the flavanones corresponding to nobiletin and 5-*O*-demethylnobiletin (Row and Sastry, 1963); in addition, hesperidin was present to the extent of 2% dry weight. These flavanones, which were named citromitin and 5-*O*-demethylcitromitin, were subsequently isolated from Assam varieties of the mandarin, together with hesperidin, neohesperidin, and tangeretin (Chaliha *et al.*, 1965b, 1967).

C. Pummelos and Related Fruits The pummelo can be placed in the same category as the grapefruit; its neohesperidoside:rutinose ratio does not appear to change with maturity, location, and variety, since the flavonoids reported as principle constituents by different workers are all neohesperidosides. Naringin has long been considered the main flavonoid, as confirmed by Khan *et al.* (1959). Recently, however, Chaliha *et al.* (1965b) reported the presence of poncirin (in addition to naringin) in the immature fruit, but like neohesperidin in the sour orange, poncirin almost completely disappears in the mature fruit.

Two Japanese varieties of pummelo contained naringin (0.1%) and rhoifolin (0.1%), while a third, the Tanikawa buntan, contained predominantly neohesperidin, some naringin and a little rhoifolin, with only slight changes in flavonoid composition with maturity (Matsuno, 1959b; Matsuno and Amano, 1968). Naringin, however, was the only flavonoid found by Nishiura *et al.* (1969) in the mature fruits of several pummelo varieties and pummelo-like fruits grown in

Japan, although some contained neohesperidin, either alone or accompanied by naringin and rhoifolin, in the immature fruits. These workers also reported a third isomer of naringin and narirutin, which they called isonaringin, as a constituent of two pummelo-like fruits, Tengu and Jagatarayu.

Paper chromatographic studies indicate that the pummelo contains the same two isomeric trisaccharides of naringenin that are present in grapefruit (Mizelle *et al.*, 1967).

D. Minor Species The trifoliate orange may be placed in the neohesperidoside category: Wan (1942) has identified neohesperidin in this fruit, in addition to the previously noted neohesperidosides, naringin and poncirin. The "citrofolioside" isolated from the trifoliate orange by Sannie and Sosa (1949) has been identified with poncirin by Kariyone and Matsuno (1954). In a recent study, Nishiura *et al.* (1969) could isolate only poncirin from immature fruits.

The Ponderosa hybrid was reported by Yamamoto and Oshima (1931) to contain "citronin," claimed to be a 2′-methoxylated flavanone otherwise unknown in citrus. However, (1) a synthetic specimen of the postulated aglycone had different properties from the natural product (Shinoda and Sato, 1931), (2) the reactions of "citronin" did not agree with the proposed structure as pointed out by Horowitz and Gentili (1960c), and (3) attempts to reproduce the isolation yielded only neohesperidin (Sarin and Seshadri, 1959; Horowitz and Gentili, 1960c). Horowitz and Gentili therefore suggested that this fruit is a relative of the grapefruit and not, as previously believed, of the lemon or citron both of which contain hesperidin as the principal flavonoid. An Asian variety of the citron has been found to contain a second rutinoside, diosmin (Matsuno, 1959a).

The rough lemon (jambhiri) varies in flavonoid content depending upon growing area, according to Chaliha *et al.* (1965a) who isolated both hesperidin and neohesperidin from fruit grown in one particular area of Assam and only hesperidin from fruit from other areas. Although no information on relative concentrations is available, it is tempting to suggest again that two physiologically different fruit of similar appearance have been confused. However, the permethoxylated flavonoids, tangeretin and 5-*O*-demethyltangeretin were present in all samples of rough lemon examined, although such compounds were not found in the common lemon (Chaliha *et al.*, 1965b).

Citrus sudachi, which is probably a hybrid papeda, contains the partially methoxylated flavones, sudachitin and 3′-demethoxysudachitin (Horie *et al.*, 1961, 1962) and thus stands closer to the lemon with its partially methoxylated flavones, limocitrol and isolimocitrol, than to the orange with its fully methoxylated flavones.

Another hybrid, the satsumelo, is unique among citrus fruits in containing rutin in amounts up to 3.2% of the dry weight of green peel (Krewson and

Couch, 1948). Among other citrus fruits quercetin glycosides have been found only in lemons (Horowitz and Gentili, 1960b) and grapefruit (Maier and Metzler, 1967b) and then only in trace amounts.

The divergence of kumquats from citrus species is reflected in the presence of another unique flavonoid, fortunellin, the 7-neohesperidoside of acacetin (Matsuno, 1958; Nakabayashi, 1961a).

VII. Biochemical Aspects

The importance of flavonoids in citrus technology has resulted in numerous studies on the factors governing their occurrence. These studies have emphasized the practical aspects of the question, but in addition there is considerable interest in the physiological role of these compounds, which may be present in amounts up to 75% of the total solids of citrus fruits (Kefford, 1959).

The hydroxylated and glycosylated flavonoids are probably stored in the vacuoles of cells distributed throughout the fruit (Walther *et al.*, 1966). Until recently it was believed that they did not occur in the secretory cavities, but Peyron (1963) obtained evidence for the presence of hesperetin in the oil sacs of the orange. On the other hand, the permethoxylated flavonoids are apparently restricted to the oil receptacles of the fruit in which they occur.

The identification of a large number of variations in the basic flavonoid nucleus among the constituents of citrus fruits, particularly the lemon from which at least twenty flavonoids have been isolated, may be the consequence of the intensive studies to which they have been subjected. On the other hand, it is recognized that the more advanced plant families contain flavonoids with more complex hydroxylation, methoxylation, and glycosylation patterns (Swain, 1965). It is therefore likely that unique biosynthetic systems are present in citrus, with some systems occurring only in certain varieties.

A. Physiological Considerations Despite their omnipresence, sometimes in high concentrations, within the citrus species, the functional importance of the flavonoids in the development of the fruit is obscure. In the last decade there have been only two reports to indicate that they play any particular roles in plant physiology.

Naringenin, which occurs almost exclusively as the glycoside, naringin, in citrus fruits, acts as a growth inhibitor in dormant peach flower buds and apparently controls the emergence of peach buds from rest (Hendershott and Walker, 1959).

Secondly, certain permethoxylated flavonoids contribute to the resistance of citrus trees to the disease "mal secco." This disorder, widespread throughout the Mediterranean region, is caused by the fungus *Deuterophoma tracheiphila,* which may attack through the roots, leaves, or branches of the tree. In 1962, Ben-Aziz *et al.* reported the ability of extracts of the bark from resistant citrus varieties to inhibit growth of the fungus on synthetic media. These workers isolated two active constituents from these extracts, and identified the least active as naringin; neither of these compounds was present in the bark of susceptible species.

Later, Ben-Aziz (1967) named nobiletin as the principal fungistat in the peel, leaves, and bark of the highly resistant Cleopatra mandarin and Dancy tangerine. Rough lemon seedlings inoculated with the fungus succumbed to the disease within 2 months, while inoculated seedlings infused with nobiletin solution remained alive for 3 months. Tangeretin was also found to be a weak inhibitor of the fungus, while hesperidin had a slight stimulatory effect. Pinkas *et al.* (1968) subsequently identified two other flavones, in addition to nobiletin and tangeretin, as potent fungistats: 4′,5-dihydroxy-3′,6,7,8-tetramethoxyflavone (not previously reported in nature) and 4′-5-dihydroxy-6,7,8-trimethoxyflavone (not previously reported in citrus), both present in mandarin varieties resistant to the disease. These findings were the subject of a patent application by Ben-Aziz (1965).

Studies by Goren and Monselise (1965a) indicated interrelations between hesperidin and other components of the developing Shamouti orange, including enzyme and hormone systems, and led to the hypothesis that hesperidin functions as a growth promoter in citrus tissue. It is noteworthy that one of the threads of their argument involves the claim that hesperidin acts as protector for ascorbic acid; however, recent work by Clementson and Andersen (1966) reveals that pure hesperidin is without such protective activity, all effects previously claimed being due to impurities. Obviously, further work is required on the physiological importance of hesperidin and other flavonoids in developing citrus fruits before their functions can be defined.

Preliminary observations (Haskell, 1965) on the biochemical differences in color sectors of a chimeral orange fruit suggested that mutation had occurred in two stages since the orange sector showed the loss of two flavonoid compounds, and the yellow sector the additional loss of a third compound and the formation of a new constituent. Unfortunately, the flavonoids were not identified.

The wide variations in the modifications of the basic 15-carbon structure shown by the flavonoids of citrus open up prospects for the application of such knowledge to problems of citrus taxonomy. The suggestion that citrus fruits may be differentiated on the basis of their neohesperidoside:rutinoside ratio (Horowitz, 1964) has already been discussed while another major difference in the flavonoid composition of citrus fruits lies in the occurrence of the permethoxylated flavones. Differences in the flavonoid components during the maturation of citrus fruits do not appear to affect these particular aspects of their chemistry, but complications arise in the variations reported in mandarins, sour oranges, and rough lemons according to the growing area. Further studies on minor varieties are needed before reliance can be placed on the assignment of taxonomic classifications to citrus fruits on the basis of their flavonoid composition.

B. Effects of Horticultural Variables Although it is generally believed that the flavonoid content of citrus fruits decreases with increasing maturity (Kefford, 1959), this is not always the case. In Morocco (Huet, 1962b), the flavonoid content of Washington navel orange and grapefruit juices fell steadily in some seasons, but remained comparatively constant in others. In certain citrus fruits, one flavonoid component, neohesperidin in the sour orange and poncirin in the pummelo, has been reported to disappear with maturity, while another component, naringin in both cases, remains at a constant concentration (Nakabayashi, 1961c; Chaliha *et al.*, 1965b).

In more extensive studies, Hagen *et al.* (1966) applied techniques developed for the estimation of individual grapefruit flavonoids to follow the seasonal variations in six glycosides in the juice sacs of Texas Ruby Red grapefruit. The most significant changes occurred in the period July-November when the concentrations of all six glycosides decreased rapidly and proportionately to roughly one-third of the initial values. The average quantities of individual and total glycosides in the juice sacs of single grapefruit showed little seasonal variation and the results suggested that the decrease in flavonoid concentration was principally but not entirely the result of dilution with increasing fruit size. The ratios of the concentrations of the six glycosides remained comparatively constant during maturation, and there was no evidence of any conversion of the bit ter neohesperidosides to the tasteless rutinosides as suggested by Horowitz (1961b, 1964), nor of any other interconversions.

Shimba and Nakayama (1963) found no appreciable differences in the naringin content of small and large Natsudaidai fruits, while Bessho *et al.* (1964) reported that immature fruit contained more naringin than fully mature fruits, all differences between fruits being explicable in terms of the maturity of the fruit at harvest. However, according to Tasaka (1965), where other factors were equal, growth conditions and hence fruit size affected the concentration of flavonoids in the fruit; young trees bore fruits with lower flavonoid contents than old trees, and high acid and sugar contents were associated with high flavonoid contents.

The juice of the Shamouti orange seemed to be saturated with hesperidin during the maturation period, and the increase in the total hesperidin content of the fruit corresponded to the increase in juice content (Goren, 1965). Even the effect of exposure to sunlight, which increased the hesperidin content of the fruit on a dry weight basis, was seen as the result of the increase in juice content in exposed fruit. More hesperidin was found in the stylar half of the peel than in the stem half.

Large differences in the hesperidin content of the peels of seven Spanish varieties of orange were reported by Tarazona *et al.* (1959); the Cadenera orange contained almost twice as much as the Comuna orange at similar maturity as measured by sugar:acid ratio. Maleki and Sarkissian (1967) also found differences in the flavonoid content of Lebanese orange juices, Valencia juice having about half the flavonoid content of other varieties.

The effect of two rootstocks on the flavonoid content of Washington Navels was studied by Huet (1962b); early in the season, fruit from trees grown on sour orange stock had flavonoid contents 50% higher than those from trees on trifoliate stock, but at the end of the season differences were slight.

Limited information has accumulated on the effect of growing region on flavonoid composition, but it is more specific than other data reported in this section. Indian workers found the sour orange to contain, in addition to auranetin, 5-hydroxyauranetin when grown at Waltair on the east coast, and 5-demethylnobiletin when grown at Nagpur in the Central Provinces (Sarin and Seshadri, 1960). Similarly the Indian rough lemon (jambhiri) was found to contain, in addition to hesperidin and tangeretin, 5-*O*-demethyltangeretin and a new flavone when grown in the hills of Assam, and neohesperidin and an unidentifiable flavonoid when grown on the plains (Chaliha *et al.*, 1965a,b). As previously pointed out, these differences may have been the result of the examination of similar-looking but physiologically different fruits. The same workers also reported that mandarin varieties in Assam were unusual for the presence of tangeretin and neohesperidin, and the absence of auranetin, while still containing hesperidin, citromitin, and 5-*O*-demethylcitromitin like mandarins grown elsewhere (Chaliha *et al.*, 1967).

The flavonoid constituents of lemons grown in Florida and on the California coast were generally similar (Vandercook and Stephenson, 1966). However, the coastal lemons contained more phenols than fruit from the inland desert area, which had relatively lower eriodictyol and higher hesperetin contents. A similar trend was observed by Herzog and Monselise (1968) in grapefruit grown in Israel, where the naringin content of fruit grown in warm, arid areas was lower than that of fruit from cooler, humid areas. Less specific analytical methods led Rouse *et al.* (1958) and Maraulja *et al.* (1963) to state that oranges maturing in a severe winter yield juice with higher hesperidin content than usual, and it is possible that all these observations may be explained in terms of climatic differences.

Despite numerous studies on the effects of horticultural sprays on citrus quality, little is known of effects on flavonoid composition. Lead arsenate sprays significantly increased the flavonoid content of grapefruit, as well as the total sugar content and pH of the juice (Deszyck and Ting, 1960b). The application of gibberellin and 2-chloroethyltrimethylammonium chloride to Shamouti oranges when approaching maturity was studied by Monselise and Goren (1965); significant differences were found between the low hesperidin content of the gibberellin-treated fruit and the relatively high hesperidin content of the fruit treated with the quaternary ammonium salt, and parallel differences in reducing sugar content.

The effect of fruit storage on flavonoid content has also received little attention; while Vandercook *et al.* (1966) reported no change in the polyphenolic content of lemon juice with storage of the fruit, their analytical methods were not specific enough to demonstrate any changes in the contents of total or individual flavonoids.

C. Biogenesis of Citrus Flavonoids Two aspects of the biogenesis of citrus flavonoids are basically interesting: the formation of the flavonoids as a group, and the variations in flavonoid structure under different conditions. The wide variety in the patterns of hydroxylation, methoxylation, and to a lesser extent glycosylation, among the citrus flavonoids may arise by modification of a preformed flavonoid nucleus, but the isolation of *p*-coumaric and sinapic acids and the coumarins, scopoletin and umbelliferone, from lemons by Horowitz and Gentili (1960b) suggests that these compounds may be precursors of the variously substituted flavonoids. With the analytical methods recently developed for citrus flavonoids some advance is now possible in studies on their biogenesis. The procedure reported by Fisher (1968) for obtaining radioactive naringin from grapefruit leaves provides another useful tool for biosynthetic investigations.

One particular biogenetic process involving citrus flavonoids is of interest to both the plant biochemist and the food technologist. This is the marked decrease in the naringin content, and hence the bitterness, of the ripening grapefruit (Hagen *et al.*, 1966); similar observations have been made for neohesperidin in the sour orange (Nakabayashi, 1961c), for naringin in the Natsudaidai (Bessho *et al.*, 1964), and for poncirin, naringin, and neohesperidin in the pummelo (Chaliha *et al.*, 1965b; Nishiura *et al.*, 1969).

Griffiths and Lime (1959) found no evidence for the formation of naringenin or prunin, the tasteless aglycone and 3-glucoside from naringin, when this bitter flavanone disappeared from ripening grapefruit. However, the co-occurrence of rhoifolin, the corresponding tasteless flavone, with naringin in grapefruit (Dunlap and Wender, 1962) led Rowell and Beisel (1963) to suggest the conversion of naringin to rhoifolin as an explanation of the decline of grapefruit bitterness with maturity. A further possibility, proposed by Horowitz (1964), is the transglycosylation of the bitter 2-rhamnoglucoside (naringin) to the tasteless 6-rhamnoglucoside. Hagen *et al.* (1966), however, reported a slight decrease in the ratio of the bitter isomers to the tasteless ones in the endocarp of the maturing grapefruit, with a continuous decrease in all types of flavanones.

When Maier and Metzler (1967a) separated dihydrokaempferol from hydrolyzates of grapefruit flavonoids, completing with naringenin and kaempferol the first trio of related flavanone, dihydroflavonol, and flavonol isolated from the same actively metabolizing tissues, they suggested that dihydrokaempferol was an intermediate in the formation of kaempferol from naringin. Extending this idea, Maier and Metzler (1967b) reported the presence in grapefruit of the flavanone, eriodictyol, and the corresponding flavonol, quercetin, together with isorhamnetin, and a number of related polyphenolics, coumarins, coumaric acids, and phloroglucinol. With one exception, the furocoumarin, bergaptol, these compounds were all present in the "bound" form and were distributed in generally similar proportions in both peel and endocarp. Each compound can be related to one or more compounds with similar hydroxylation and methoxylation patterns in the "metabolic grid" previously mentioned with *p*-coumaric acid at its center. The absence of precursors with similar methylation patterns to hesperetin and isosakuranetin suggested that they are formed by direct enzymic methylation of eriodictyol and naringenin, respectively; while the transformation of the bitter glycosides to nonbitter glycosides of apigenin, dihydrokaempferol, or kaempferol, appears to be dependent on enzymes capable of acting on forms of naringenin (e.g., one particular diastereoisomer of free naringenin) that are present only in small amounts. No doubt these authors will extend their studies to variations in the polyphenolic compounds with maturity. The results of their work will be awaited with interest.

In tissue culture studies by Kordan and Morgenstern (1962) and Kordan (1965), flavonoids were found in the exudates from proliferating vesicle stalks from mature lemon fruits, and the continuous synthesis of these compounds by the growing tissues, and not by living but nonproliferating tissues, provided evidence that flavonoid synthesis occurs only in dividing or growing cells.

VII. Technological Aspects

A. Bitterness Problems The characteristic of citrus flavonoids of prime importance in food technology is their bitterness, and the relation between their structure and taste has been discussed in detail by Horowitz (1964) and Horowitz and Gentili (1969). The flavonoid neohesperidosides are generally, but not always, bitter, and a few other flavonoid types also show this property. However, as elaborated by Horowitz and Gentili (1959), Horowitz (1964), Maier and Dreyer (1965), and Hagen *et al.* (1966) among others, there are problems in assessing the bitterness of citrus products in terms of the content of flavonoids, either individually or as a group, because complications are introduced by the presence of the bitter limonoids and bitterness suppressors.

Only at high concentrations is the bitterness due to citrus flavonoids objectionable, and at moderate concentrations they may be desirable constituents. Thus, Buffa and Bellenot (1962b) suggest that a level of 0.03-0.07% naringin is required to give grapefruit juice its characteristic bitterness, juices with naringin contents below 0.03% being of poor quality. Naringin finds some use in the preparation of mildly bitter beverages. It was one of the bitter principles used by Dastoli *et al.* (1968) in a study of the relation between bitterness and absorption by a "bitter-sensitive" protein isolated from porcine taste buds.

The effect of variations in the extraction process on the naringin content and hence the bitterness of grapefruit juice was studied by Hendrickson *et al.* (1958a) using direct ultraviolet absorption measurements to determine naringin. The most important factors were found to be the amount of rag and peel incorporated into the juice, and the pressure used in juice extraction; there was also an unavoidable increase in naringin content during the heat process.

Huet (1961b) reported that in normal commercial operations naringin passed so rapidly into the juice from the albedo and membranes that bitterness could be decreased only by limiting the amount of pulp incorporated into the juice, e.g., by suitable settings of the juice extractor. Straining the juice almost at the moment of extraction reduced the naringin content by 20%, but the usual methods of pulp removal resulted in only 10% decrease and did not appreciably improve juice quality. Huet emphasized the importance of variety and maturity in controlling the quality of grapefruit products. Similar results were obtained by Buffa and Bellenot (1962b) for pummelo and sour orange juices as well as grapefruit, while with lemon juice Vandercook *et al.* (1966) found a significant increase in polyphenolic (principally flavonoid) content with increasing extraction pressure, although different sieving procedures had little effect.

Japanese workers have been particularly concerned with the Natsudaidai, an important commercial variety in that country. Nakabayashi (1962a) reported that 30-60% of the naringin in the peeled Natsudaidai could be readily removed by washing. Heating peeled fruit in water for 8 minutes at 60°-65°C solubilized 90% of the naringin without loss of texture, and in this way less bitter products, including canned segments, could be obtained from these fruit.

Bessho *et al.* (1964) found the distribution of naringin within the Natusdaidai to be similar to that in grapefruit (Kefford, 1959): the albedo contained most of the naringin in amounts up to 1.9%, while only 0.02 to 0.03% was present in the juice sacs. Most of the naringin in the endocarp was readily solubilized during extraction of the juice (Manabe *et al.*, 1964), while treatment with heat or with hemicellulase only slightly increased the amount dissolved. The juice dissolved far more naringin (0.02%) than did water (0.04%), but no explanation of this solubility difference was apparent from the composition of the juice. In processing the Natsudaidai as segments or whole fruit, retention of naringin within the core presents a problem, only 70% of its naringin content being removed in the hot-water treatment recommended earlier by Nakabayashi (1962a). Attempts to accelerate the removal of naringin by treatment with pectinase, cellulase, and hemicellulase were unsuccessful and also resulted in texture loss (Kodama *et al.*, 1964).

B. Crystallization Problems The unusual solubility characteristics of the flavonoids cause problems in citrus fruits and processed products (Kefford, 1959). They may separate as crystals in frosted oranges and grapefruits, they readily form supersaturated solutions on heating and may slowly crystallize out of citrus products during storage, and they may appear as deposits on the surfaces of processing equipment. These properties were generally reviewed by Braverman (1959) who, however, considered the precipitates in stored citrus oils to be degraded and oxidized flavonoids.

Naringin crystallization, a common defect in stored grapefruit products, has been the subject of two recent papers. Koeppen and Smit (1960) described a simple procedure for the identification as naringin of the crystalline deposits in canned grapefruit segments, and Basker (1965) determined the solubility of naringin in model systems in order to specify the conditions under which crystallization might occur in concentrated grapefruit juice. Basker found that crystal formation, which depended on the original naringin content and the level of juice concentration, could be reversed simply by heating the reconstituted juice, whereas deposits of hesperidin which sometimes form in orange juice concentrates could not be removed by this means.

In other citrus products hesperidin often separates as a cloud to give a muddy appearance. From recent studies on canned mandarin oranges Nakabayashi *et al.* (1959) concluded that there is a factor in the juice that keeps hesperidin in supersaturated solution for considerable storage periods. This observation is analogous to the report already mentioned on the increased solubility of naringin in Natsudaidai juice.

C. Flavonoid-Hydrolyzing Enzymes Because of the importance of bitterness in the consumer acceptance of many processed citrus products considerable interest has been aroused in enzymes of fungal origin which completely hydrolyze naringin to nonbitter products.

A Japanese worker, Kishi (1955), was the first to study naringinase enzymes, which he isolated from strains of *Aspergillus niger* grown on a citrus medium. Production of the enzyme was promoted by addition to the medium of naringin (0.05%) or rhamnose (Kishi, 1957), or soybean flour hydrolyzates (Kishi, 1958). However, from studies on the products of the action of the enzyme on naringin isolated from the shaddock, Kishi (1959) mistakenly concluded that naringenin and rutinose were first produced, the rutinose being then hydrolyzed to rhamnose and glucose.

The debittering of grapefruit products by naringinase was confirmed by Ting (1958) who also believed that naringenin was the final product. However, D. W. Thomas *et al.* (1958) demonstrated that naringin was only partially hydrolyzed, the resulting product being not naringin but prunin. These workers screened a number of microorganisms for their ability to produce the naringin-hydrolyzing enzymes and obtained a highly active preparation, Naringinase C-100, which rapidly debittered natural grapefruit juice in the pH range 3.5-5.0, at temperatures between 20° and 50°C. This preparation was the subject of a patent granted to Smythe and Thomas (1960) covering the production of naringinase from *Aspergillus* species, its use in the debittering of naringin-containing foods, and its use in the preparation of prunin from naringin. More efficient methods of naringinase production by *Aspergillus niger* were reported by Bram and Solomons (1965) who obtained a preparation having five times the activity of previous samples by keeping the pH of the growth medium above 4.5 and using 0.1% naringin as a stimulant, as indicated earlier by Kishi (1957).

Japanese workers have investigated many sources of naringinase. Takiguchi (1962) obtained a highly potent preparation from *Coniella diplodiella* with properties similar to enzymes produced from *Aspergillus.* He later investigated 132 strains of molds belonging to 39 genera for the production of naringinase (Takiguchi, 1965). Japanese patents for naringinase production have been issued to Iizuka *et al.* (1964), Sumitani *et al.* (1964), and Hara and Koaze (1965).

Griffiths and Lime (1959) demonstrated that Naringinase C-100 contained a number of enzymes, mainly rhamnosidase and glucosidase, as well as significant amounts of pectinesterase which must be deactivated before addition to grapefruit juice. Subsequently, Dunlap *et al.* (1962a) separated Naringinase C-100 into rhamnosidase and glucosidase fractions and studied their specificities and properties using a micromethod for the analysis of naringin, prunin, and naringenin specially developed for the purpose. To overcome the problem of the pectinase activity of naringinase preparations, Omura and his colleagues (1963a,b,c) recommended that naringinase enzyme solutions be brought to pH 8.0 and warmed at 37°C for 2 hours to inactivate the pectinase without affecting naringinase activity, while Morikawa *et al.* (1968) made use of the different solubilities of the two enzyme systems in alcohol.

Inactivation of pectinase at high pH levels, and also the inhibition of naringinase by sugars, as first reported by Griffiths and Lime (1959) and Nakabayashi (1962c), were confirmed by Nomura (1965). Among the sugars, rhamnose and glucose were the most powerful inhibitors, while sorbitol and ascorbic acid had little effect in studies reported by Omura *et al.* (1966), who suggested that the natural glucose of the Natsudaidai would have a significant inhibitory effect on the enzyme. Sulfhydryl reagents generally depressed enzyme action even at low concentrations, but reducing agents such as hydroxylamine and hydrazine had no effect.

A crystalline naringinase was prepared from *Aspergillus niger* by S. Okada *et al.* (1963a) who also purified a crystalline hesperidinase from the same source. The two enzymes, which differed slightly in temperature and pH requirements for optimum activity, hydrolyzed the two different types of rhamnoglucosides (S. Okada *et al.* 1963b). Naringinase hydrolyzed the rhamnose from the bitter neohesperidosides such as naringin, poncirin, and neohesperidin, while hesperidinase hydrolyzed the rhamnose from tasteless rutinosides such as hesperidin. Oddly enough, hesperidinase was reported to hydrolyze rutinose, but not rutin.

Also isolated from the same source was a flavonoid glucosidase capable of completings the hydrolysis of the products of naringinase and hesperidinase activity (S. Okada *et al.*, 1963c). All sugars had a strong inhibitory action on the flavonoid glucosidase, but, contrary to other reports (Omura *et al.*, 1966; Nomura, 1965), only rhamnose strongly inhibited naringinase activity (S. Okada *et al.*, 1964a). The glucosidase component of crude "naringinase" preparations was inactivated by heating for 1 hour at 60°C at pH 6.4-6.8, while the naringinase retained its activity; the resultant preparation was used to obtain several flavonoid glucosides not previously prepared (Kamiya *et al.*, 1967).

D. Enzyme Treatment of Grapefruit Products In their early work on the use of naringinase for debittering grapefruit products, D. W. Thomas *et al.* (1958) showed that only partial hydrolysis of naringin (naringenin-7-rhamnoglucoside) to prunin (naringenin-7-glucoside) was necessary for successful debittering. The Davis colorimetric test for naringin could not, therefore, be used to follow debittering by the enzyme since it did not distinguish between naringin and prunin. Accordingly, these workers developed a specific method for naringin analysis; other suitable procedures have been suggested by Dunlap *et al.* (1962a), S. Okada *et al.* (1965), and Hagen *et al.* (1965).

The naringinase system was investigated by Griffiths and Lime (1959) with particular reference to its use in debittering Texas Ruby Red grapefruit juice which had been brought to an acceptable color by the addition of pigmented pulp. With an enzyme concentration of 0.01-0.05% and a natural juice pH of 3.1, incubation periods of 1 to 4 hours at 50°C or 44 hours at 4°C were required.

The application of naringinase to the debittering of concentrated grapefruit juice was studied by Olsen and Hill (1964), who found that hydrolysis occurred in reconstituted juices much more rapidly than in concentrates. Enzyme preparations from *Aspergillus niger* have recently been used in Russia for the debittering of grapefruit juices (Fishman and Gumanitskaya, 1967).

E. Enzyme Treatment of Japanese Citrus Products After the Satsuma mandarin, the next most important citrus fruit in Japan is the Natsudaidai which contains both bitter flavonoids and bitter limonoids. Much work has therefore been done to obtain nonbitter products from it (Fukomoto and Okada, 1963).

The first application of naringinase to Japanese citrus products was reported by Nakabayashi (1962c) at a stage when the purity of the preparations was questionable; the activity of his sample was reduced in proportion to the square root of the sugar concentration of the juices tested. Nomura *et al.* (1963) and Nomura and Akiyama (1964) suggested that this inhibition could be overcome by soaking the peeled fruits in solutions of pectinase-containing naringinase at 20°C for 16 hours prior to processing. However, Kodama *et al.* (1964) found the naringin in the central bundle to be resistant to enzyme action although most of the naringin was removed from the membranes. Tsusaka (1965), on the other hand, obtained satisfactory results with a naringinase preparation of low pectinase activity and an even better product by adding the enzyme to the fruit in the can.

Naringinase retained its activity after heating 10 min. at 80°C in sugar syrup at pH 3.0-3.4 (Kubo *et al.*, 1966a,b). The one-step process debittered canned Natsudaidai almost completely after storage at 30°C for 2 weeks or at room temperature for 2-3 months; the same process could be used for production of an acceptable marmalade, while incubation periods of about 2 hours were effective for canned juice (Shimoda *et al.*, 1966b, 1968a). The process was ineffective in cyclamate-sweetened products, owing to inactivation of the enzyme by the artificial sweetener; saccharin and dulcin however did not significantly inhibit naringinase activity (Shimoda *et al.*, 1968b).

Shimoda and his group also studied the application of hesperidinase for preventing turbidity in canned mandarin orange resulting from crystallization of hesperidin during storage (Sawayama *et al.*, 1966). Hydrolysis of hesperidin to the more soluble hesperetin glucoside by hesperidinase added to the canned product was more effective than the previously used method involving addition of methylcellulose (Anonymous, 1963), giving a satisfactory product after 1 month at room temperature. Again, cyclamate inhibited the enzyme but sufficient activity was retained when 50% or more of the cyclamate was replaced by saccharin (Shimoda *et al.*, 1968c).

F. Citrus Flavonoids as Antioxidants The general antioxidant activity of citrus flavonoids has long been recognized, particularly with respect to protection of ascorbic acid, but conflicting results have been reported in recent studies of this property, arising in some cases from the fact that different types of antioxidant activity were investigated.

Di Giacomo and Rispoli (1962b) reported that component flavonoids afforded considerable protection of citrus beverages against ultraviolet-catalyzed oxidations which caused loss of color, flavor, and ascorbic acid, with the result that products containing pulp had longer storage life than clear beverages. Thus, the color of nonpulpy fruit drinks, as measured by carotenoid content, was maintained by the addition of ascorbic acid (40 ppm) and citrus flavonoids (250 ppm) (Rispoli and Di Giacomo, 1962).

On the other hand, Gutfinger and Zimmermann (1962) reported that hesperidin, naringin, and their aglycones did not retard the oxidative spoilage of several vegetable oils, although crude citrus flavonoids were satisfactory antioxidants for animal fats, vegetable oils, and ascorbic acid. Similarly Davidek (1963) tested a number of flavonoids for their inhibition of ascorbic acid oxidation and found that pure naringin had only slight activity while crude orange flavonoids showed the greatest stabilizing activity.

In 1965, Ting and Newhall demonstrated that the principal natural antioxidant of citrus fruits, as measured by inhibition of limonene oxidation, was probably a tocopherol and certainly not a common flavonoid. Oranges yielded highly active preparations, while lemons contained little of the antioxidant. Pure hesperidin and naringin were without effect, and the antioxidant was later identified as the vitamin-E factor, α-tocopherol. This compound is present to the extent of about 0.01% in orange flavedo and 0.4% in crude dewaxed orange oil, but could not be detected in screened samples of juice (Newhall and Ting, 1965). In Satsuma mandarin flavedo Inagaki *et al.* (1968) found similar concentrations of both α- and γ-tocopherols.

Clementson and Andersen (1966) confirmed that hesperidin had no antioxidant activity with respect to ascorbic acid oxidation and pointed out that neither hesperidin nor naringin contains the chelating functions associated with strong inhibition of the metal-catalyzed oxidation of ascorbic acid, such as exist in quercetin.

The position appears to be, therefore, that lipid oxidations, such as those associated with some color and flavor changes in citrus juices, may be inhibited by α-tocopherol, while inhibition of ascorbic acid oxidation is the result of complex formation between certain minor citrus flavonoids and metals which catalyze this oxidation. In the latter connection it is interesting to note that complex formation between contaminating iron and a flavonoid has been blamed for darkening of Natsudaidai peel oil (Hattori and Konishi, 1963).

G. Flavonoids as Evidence of Authenticity Measurements of the flavonoid content have been suggested for testing the authenticity of citrus products (Kefford, 1959). Thus Vandercook and Rolle (1963) used ultraviolet spectrophotometric measurements of the flavonoid content, together with data on total amino acids and *l*-malic acid, to characterize the authenticity of lemon juice samples and to check the fruit content of lemon juice products (see Chapter 4); the method gave satisfactory results despite the variations in the concentration of flavonoids and other constituents in lemon juice that arise from varying extraction pressures (Rolle and Vandercook, 1963). This procedure was later examined in collaborative studies and was approved as official, first action, by the Association of Official Analytical Chemists (H. Yokoyama, 1965). Wagner and Monselise (1963) also considered the flavonoid content of Israel citrus juices in relation to adulteration and found no useful correlations, but the formula applied by Rolle and Vandercook involved factors not considered by the workers in Israel. Koch and Haase-Sajak (1965a) reported that hesperidin content, alongswith crude fiber and pectin contents, provided a suitable means of identifyingsthe presence of orange juices in fruit drinks.

IX. Flavonoids as By-Products

A. Recovery of Hesperidin and Naringin Because of the comparatively high concentrations in which they occur in citrus fruits and the ease with which they can be isolated, the flavonoids have often attracted attention as a means of increasing the profitability of citrus processing (Vincent, 1962; Hendrickson and Kesterson, 1965). It has been calculated that 16 million pounds of hesperidin could be recovered annually from the residues of Florida processors alone, while the present production in the United States of this chemical amounts to only about 100,000 pounds. Similarly the annual wastes of the Florida grapefruit industry contain several million pounds of naringin, little of which is recovered despite its immediate use as a bittering agent for beverages.

Comprehensive studies of the recovery and utilization of citrus flavonoids have been made by the research team of the Florida Citrus Commission at the Citrus Experiment Station, Lake Alfred, and bulletins have been published on naringin (Kesterson and Hendrickson, 1957) and hesperidin (Hendrickson and Kesterson, 1964b). Assessing the value of lemons as a commercial crop, these workers (Kesterson and Hendrickson, 1958) concluded that they had no competitive advantage over oranges as a source of flavonoids. When oranges were looked at solely as a source of hesperidin, on the assumption that it might be profitable to dispose of surplus crops in this way, peak yields in terms of pounds of hesperidin per acre were obtained when the fruit was about 2 inches in diameter (Hendrickson and Kesterson, 1962, 1964b).

The extraction of hesperidin from citrus wastes was the subject of two Japanese patents (Aratake, 1960; Nakabayashi, 1960) and of two papers in which tangerines (Di Giacomo and Lo Presti, 1959) and mandarins (Hirota, 1962b) were the starting materials. Tarazona *et al.* (1959) have described its recovery from sediments formed in orange juice evaporators. Most of these methods employed caustic soda or lime to extract the hesperidin, but Bonnell (1958) recommended the use of liquid ammonia to give a more marketable residue.

The preparation of "bioflavonoids" from citrus wastes is the subject of patents in the United States (Horowitz, 1958; Freedman *et al.*, 1959, 1960; Sudarsky and Fisher, 1961), and France (Arnaud, 1962a,b), and of a paper by Romanenko (1964). In addition, procedures have been described for the preparation from citrus peel of highly purified flavonoids suitable for biological testing (Dunlap *et al.*, 1962b).

B. Citrus Flavonoids as Pharmaceutical Chemicals Claims continue to be made for the pharmacological activity of citrus flavonoids, which have been grouped with flavonoids from other sources under the labels "vitamin P" and "bioflavonoids." It is, however, noteworthy that the number of references to "vitamin P" in *Chemical Abstracts* has markedly declined over the period under review, e.g., from 28 in 1961, 34 in 1962, and 27 in 1963 to only 12 in 1965 and 11 in 1966. There is still considerable confusion in the literature arising from the failure of many investigators to indicate the nature or source of the bioflavonoids used in their tests, thereby compounding the difficulties involved in pharmacological studies by hindering attempts to confirm their results.

Several books (Böehm, 1967; Harborne, 1967; U.S. Department of Commerce, 1963) and reviews have discussed the general and specific aspects of flavonoid pharmacology as follows: the physiological effect of flavonoids and their metabolic fate (De Eds, 1959), the action of flavonoids on capillary structure (Lockett, 1959), the activity of flavonoids on vascular resistance (Lavollay and Neumann, 1959), the pharmacology of bioflavonoids (Babin *et al.*, 1959; Vogin, 1960), the promotion of vitamin C activity by bioflavonoids (Fabianek, 1961), the use of flavonoids in cosmetic preparations (Avalle, 1961), the significance of vitamin P in food plants (Heintze, 1965), and the role of flavonoids in human physiology and nutrition (Charley, 1966).

Claims made specifically for citrus flavonoids include protective activity against adrenergic poisons (Call and Patterson, 1958), antitumor activity (L. Cohen and Cohen, 1959), antiinflammatory and capillary-strengthening activity (Menkin, 1959; Leger *et al.*, 1960; Robbins, 1966), activity in accelerating hydrocortisone metabolism (Weichselbaum and Margraf, 1961), effects on lipid and protein metabolism during senescence (Ravina, 1964), cytostatic activity against Zebra-fish embryos (R. W. Jones *et al.*, 1964), activity in promoting the retention of natural teeth (El Ashiry *et al.*, 1964), choleretic activity (Bezanger-Beauquesne *et al.*, 1964), activity in reducing the effects of bruising during "knockabout" sports (Woods, 1966), activity in blood cell aggregation (Robbins, 1966), and activity in promoting the transplacental passage of fetal erythrocytes during normal pregnancy (Clayton *et al.*, 1967). In addition, the consumption of citrus fruits has been suggested as a preventative of "little stroke" through the intake of bioflavonoids (McConnell and Sokoloff, 1959). A brief summary of the use of citrus flavonoids in pharmacology was made by Huet (1962a).

In many cases where positive results were reported after treatments with citrus flavonoids, the investigators have pointed out the difficulties that may arise in the testing procedures and in the interpretation of the results. In particular no method for assessing bioflavonoid activity has received general acceptance although the 20-year old procedure of Bacharach and Coates (1942,1943) was still recommended in a recent review (Charley, 1966).

A standard procedure involving direct measurement of edema reduction was suggested by Freedman and Merritt (1963), who found that naringin and hesperidin were inactive in reducing edema. However, nobiletin and a pentamethoxyflavone isomeric with tangeretin were almost as active as hydrocortisone phosphate, while an unidentified citrus constituent with flavonoid properties was more active still.

It is difficult to obtain pure preparations of citrus flavonoids, particularly the minor ones, for assessment of specific pharmacological properties, and such preparations when obtained tend to be more insoluble than the impure products. For these reasons, Aichinger *et al.* (1964) resorted to using commercial citrus flavonoid mixtures, which they found to inhibit the penetration of polyvinylpyrrolidone through vascular walls in rats by reducing the number or diameter of large pores.

There is considerable evidence that the results obtained in the administration of flavonoids in antiinflammatory studies vary considerably according to the circumstances of use. In the studies of Bavetta *et al.* (1964), various bioflavonoids, administered either orally or locally, showed no antiinflammatory activity in wounded rats, but completely counteracted the inhibitory effects on collagen synthesis produced by treatment with steroids. Different results were obtained in feeding trials with rats according to whether the animals were on a thrombogenic or an atherogenic diet (Robbins, 1967); crude citrus bioflavonoids and pure hesperidin, naringin, and tangeretin, when fed at the rate of 1-5 g/kg of the diet, increased the survival time of rats on a thrombogenic diet, while rats on an atherogenic diet gave variable results from which no conclusion could be drawn.

The claims that administraiton of "citrus bioflavonoids" reduced capillary fragility in human subjects lead Deyoe *et al.* (1962) to study their use in the control of bruising during the handling of broiling chickens. No apparent differences were observed in the dressed birds at the conclusion of growth studies in which they were fed citrus bioflavonoids up to a level of 5% of the diet; there was, however, no attempt to subject the birds to commercial handling, processing, or marketing. The flavonoids had no detrimental effect on the birds up to a level of 2½% of the diet; at 5% there was a marked reduction in growth and feed utilization which could have arisen from the substitution of the flavonoids for nutrients in the diet.

Many negative results have been reported from pharmacological tests involving citrus flavonoids: large doses had no effect on the severity of anaphylactic shock in rabbits (Lecomte, 1959); the fertility of male and female rabbits was not decreased (Friz, 1959); no important effects were observed on growth and capillary resistance of guinea pigs (McGraw, 1960); the survival time of mice and rats exposed to X-rays was not increased (Ershoff and Steers, 1960; Mucci *et al.*, 1966); there were no gross signs of dietary deficiency, nor any peripheral vascular phenomena observed in rats under several dietary regimens that could be explained by a lack of hesperidin in the diet or corrected by its supplementation (Lee, 1960); there was no acceleration of healing in rats (Vogin and Rossi, 1961), no estrogenic activity (Frank *et al.*, 1963), and no effect on lymph flow or on the permeability of the intact plasma-lymph barrier (Vogel and Stroecker, 1966).

This catalog of negative findings points strongly to the conclusion that the major citrus flavonoids have no beneficial pharmacological properties (Hänsel, 1965; Matusis, 1965). The positive therapeutic effects occasionally demonstrated were probably due to trace constituents which may have been permethoxylated flavonoids.

C. Citrus Flavonoids as Fine Chemicals A further outlet for hesperidin and naringin lies in the field of intermediates for preparation of other organic chemicals. A patent has been granted for the production of fluorescent wood dyes by the reaction of chlorinated hesperetin, naringenin, and eriodictyol with triazine compounds (Drummond, 1960). Greater therapeutic activity is claimed for flavonoid derivatives of higher water solubility, and a patent (B. F. Hart, 1960) covers the production of soluble hesperidin carboxyacid chalcone for the treatment of chronic kidney diseases, diseases of the eye, and rheumatoid diseases such as bursitis and osteoarthritis. Patents have been granted on the preparation of artificial sweeteners from the bitter flavonoid neohesperidosides (Horowitz and Gentili, 1963b, 1968; Feldman *et al.*, 1968), and it has been estimated that about 6 million pounds of citrus flavonoids could be absorbed annually in the production of such products.

Hesperidin and naringin are readily converted to the aglycones, hesperetin and naringenin, which give on fusion the corresponding benzoic acids in 80-85% yield and phloroglucinol in 65% yield. Newhall and Ting (1967) studied the optimum conditions for this reaction which offers a profitable new field for flavonoid utilization.

Finally two discoveries, already referred to, open up possibilities for the use of citrus flavonoids as agricultural chemicals: the inhibition of the citrus disease "mal secco" by permethoxylated flavones from the mandarin, and the inhibition of the growth of peach buds and flowers by nariningin (Chapter 13, Section VII).

Chapter 14 Triterpenoids and Derivatives

I. Steroids and Triterpenoids

β-Sitosterol, previously found in the oil, juice, and peel of the grapefruit (Kefford, 1959), is apparently the principle citrus steroid, occurring as the β-D-glucoside in grapefruit juice (Maier and Dreyer, 1965), in orange juice (Tatum *et al.*, 1965) and in lemon juice (Vandercook and Yokoyama, 1965) at about 80 ppm. Sitosterols, mainly β-sitosterol, have been isolated from the peels of the lemon (Chopin *et al.*, 1964), the mandarin (Chaliha *et al.*, 1967), and two mandarinlike fruit, the Rangpur (Chaliha *et al.*, 1964a) and the Calamondin (Row and Sastry, 1962b). Smaller amounts of cholesterol and γ-sitosterol accompany β-sitosterol in the nonvolatiles of orange peel oil (Böhme and Volcker, 1959), while ergosterol, which is closely related to β-sitosterol, has recently been isolated from the peel of the Rangpur (H. Yokoyama and White, 1968a). The amount of sitosterol in lemon juice remained constant under variations in storage of the fruit or processing of the juice which produced changes in the concentration of some other juice constituents, but the sterol content could not be used as an index of the authenticity of juice samples (Vandercook *et al.*, 1966).

The co-occurrence of the 4-α-methyl sterol, citrostadienol, with β-sitosterol in grapefruit peel (Mazur *et al.*, 1958) led to the suggestion that such compounds are intermediates in the biosynthesis of phytosterols from squalene cyclization products, fulfilling the same role as lanosterol in animals and fungi. More recently 4-α-methyl sterols have been shown to occur in other higher plants, and a closer examination of grapefruit sterols (Williams *et al.*, 1967; Goad *et al.*, 1967) revealed the presence of the following compounds which may be regarded as sterol intermediates: two 4,4-dimethylsterols, cycloartenol (21), and 24-methylene cycloartenol; four 4-methyl-sterols, cycloeucalenol (22), obtusifoliol (23), 24-methylene lophenol (24), and citrostadienol, together with four other unidentified 4-α-methyl sterols, two of which are probably 2-ξmethyl lophenol and 2-ξ-ethyl lophenol; and three sterols, stigmasterol, campesterol (25), and sitosterol. The biosynthetic pathway involves loss of a methyl group [e.g., (21)→(22)], opening of a cyclopropane ring [e.g., (22)→(23)], then loss of a second methyl group and migration of double bonds [(24)→(25)]. Investigations now center upon the detection of enzyme systems capable of carrying out these transformations.

The pentacyclic triterpenoid, friedelin (26), has not been mentioned as a citrus constituent since the first report of its isolation from grapefruit (Weizman *et al.*, 1955). Confirmation of its occurrence is desirable in the light of recent studies since it does not fit in with biogenetic patterns established for citrus triterpenoids, steroids, and limonoids, and since it is normally restricted to the bark and leaves of trees.

II. Limonoids

About 120 years after its first isolation in 1841, limonin, the main bitter principle of Navel oranges, was identified as a triterpenoid derivative (27) following the collaborative efforts of four leading research teams (Arigoni *et al.*, 1960; Arnott *et al.*, 1960), and it is noteworthy that its complicated structure[2] was elucidated completely only with the aid of X-ray crystallography. Limonin is thus a dicarbocyclic (B and C) compound with two lactone rings (A and D), a cyclic ether ring (A′), an epoxide group (E), a furan ring (F), and a ketone group. The arguments for this structure will not be given here, nor will the complex chemistry of the limonoids be discussed except where it is related to their occurrence in citrus fruits and products. The chemical and physical properties of limonin have been reported in detail by Barton *et al.* (1961), Arnott and Robertson (1959), and Arnott *et al.* (1961). Application of nuclear magnetic resonance, optical rotatory dispersion, and circular dichroism to structural and stereochemical problems in the limonoid series has been fully discussed by Dreyer (1965a, 1968a), who has also reviewed (Dreyer, 1968c) the structure determination and chemistry of natural products related to limonin including the meliacins and mexicanolides.

Early reports on various physiological activities of limonin have since been almost completely discounted, e.g., Emerson (1949), although according to Nikonov and Molodozhnikov (1964) it possesses hypoglycemic activity.

[2] *Chemical Abstracts* name: 8-(3-furyl)decahydro-2,2,4a,8a-tetramethyl-11H,13H-oxireno [*d*] pyrano[4′,3′:3,3a] isobenzofuro[5,4-*f*] [2] benzopyran-4,6,13(2H,5aH)-trione.

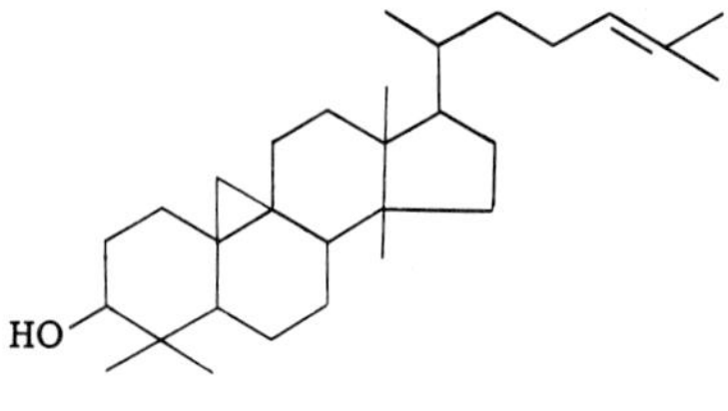

HO

Cycloartenol
(21)

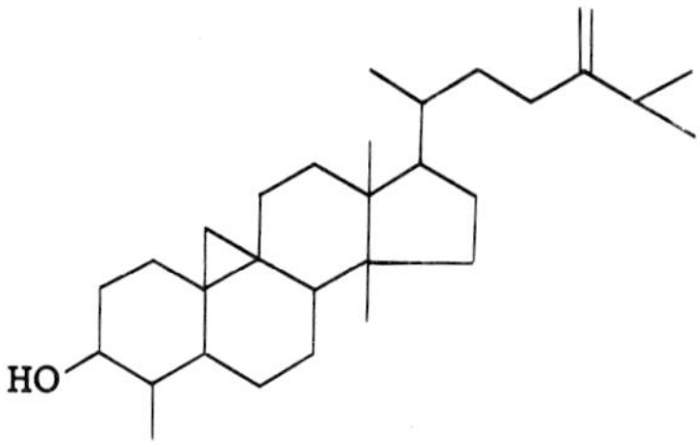

HO

Cycloeucalenol
(22)

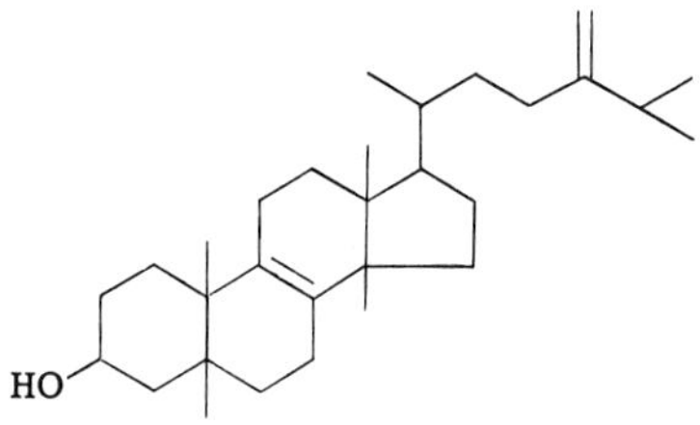

HO

Obtusifoliol
(23)

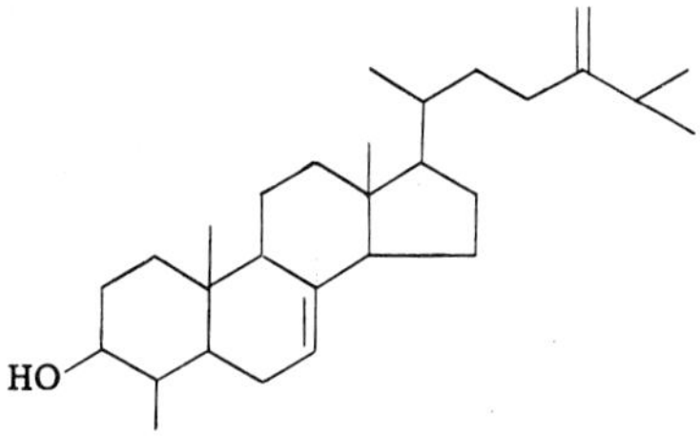

HO

24-Methylene lophenol
(24)

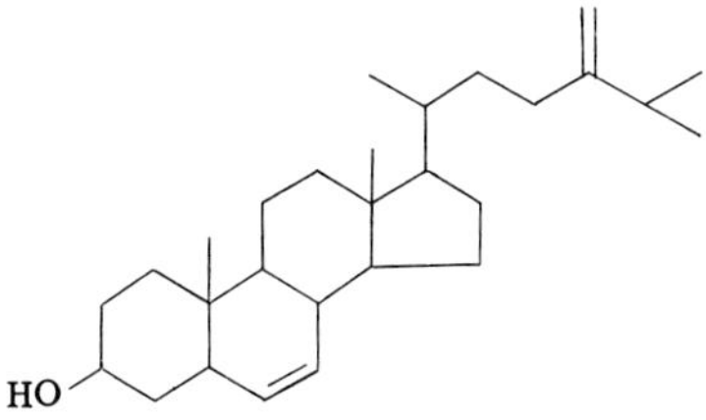

HO

Campesterol
(25)

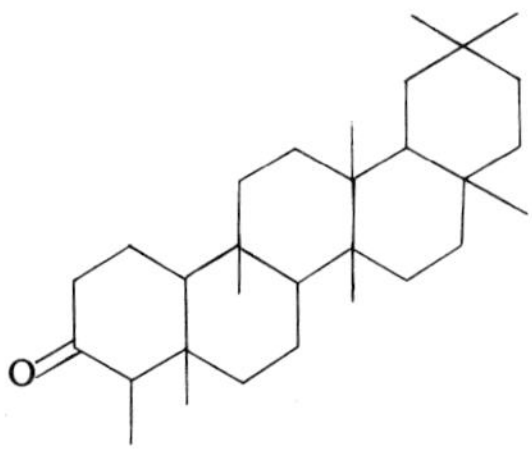

O

Friedelin
(26)

A. Biosynthesis and Biodegradation In the theory of limonin biogenesis proposed by Arigoni *et al.* (1960) and elaborated by Barton *et al.* (1961), a triterpenoid similar to euphol was assumed to be the starting material. Identification of the closely related cycloartenol (21) as a constituent of citrus fruits (Williams *et al.*, 1967) suggests this compound as an intermediate in limonoid formation by oxidative opening of the terminal dimethylcyclohexane ring, instead of loss of these methyl groups leading to the sterols.

Possible subsequent steps in passing from a triterpenoid to the complicated limonin molecule have been reviewed by Dreyer (1964) and Moss (1966), with particular reference to natural products from other plants in the Rutaceae and the related Meliaceae families having partial structures similar to the postulated limonin intermediates. These compounds present stages in the elaboration of the furan (F) and epoxylactone (DE) rings of limonin which do not occur in the citrus limonoids; however, various modifications of the remainder of the molecule that occur in recently isolated citrus constituents make it possible to formulate more complete schemes for limonin biogenesis.

A probable precursor of limonin (27), obacunone, is the second most prominent citrus limonoid, occurring in quantity in all fruits except sour oranges and limes. The structure (28) postulated for this compound by Arigoni *et al.* (1960) was shown to be correct from the studies of Tokoroyama (1958a,b), Kubota *et al.* (1960, 1961), Kamikawa and Kubota (1961), T. Matsuura *et al.* (1961), Tokoroyama *et al.* (1961), and Kamikawa (1962), and the conversion of obacunone to a limonin derivative (Tokoroyama and Matsuura, 1962) confirmed the relationship between the two compounds.

The third most prominent citrus limonoid, nomilin, had been identified from early work as acetoxydihydroobacunone (29), and its formation seems to be essential for limonin biogenesis since it readily loses acetic acid to give obacunone. On the other hand, the corresponding unacetylated compound, deacetylnomilin, recently isolated from the seeds of several citrus fruits but most readily from the Ichang lemon (Dreyer, 1965b), cannot be dehydrated to obacunone. Dreyer suggested that deacetylnomilin was identical with the "isolimonin" previously reported as a constituent of various citrus fruits, and Nomura (1966a) has confirmed this identity for "isolimonin" from the Natsudaidai.

The alcohol corresponding to obacunone, obacunol, isolated by Dreyer (1968b) from *Casimiroa edulis* (also a member of the Rutaceae family), represents the only limonoid without a ketone group in the B ring; it should be regarded as the result either of biological reduction of obacunone or of incomplete oxidation of a precursor to deacetylnomilin, rather than as evidence that the ketone group is formed after elaboration of the AA′ system.

An alternative to the postulated sequence–deacetylnomilin, nomilin, obacunone, limonin–is suggested by the isolation from citrus fruits of a compound corresponding to limonin but lacking the cyclic ether (A′) ring. This compound,

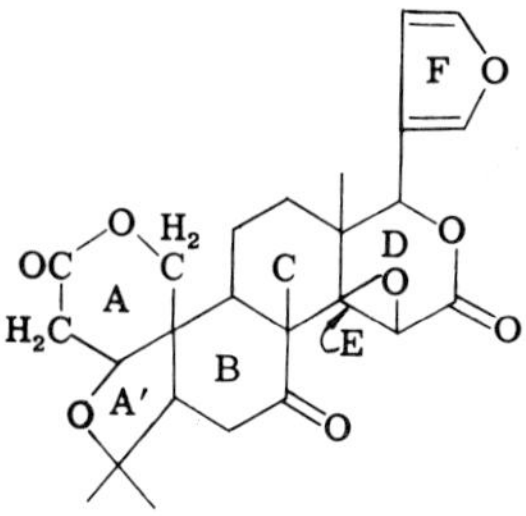

Limonin
(27)

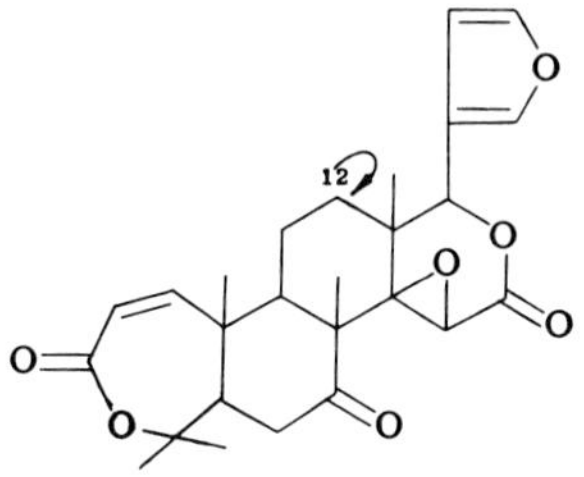

Obacunone
(28)

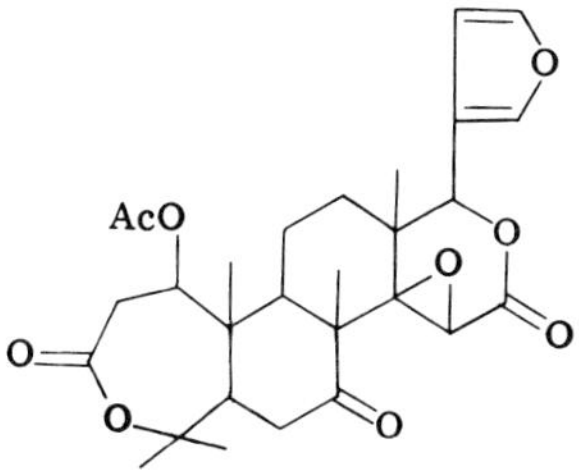

Nomilin
(29)

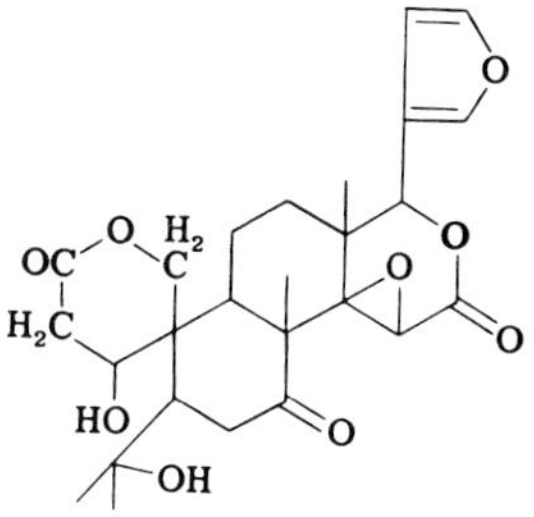

Ichangin
(30)

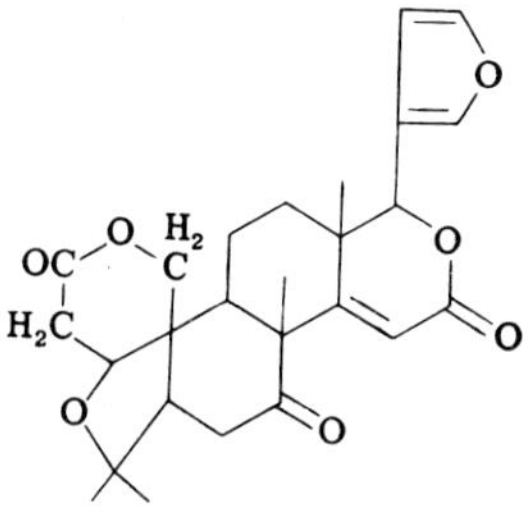

Deoxylimonin
(31)

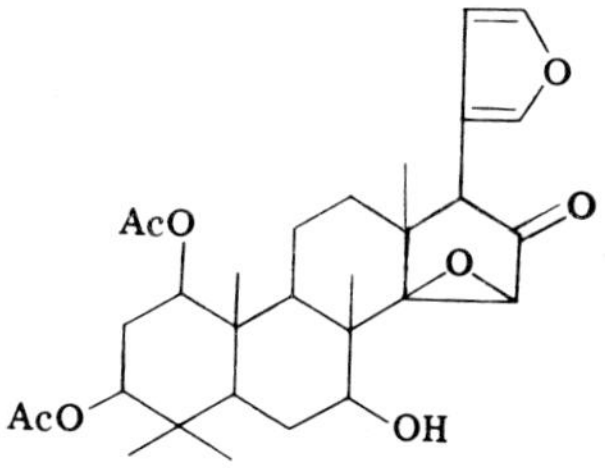

Grandifolione
(32)

ichangin (30), has been isolated only from the seeds of the Ichang lemon (Dreyer, 1966b), which seems to be notable among citrus fruits for its ability to accumulate large amounts of limonin precursors. Ichangin, which can be readily converted chemically to limonin, presents an alternative to obacunone as a limonin precursor, since its formation from deacetylnomilin could proceed by a similar series of reactions to those by which obacunone is converted to limonin. Since, however, ichangin has been detected in only one citrus species whereas nomilin and obacunone occur in most fruits containing limonin, greater credence must be given to the biosynthetic sequence involving these two compounds.

The biogenetic events so far discussed occur after the elaboration of the furan (F) and epoxylactone (DE) rings of the limonoids, since citrus fruits are noted for the absence of compounds representing stepwise elaborations of these parts of the limonoid structure. One compound, deoxylimonin (31), with the partially developed DEF system, has been isolated from grapefruit (Dreyer, 1965b), but its formation should be regarded as the result of incomplete epoxidation of a precursor to deacetylnomilin, or even of biological reduction of limonin, rather than as evidence that the epoxide ring (E) is the final stage in limonin synthesis. Support for considering deoxylimonin as a side-track in the biogenesis of limonin lies in its very limited occurrence, and in the variety of compounds found in other plants which may be regarded as representing steps in the formation of the DEF system in the limonoids although they do not possess the AA′ system. Grandifolione (32), for instance, from *Khaya grandifolia* contains the epoxide group, but the D ring is a cyclopentanone which has not yet been oxidized to the lactone (Connolly *et al.*, 1968a).

It is generally accepted that limonin develops in the albedo of citrus fruits and then disappears with advancing maturity. Recently, enzymic systems degrading limonin have been found in the albedo of Shamouti (Flavian and Levi, 1970), and Valencia and Navel oranges (Chandler and Kefford, 1969). The latter workers have demonstrated that the system is enzymic in nature since (1) it is inactivated by boiling, (2) it is inhibited by copper and cyanide, and (3) it shows a pH optimum at 5.3-5.5. Moreover, the degradation is evidently not an atmospheric oxidation of limonin, such as those leading to evodol (Barton *et al.*, 1961) or limonexic acid (Melera *et al.*, 1957), since it occurred in the absence of oxygen and was not inhibited by antioxidants. The most active systems were found in Navel oranges on trifoliate orange rootstock and in Valencia oranges on rough lemon rootstock and the activity increased to a maximum during maturation of the fruit.

So far, however, no authentic degradation product of limonin has been isolated from citrus fruits, although several possibly related compounds occur in other members of the Rutaceae, particularly in plants of the Rutoideae. As Dreyer (1966a) has pointed out, this subfamily of the Rutaceae is remarkable for the presence of appreciable amounts of what may be regarded as limonin oxidation products. Evodol, long recognized as the main bitter principle of *Evodia rutaecarpa*, was identified (Hirose, 1963) as limonin diosphenol (33), which can be prepared by chemical oxidation of limonin (Barton *et al.*, 1961). Rutaevin, which occurs together with evodol and limonin in *Evodia*, was shown by Dreyer (1967) to have the structure (34); it can be converted to the diosphenol by atmospheric oxidation and probably represents the most stable form of the α-hydroxyketone-enediol system, intermediate in the oxidation of limonin to the diosphenol.

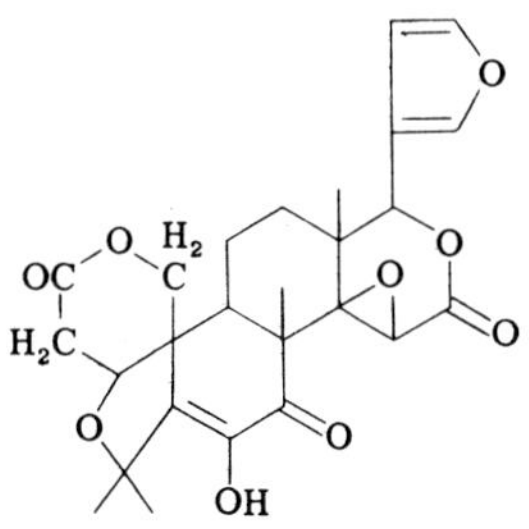

Evodol

(33)

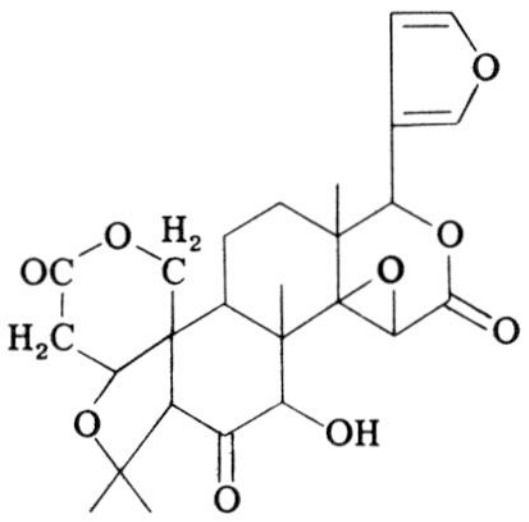

Rutaevin

(34)

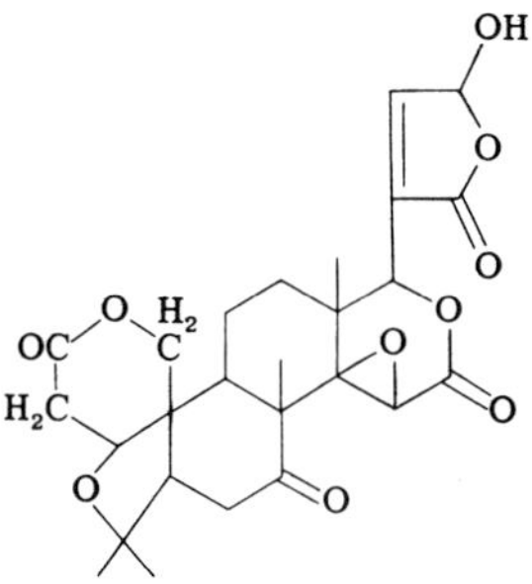

Limonexic acid

(35)

Confirmation of the conversion of limonin to rutaevin in maturing fruits was provided by Hirose *et al.* (1967) who found that fruit of *Evodia rutaecarpa* contained mainly limonin during the early stages of maturation, rutaevin and evodol in the intermediate period, and rutaevin when fully mature. Although these results seem to conflict with the role of rutaevin as an intermediate between limonin and evodol, it is possible that evodol cannot be accumulated in the mature fruit because very efficient enzyme systems are present for its transformation to other products. No higher oxidation products of limonin itself are known but several compounds (meliacins) that have been isolated from plants of the Rutaceae have structures representing oxidative opening of the B ring of limonin, although in these compounds the intact A ring of the triterpenoids is present instead of the fully elaborated AA′ system of the limonoids.

Limonin can be oxidized chemically to limonexic acid, most readily by atmospheric oxygen in a light-catalyzed reaction at pH 7.0 (Melera *et al.*, 1957); this compound was assigned the structure (35) by Arigoni *et al.* (1960) and Dreyer (1965a), with the furan ring of limonin oxidized to a hydroxylactone in a reaction with well-established analogies (Foote *et al.*, 1967). A product with similar properties can be isolated from extracts of orange seeds (Chandler and Kefford, 1953), but in view of the ready autoxidation of limonin to limonexic acid the product from seed extracts may be an artifact although it can be obtained in high yield under conditions that minimize autoxidation. Photogedunin in which the furan ring of the limonoid gedunin is replaced by a hydroxylactone has been obtained both as a natural plant constituent and as a photooxidation product of gedunin (Burke *et al.*, 1969). In another limonoid, nimbin, however, the furan ring undergoes autoxidation to an unhydroxylated lactone ring (Connolly *et al.*, 1968b) of the kind most frequently encountered in natural products. Clearly, further study of the occurrence of limonexic acid is required before it is established as an authentic biooxidation product of limonin.

The postulated relationship (Dreyer, 1964; Moss, 1966) between limonoids and the simarubolides of the Simarubaceae, taxonomically closely related to the Rutaceae, suggests another mechanism for removal of limonoids from maturing citrus fruits. The simarubolides may be regarded as relatives of the limonoids in which elaboration of the AA′ system has stopped at the obacunone (28) stage, while ring C has been oxidized and the furan ring F removed. Support for the conversion of limonoids into simarubolides may be found in the constituents of *Casimiroa edulis,* a member of the Rutaceae family remarkable for its inability to complete the elaboration of the AA′ system. Seeds of *C. edulis* contain nomilin, obacunone, and deacetylnomilin, together with zapoterin, which was shown by Dreyer (1968b) and Murphy *et al.* (1968) to be 12-hydroxyobacunone cf. (28), and a possible link between the limonoids and the simarubolides. The simarubolides, it should be noted, appear to be far more susceptible to biological oxidation than the limonoids, and some difficulty may be expected in their isolation and detection in citrus fruits.

At the moment, therefore, no single scheme of biodegradation can be offered to explain with certainty the decrease in the limonin content of citrus fruits with maturity. Any such explanation must conclude with the formation of nonbitter products, but considerations of this factor do not distinguish between the three mechanisms available. Rutaevin and evodol are reported by various workers to be very slightly bitter or tasteless, and limonexic acid to be tasteless, while the meliacins and simarubolides, to which the limonin degradation products may be related, vary in bitterness according to their individual structures. Such variations are similar to those occurring in the bitterness of the limonin "precursors": only nomilin and ichangin approach limonin in bitterness–deoxylimonin, deacetylnomilin, and obacunone being tasteless. The variations occurring in the bitterness of processed orange juice with maturation of the fruit therefore depend fairly specifically on the elaboration and degradation of the nomilin-limonin system.

Attempts to follow the biogenesis and biodegradation of limonin using radioactive tracers have met with difficulties in introducing a labeled atom into the molecule. The only method found by Datta and Nicholas (1968) to give a satisfactory labeled product was to soak germinated seeds of Valencia oranges in labeled mevalonic acid solution for several hours with subsequent incubation of the seeds for 2 to 6 days. Effectively labeled β-sitosterol and citrostadienol were obtained by injecting mevalonic acid directly into oranges or by standing cut stems bearing fruit in a solution of the acid for several hours; the two labeled steroids could be isolated from the fruit, stems, and leaves, but the limonin in the seeds was only weakly radioactive. These observations indicate that the isolated orange seed is capable of limonin synthesis, but leave it uncertain as to whether limonin also accumulates in the seeds by migration from the albedo during maturation of the fruit.

The report of Nomura (1963) that the limonin content of the seeds decreased in the maturing Natsudaidai while that of the membrane increased suggests that a closer study should be made of the distribution of limonin in the maturing orange.

B. Horticultural Considerations The large number of enzyme systems that must be involved in the processes of limonin synthesis and degradation help to explain the wide variations in limonin content that occur in citrus fruits under the influence of horticultural factors. It is impossible at the moment to define the relations between these factors and the enzyme systems, but future studies may enable the varietal factor, for instance, to be associated with the presence or absence of certain enzymes, as suggested 20 years ago by Balls (1949).

The rootstock is probably as important as the variety in determining the bitterness of processed orange juices. For instance, Kefford and Chandler (1961) demonstrated that trifoliate orange stock promoted the disappearance of bitter principles from all parts of Navel and Valencia oranges with advancing maturity, but rough lemon stock retarded this process. Sweet orange stock was intermediate in its effect, while tangelo and Cleopatra mandarin stocks resembled trifoliate orange, and sweet lime, Kusaie lime, and East India lime stocks resembled rough lemon. These results were confirmed in part by Bowden (1968), and the importance of the rootstock factor was further emphasized when Chandler *et al.* (1966) demonstrated the comparatively low concentrations of limonin in Navel oranges from rooted cuttings. Thus juices from comparable fruits borne by cuttings grown on their own roots, and on trifoliate orange, sweet orange, and rough lemon stocks contained 7.5, 8.0, 11.0, and 17.0 ppm of limonin, respectively.

The remarkable effect of rootstock on the limonin content of oranges makes it difficult to compare the few results that have been reported on the limonin content of citrus fruits. No doubt variety is a dominating factor in controlling limonoid biogenesis, as mentioned above, but it seems unrealistic to compare the limonin contents of commercial fruits, for instance, without reference to the original rootstocks. On present evidence, the grapefruit seems to be a better source of limonin than any other readily available fruit: the seeds are reported to contain up to 0.86% limonin, compared with 0.54%, 0.21%, and 0.08%, respectively, in orange, lime, and lemon seeds (Barton *et al.*, 1961), 0.07% in mandarin seeds (Row and Sastry, 1962b), and 0.09% in trifoliate orange seeds (Nikonov and Molodozhnikov, 1964). The highest limonin yield (16%) so far reported, however, is from the seeds of the Meyer lemon (Nikonov and Molodozhnikov, 1964), a figure so remarkably high that confirmation is desirable.

The endocarp of the grapefruit is reported to retain limonin (in amounts up to 140 ppm on a wet weight basis) far longer after commercial maturity than does even the Navel orange, with the result that the limonin content (up to 10 ppm) can make a significant contribution to the bitterness of processed grapefruit juice (Maier and Dreyer, 1965). The Natsudaidai is also reported to be a good source of limonin, with 0.25-0.56% in the dried membranes (Nomura, 1963).

The effect of variety and maturity on the bitterness of processed orange juice has been studied by Kefford and Chandler (1961) and Bowden (1968) in Australia, and by Exarchos and Aspridis (1962) in Greece. The greater bitterness in juice from Navel oranges compared with seeded varieties, and the lower bitterness in juices from mature fruits were again noted. In addition, the Australian workers observed an important effect of growing area which may be due to climatic factors. Bitterness in mandarin products, also due to limonin, is reported to vary according to growing area: fruit from Central India gave bitter juice and segments after pasteurization, while fruit from Assam gave nonbitter products (Siddappa and Bhatia, 1954).

One of the defects noted by Rouse *et al.* (1963) in Pineapple oranges harvested after the disastrous Florida freeze of 1962 was the development of bitterness in the extracted juice, whereas this variety is normally regarded as free from bitterness. Solvent extraction of the membranes, juice sacs, pulp and juice, and seeds yielded crystals which may be identified from the mode of isolation and crystalline form as limonin. This finding emphasizes facts frequently forgotten: that limonin is a normal constituent of immature oranges, disappearing at varying rates from the maturing fruit, and that bitterness can be a problem in juices from oranges other than Navel oranges (Kefford and Chandler, 1961; McDuff, 1967; Bowden, 1968). It is unfortunate that the wording in the paper by Rouse *et al.* (1963) has been interpreted to mean that limonin crystals were present as such in the intact fruit; reference to the isolation procedure makes it clear that the crystals were obtained by the normal process of extraction and crystallization. The observation of Rouse *et al.* (1963) can be explained simply as a further example of the disturbing effect of climate on fruit maturation. Bitterness may also occur in juice from "late-bloom" fruit of varieties not normally susceptible to this problem (McDuff, 1967), while physiologically abnormal fruit, such as Valencia oranges affected by the "greening" disease of South Africa, may yield a bitter juice when normal fruit is free from bitterness (Reuther, 1962).

Two apparent mutations of the Washington Navel orange originating in Spanish orchards are reported to yield juice similar in quality to the Washington Navel except that bitterness is not detectable on processing unless immature fruit is used; in each case the fruit is described as quick maturing, losing limonin rapidly during ripening (Primo *et al.*, 1962, 1964a, 1965b; Feliu and Bernal, 1966).

There is wide variability in the limonin content of Valencia oranges, even when harvested from trees of identical origin and orchard treatment (Chandler and Kefford, 1966); no correlations could be found with fruit size or seed number, two factors which might be expected to influence limonin content.

C. Technological Considerations The importance of limonoids in the processing of citrus fruit lies in the fact that limonin is responsible for the bitterness that develops on pasteurization or storage of many citrus products, most notably Navel orange juice and concentrates, as reviewed by Huet (1961a). Other oranges may also yield bitter juices under certain abnormal conditions of growth (Reuther, 1962; Rouse *et al.*, 1963), and so may normally developed Valencia oranges under certain horticultural conditions (Kefford and Chandler, 1961; McDuff, 1967; Bowden, 1968). The bitterness of grapefruit products also may be partly due to limonin (Maier and Dreyer, 1965) as well as to such bitter flavonoids as naringin. Although such flavonoids have been reported in mandarins grown in certain areas (Chaliha, *et al.* 1965b, 1967), they are generally absent from this fruit, and the bitterness in canned mandarin products (Siddappa and Bhatia, 1954, 1959) is due to limonin which may be present in the juice in concentrations up to 25 ppm (Chandler, 1958b).

The amount of limonin required in a juice before bitterness becomes detectable varies with the sweetness and acidity of the juice as well as the sensitivity of the taster (Chandler and Kefford, 1966). As a general rule a juice containing less than 6 ppm is unlikely to taste bitter, but a juice with more than 9 ppm will seem bitter to most tasters. Limonin contents higher than 30 ppm are comparatively rare, even in Navel orange juice.

In contrast to bitterness due to flavonoids, the bitterness in citrus products containing the comparatively insoluble limonin appears only after storage or heat treatment. In his early work on limonin, R. H. Higby (1941) explained this phenomenon of delayed bitterness in terms of a nonbitter precursor, such as a glycoside of the hydroxy-acid form of limonin, which was assumed to be hydrolyzed to the bitter principle under the acid conditions of the juice. Support for this theory was found in the results of Emerson (1948) who extracted Navel orange juice with benzene immediately after expression and again after 4 hours; when the second extract contained most of the limonin it was postulated that the nonbitter precursor was insoluble in benzene and that some time was required to convert it to limonin. However, in a later paper Emerson (1949) reported that any limonin precursor present in immature Valencia peel amounted to not more than 10% of the total limonin. Emerson also believed that the development of bitterness in Navel juice was a different process from the development of bitterness in Valencia juice to which limonin in the nonbitter hydroxy-acid form had been added.

In the precursor theory, the delay in bitterness development is ascribed to the time taken to convert the nonbitter precursor to limonin, and the assumption is made that the precursor, the free hydroxy-acid or its ion or glycoside, passes rapidly into solution, at least as quickly as, say, naringin from grapefruit albedo. However, although they subscribed to the precursor theory, Samish and Ganz (1950) concluded from filtration experiments on albedo suspensions and extracted juices that the bitter principle (or its precursor) diffused only slowly from the tissue particles in which it was located. This observation provides an explanation for Emerson's results with benzene extracts of Navel orange juice; immediate extraction was ineffective not because the nonbitter precursor had not been converted to limonin but because limonin (or its precursor) had not had time to diffuse into the juice, and because, as Samish and Ganz (1950) also demonstrated, hydrophobic solvents are inefficient for the extraction of limonin from moist albedo. As discussed in the previous review (Kefford, 1959), observations such as these led to an explanation of delayed bitterness in terms of the diffusion of insoluble limonin from the albedo particles, a slow process unless heat is applied.

Because limonin is unstable in aqueous solutions at the pH of the albedo, and the hydroxy acid at the pH of the juice, and because no distinction can be made between the heat catalyzed physical and chemical reactions suggested in explanations of delayed bitterness, studies on this problem have stagnated for many years. Recently, however, Maier and his co-workers at the U.S. Department of Agriculture Fruit and Vegetable Chemistry Laboratory, Pasadena, have obtained evidence in favor of the precursor theory by the application of high-voltage paper electrophoresis. They reported the presence in early season Navel oranges and grapefruit of a nonbitter limonin monolactone which was stable at the pH of the tissues (5.1), but was converted to limonin at the pH of the juice (3.0) (Maier and Beverly, 1968). Their evidence suggested that all the limonin was in the monolactone form in Navel orange tissues, contrary to the finding of Emerson (1949) that at least 90% of the limonin in Valencia peel was preformed limonin.

Subsequently, the Pasadena workers identified the limonin monolactone in their extracts as the A-ring lactone (36), which they were able to prepare as an impure solid for the first time, and they speculated that this compound may be a biosynthetic precursor of limonin in those tissues, such as the seeds, in which it accumulates (Maier and Margileth, 1969). In the biogenesis of limonin from obacunone (Barton *et al.*, 1961; Dreyer, 1965b), the D-ring monolactone (37), i.e., the A-ring hydroxy acid, is a necessary intermediate, but the A-ring monolactone, i.e., the D-ring hydroxy acid, is not. It is therefore unexpected to find limonin existing in the tissues as the A-ring monolactone. The explanation of this apparent anomaly may lie in the presence in tissue extracts of an enzyme which catalyzes the opening and closing of the D lactone ring, but not the A lactone ring. The purified enzyme, limonin D-ring lactone hydrolase, was isolated from orange seeds by Maier *et al.* (1969). The function of this enzyme and the two nonbitter limonin hydroxy acids in the biogenesis of limonin is probably of less significance than their role in bitterness development. The Pasadena workers have therefore indicated that their findings may have commercial application in the production of nonbitter juices from Navel oranges.

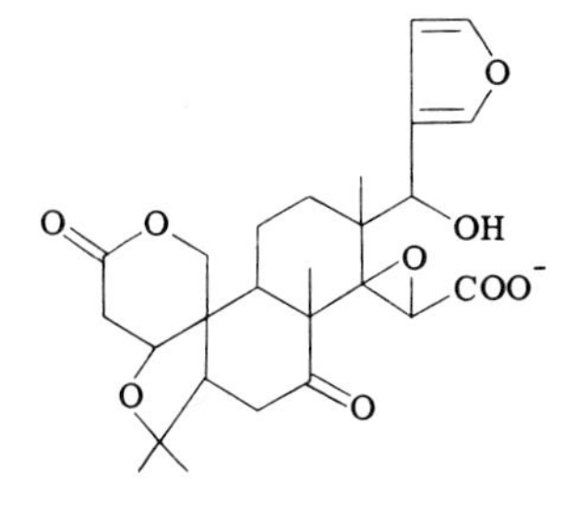

Limonin A-ring lactone
(36)

HO
⁻OOC
O
O
O
O
O
O

Limonin D-ring monolactone
(37)

The problem of preparing nonbitter juices from fruit of high limonin content has long attracted the attention of food chemists and technologists. Several procedures were tested by R. H. Higby (1941) who found that a nonbitter juice could be prepared from Navel oranges only by removing the navels by hand, extracting the juice without gross tissue maceration, and raising the pH of the juice to 3.8-4.0, to prevent hydrolysis of the postulated nonbitter precursor. Control of the extraction process to minimize the inclusion of albedo particles was recommended for the preparation of nonbitter juices from Shamouti oranges (Samish and Ganz, 1950) and mandarins (Siddappa and Bhatia, 1959), while Siddappa and Bhatia (1954) had previously suggested boiling mandarin segments in 1% sodium hydroxide solution to remove albedo before expressing the juice.

Juices from fully mature fruit are less susceptible to bitterness than those from partially mature fruits, but attempts to promote the ripening process by treating oranges on the tree with ethylene proved unsuccessful (Balls, 1949; Emerson, 1949). When oranges were stored in an atmosphere containing ethylene, Emerson (1949) reported a decrease in bitterness in the juice, but an increase in other off-flavors. However, according to Samish and Ganz (1950), ethylene treatment showed no practical advantage over ordinary storage, even though the amount of bitter principle in the albedo and segment walls was significantly decreased; they therefore recommended off-tree ripening to extend the processing season for Shamouti oranges. Subsequently American workers were granted a patent for the removal of bitter principles from citrus fruits by storage for 7 to 20 days under warm, moist conditions (80°-90°F, 70-90% relative humidity) intended to promote metabolic changes in the fruit (Rockland *et al.*, 1957).

Although unaware that the bitter principle involved was limonin and not naringin, Russian workers reported the removal of bitterness from tangerine and other citrus products by the simultaneous addition of peroxide and peroxidase preparations from such sources as cabbages and apples (Markh and Fel'dman, 1949, 1950; Markh, 1953). In American studies (Anonymous, 1950), pectin-destroying enzymes from fruits or fungi were found to debitter orange juice after several hours' treatment at 4°-10°C; the process was associated with off-flavor development and loss of cloud, however, and apparently involved the breakdown of colloidal systems stabilized by pectins without chemical alteration of the limonin molecule. This explanation is supported by observations that the presence of pectin is necessary to maintain limonin in aqueous solution in concentrations high enough to taste bitter (Chandler, 1958b). It seems likely that pectinase activity was responsible for the slightly positive results obtained by Nomura (1966b) in the use of preparations from *Aspergillus* and *Penicillium* species to debitter Natsudaidai juice.

Selective extraction of limonin was the basis of a patent for the preparation of a nonbitter product from Navel orange juice (Pritchett, 1957). The bitter principle is first extracted with a solvent such as isopropanol and removed from this extract by absorption onto charcoal; the extract, which still contains some desirable juice constituents, is then added back to the juice, and the solvent is removed in the concentration process. Another patented method for the preparation of acceptable orange drinks from Navel orange juice (Swisher, 1958) involves adjustment of the pH of the juice to 6.0 to convert limonin into its nonbitter hydroxy-acid form; then, as the product is dispensed to the consumer, a concentrated citric acid solution is automatically added to bring the pH down to a palatable level. Since limonin is not immediately reformed on acidification, the drink is not bitter if consumed in a reasonably short time.

Processes based on the use of solvents or chemical modification of the juice must involve the risk of changing the fresh character of the juice to an unacceptable extent. A simpler procedure developed by Chandler *et al.* (1968) for the removal of limonin from orange juice depends upon selective absorption of the limonin by polyamide, which has been used in the preparation of other beverages for reducing astringency and haze. Orange juice is first given a minimum pasteurization to bring the limonin into solution; the insoluble solids are centrifuged off and the clear serum shaken gently for about 10 minutes with the polyamide absorbent, which is then removed by centrifugation; finally, the orange cloud is resuspended in the debittered serum. A process recently patented by Sperti (1968) is somewhat similar in operation in that the juice is separated into serum and pulp immediately after extraction, but in this case pulp containing no bitter principle is incorporated into the serum.

In a later patent (Hanson, 1968), the juice is extracted at low temperature to inhibit the physical and/or chemical changes involved in bitterness development and is treated with pectinesterase before pasteurization. The latter step is claimed to inhibit the formation of limonin from its precursors, but it is here suggested that it lowers the solubility of limonin by reducing the pectin content of the juice.

D. Analytical Methods Limonoids lend themselves particularly well to separation by thin-layer chromatography on silica gel (Dreyer, 1965b), and Maier and Beverley (1968) record R_f values for most citrus limonoids in two solvent systems. High-voltage paper electrophoresis has also been applied to limonin studies (Maier and Beverley, 1968; Maier and Margileth, 1969). The chromogenic spray recommended by these workers is Ehrlich's reagent (*p*-dimethylaminobenzaldehyde) for detection of furans, followed by fuming with acid. This reagent had been used by Nomura and Santo (1965) for colorimetric estimation of limonoids in citrus juices. A quantitative procedure claimed to be capable of measuring limonoid concentrations down to 0.5 ppm has recently been described by Maier and Grant (1970). After extraction with chloroform and concentration in acetonitrile the limonoids are separated by thin-layer chromatography on silica gel. The spots are sprayed with Ehrlich's reagent, fumed with acid, and compared with standard spots by visual evaluation of spot size and density to provide a measure of limonoid content. The total time required to carry out this analysis is about 4 hours.

A procedure developed by Chandler and Kefford (1966) for limonin assay involved selective extraction and reaction of the ketone group with dinitrophenylhydrazine; the limonin dinitrophenylhydrazone is separated by thin-layer chromatography on silica gel and the analysis completed by spectrophotometry. The method is long but can be adapted to routine work by carrying out the most time-consuming operation, the reaction with dinitrophenylhydrazine, overnight, thus completing the procedure in a 2-hour working period.

More recently, Chandler and Kefford (1969) have used the production of reducing agents by bromination of the furan ring as the basis for a rapid and selective method for determining limonoids (except limonexic acid). Following thin-layer chromatography of an extract, the plates are exposed to bromine vapor under controlled conditions. Although the bromination is nonstoichiometric, the amounts of reducing substances are proportional to the limonoid content. After spraying with silver nitrate, the intensity of the spots is measured by densitometry in comparison with standard spots to estimate the individual limonoids. Natural reducing substances do not interfere, while furanocoumarins, which also react with bromine, have much higher R_f values than the limonoids.

The method of Wilson and Crutchfield (1968) is somewhat more direct but employs a less specific reaction–formation of ferric hydroxamates from the lactone groups–and cannot be used in products where coumarins are likely to interfere. Moreover, the method does not differentiate between individual limonoids. For analyses of orange peel, grapefruit juice, and parts of citrus trees, the hydroxamate method should not be used.

A simple empirical method was devised by R. H. Higby (1941) to test oranges for the potential bitterness of their juice. The procedure involved the hot acetone extraction of twelve cores from a representative sample of the fruit, followed by evaporation of the acetone after dilution with water. If the final solution, when diluted to 100 ml with water, tasted only slightly bitter, then the oranges were acceptable for processing.

Chapter 15 Research Needs

There still remain constituents of citrus fruits known to be present by their biochemical actions but as yet not identified chemically. Notable among these compounds are several plant growth regulators evidently involved in the control of growth and maturation in citrus fruits. An auxin active in the *Avena* coleoptile curvature test was reported by Khalifah *et al.* (1963) in immature Valencia and Navel oranges, and later in lemons and other noncitrus fruits (L. Lewis *et al.*, 1965a). This auxin showed paper chromatographic behavior similar to indoleacetic acid, but the two compounds moved independently on silicic acid columns and differed in other chemical and physical properties (Khalifah *et al.*, 1966). The citrus auxin was accompanied by at least two auxin inhibitors (L. Lewis *et al.*, 1965b). The same workers have also extracted from Navel oranges and lemons three different gibberellinlike substances (Khalifah *et al.*, 1965), and from lemon seeds a purine derivative resembling kinetin in its effect on cell division in tobacco callus tissue and its reaction with xanthine oxidase (Khalifah and Lewis, 1966). In contrast to these findings, when Monselise *et al.* (1967) subjected aqueous extracts of developing and mature Shamouti oranges to the wheat coleoptile test, they observed strong inhibition but little promotion of growth. A substance inhibiting the growth of *Phomopsis citri* was reported to be present in aqueous extracts of orange peel (El Tobsky and Sinclair, 1964).

It would be true to say, however, that the principal needs for research in the field under review lie outside of chemistry in the narrow sense.

There is no doubt that our understanding of the factors influencing the composition of citrus fruits has advanced in 10 years, but progress towards prediction and control of the composition of a crop is inherently slow. For instance in one particular aspect—the nitrogen nutrition of citrus trees—accumulated experience at the University of California Citrus Research Center has led no further than to the conclusion that the effects of nitrogen level on the composition of orange juice are inconsistent and hence unpredictable (W. W. Jones *et al.*, 1968)! Although the specific effects of rootstocks on citrus composition have been clarified, little else is known about the mechanism of the influence of rootstocks on a scion, an influence that seems too subtle to be explained solely in terms of different capacities to extract nutrients from the soil, of which one example is the differential uptake and translocation of iron and zinc by trifoliate orange and rough lemon stocks (Khadr and Wallace, 1964).

This field of study, the borderland between food science and horticulture, has been explored mainly by empirical trials attended by all the difficulties of planned experimentation with perennial crops. There is a pressing need for more basic studies of the enzymic and plant hormonal processes contributing to the composition of the mature fruit. Except perhaps for the acids in citrus fruits, too little is known about the sites and mechanisms of biosynthesis of specific constituents. The potential value of tissue culture as a technique for studying metabolic processes in citrus fruits has been demonstrated by Oberbacher (1967) who used preparations of grapefruit albedo tissue to study the effects of growth regulators and protein inhibitors on the synthesis of carotenoid pigments.

Several members of particular chemical families, the bitter flavonoids and limonoids, for example, contribute significantly to the quality of citrus products, and studies on the basic factors controlling their occurrence, such as the pioneering work of Maier and Metzler (1967b) on grapefruit phenolics, should ultimately result in product improvement. Moreover, the minor constituents–flavonoid *C*-glycosides, permethoxylated flavones, coumarins, carotenoids, terpenoids, and fatty acids–often show marked differences in distribution among the kinds and varieties of citrus fruits. Although Dreyer (1966a) considered that differences in limonoid components had taxonomic significance only between, and not within, genera of the Rutaceae, a recent paper on citrus flavanones by Albach and Redman (1969) demonstrates the value to taxonomists and plant breeders of a knowledge of the distribution of minor constituents between species and varieties.

Another borderland into which the citrus chemist must move if his findings are to have practical meaning is the field of food acceptance and the psychometrics of flavor. This message has been frequently preached to flavor chemists but it applies to citrus products more than to many others because present chemical information so far outstrips organoleptic understanding. The citrus flavor system consists of straightforward primary tastes: sweet, sour, and bitter, plus a superstructure of more than 100 volatile flavors, the respective roles of which are hardly known. This simple yet complex system should be amenable to the experimental approach. There has been no attempt as yet to apply the subtractive techniques that Jennings (1962) and Guadagni *et al.* (1966) have applied so successfully to pinpoint the important volatile flavor constituents of pears and apples.

An attempt to establish an objective method for the assessment of flavor in concentrated orange juice was made by Moore *et al.* (1967) on the basis of 15 chemical and physical characteristics for which the following data had been collected–Brix:acid ratio, pH, recoverable oil, flavonoids, ascorbic acid, diacetyl, pulp, pectin, water-insoluble solids, viscosity, and color. Multiple regression equations based on these variables could account, however, for only about 50% of the variation in flavor scores. This means that there remain many unexplained factors contributing to the reaction of the human palate to orange juice, and there is little doubt that some of these factors are associated with volatile flavoring constituents.

To the food scientist it is a barren exercise merely to explore the biogenetic virtuosity of the citrus species; to catalog the chemical composition of citrus fruits is not enough, he also seeks to control it and to know what it means to consumers of citrus fruits as food.

REFERENCES

Abraham, E. P., Newton, G.G.F., and Jeffery, J. D'A. 1964. Deacetylcephalosporin C derivatives. German Patent 1,161,276; *Chem. Abstr.* **60**, 12310a (1964).

Agricultural Research Service. 1962. Chemistry and technology of citrus, citrus products, and by-products. *U.S. Dept. Agr., Agr. Handbook* **98.**

Ahmad, M. N., Bhatty, M. K., and Karimullah. 1962a. Some essential oil sources of West Pakistan. *Pakistan J. Sci.* **14,** 12-15; *Chem. Abstr.* **58,** 1297e (1963).

Ahmad, M., Bhatty, M. K., and Karimullah. 1962b. Deterpenation of Pakistani lemon and orange oils. *Pakistan J. Sci. Ind. Res.* **5,** 153-155; *Chem. Abstr.* **58,** 13705e (1963).

Aichinger, F., Giss, G., and Vogel, G. 1964. New pharmacodynamic results concerning bioflavonoids and the horse chestnut saponin escin as basis for therapeutic use. *Arzneimittel-Forsch.* **14,** 892-896; *Chem. Abstr.* **62,** 2119a (1965).

Albach, R. F., and Redman, G. H. 1969. Composition and inheritance of flavanones in citrus fruit. *Phytochemistry* **8,** 127-143.

Alberola, J., Casas, A., and Primo, E. 1967. Detection of adulteration in citrus juices. X. Identification of sugars in orange juices and in commercial sucroses by gas-liquid chromatography. *Rev. Agroquim. Tecnol. Alimentos* **7,** 476-482.

Alberola, J., Casas, A., and Primo, E. 1968. Determination of adulteration in citrus juices. XI. Direct method for the determination of sugars in commercial orange juices, sucrose, and citric acid by gas-liquid chromatography. *Rev. Agroquim. Tecnol. Alimentos* **8,** 127-132.

Alvarez, B. M. 1967. Detection of adulteration of fruit juices by thin-layer chromatography. *Analyst* **92,** 176-179.

Ammerman, C. B., Van Walleghem, P. A., Easley, J. F., Arrington, L. F., and Shirley, R. L. 1963. Dried citrus seeds–nutrient composition and nutritive value of protein. *Proc. Florida State Hort. Soc.* **76,** 245-249.

Anderson, C. A. 1966. Effects of phosphate fertilizer on yield and quality of Valencia oranges. *Proc. Florida State Hort. Soc.* **79,** 36-40.

Anonymous. 1950. Improved products made from Navel oranges. *U.S. Dept. Agr., Bur. Agr. Ind. Chem., Rept.* pp. 65-66.

Anonymous. 1963. Prevents turbidity. *Food Eng.* **35** (8), 43.

Anonymous. 1967a. "The State of Food and Agriculture 1967." Food Agr. Organ. U.N., Rome.

Anonymous. 1967b. Composition of Israeli citrus juices. *J. Assoc. Public Analysts* **5,** 68-71.

Anonymous. 1968. Standards for grades for various citrus juices. *Federal Register* **33,** 11881-11887; *Chem. Abstr.* **69,** 95229t (1968).

Arakawa, H., and Nakazaki, M. 1960. Absolute configuration of (–) hesperetin and (–) liquiritigenin. *Chem. & Ind. (London)* p. 73.

Aratake, S. 1960. Hesperidin. Japanese Patent 12,598; *Chem. Abstr.* **55,** 9799e (1961).

Arigoni, D., Barton, D.H.R., Corey, E. J., Jeger, O., Caglioti, L., Dev, S., Ferrini, P. G., Glazier, E. R., Melera, A., Pradhan, S. K., Schaffner, K., Sternhell, S., Templeton, J. F., and Tobinaga, S. 1960. Constitution of limonin. *Experientia* **16,** 41-49.

Arnaud, R. 1962a. Extraction of pure flavones from citrus-fruit peels with vitamin-P activity. French Patent 1,271,278; *Chem. Abstr.* **56,** 13019g (1962).

Arnaud, R. 1962b. Extraction of citrus fruit peels with high flavonoid mixture content. French Patent 1,273,399; *Chem. Abstr.* **57,** 15259c (1962).

Arnott, S., and Robertson, J. M. (1959). The lattice constants of limonin and some of its derivatives. *Acta Cryst.* **12,** 75.

Arnott, S., Davie, A. W., Robertson, J. M., Sim, G. A., and Watson, D. G. 1960. The structure of limonin. *Experientia* **16,** 49-52.

Arnott, S., Davie, A. W., Robertson, J. M., Sim, G. A., and Watson, D. G. 1961. The structure of limonin: X-ray analysis of epilimonol iodoacetate. *J. Chem. Soc.*, pp. 4183-4200.

Ashoor, S. H. M., and Bernhard, R. A. 1967. Isolation and characterization of terpenes from *Citrus reticulata* Blanco and their comparative distribution among other citrus species. *J. Agr. Food Chem.* **15**, 1044-1047.

Aspinall, G. O., Craig, J. W. T., and Whyte, J. L. 1968. Lemon-peel pectin. I. Fractionation and partial hydrolysis of water-soluble pectin. *Carbohydrate Res.* **7**, 442-452.

Attaway, J. A., and Oberbacher, M. F. 1968. Studies on the aroma of intact Hamlin oranges. *J. Food Sci.* **33**, 287-289.

Attaway, J. A., Wolford, R. W., and Edwards, G. J. 1962. Isolation and identification of some volatile carbonyl components from orange essence. *J. Agr. Food Chem.* **10**, 102-104.

Attaway, J. A., Wolford, R. W., Alberding, G. E., and Edwards, G. J. 1964a. Identification of alcohols and volatile organic acids from natural orange essence. *J. Agr. Food Chem.* **12**, 118-121.

Attaway, J. A., Hendrick, D. V., and Wolford, R. W. 1964b. Use of gas-liquid and thin layer chromatographic techniques in orange essence analysis. *Proc. Florida State Hort. Soc.* **77**, 305-307.

Attaway, J. A., Pieringer, A. P., and Barabas, L. J. 1966a. Origin of citrus flavor components. I. Analysis of citrus leaf oils using gas-liquid chromatography, thin-layer chromatography, and mass spectrometry. *Phytochemistry* **5**, 141-151.

Attaway, J. A., Pieringer, A. P., and Barabas, L. J. 1966b. Origin of citrus flavor components. II. Identification of volatile components from citrus blossoms. *Phytochemistry* **5**, 1273-1279.

Attaway, J. A., Pieringer, A. P., and Barabas, L. J. 1967a. Origin of citrus flavor components. III. A study of percentage variation in peel and leaf oil terpenes during one season. *Phytochemistry* **6**, 25-32.

Attaway, J. A., Wolford, R. W., Dougherty, M. H., and Edwards, G. J. 1967b. Methods for the determination of oxygenated terpene, aldehyde, and ester concentrations in aqueous citrus essences. *J. Agr. Food Chem.* **15**, 688-692.

Attaway, J. A., Pieringer, A. P., and Buslig, B. S. 1968. Origin of citrus flavor components. IV. The terpenes of Valencia orange leaf, peel, and blossom oils. *Phytochemistry* **7**, 1695-1698.

Avalle, N. 1961. Bioflavonoids and their use in cosmetics. *Parfums, Cosmet., Savons* **4**, 495-497; *Chem. Abstr.* **56**, 10302g (1962).

Averna, V. 1960. Nitrogen metabolism in vegetables. I. Free amino acids in various parts of citrus fruits. *Conserve Deriv. Agrumari (Palermo)* **9**(2), 83-93.

Axelrod, B. 1947. Citrus fruit phosphatase. *J. Biol. Chem.* **167**, 57-72.

Babin, R. 1957. Determination of citrus flavonoids. *Bull. Soc. Pharm. Bordeaux* **96**, 61-64; *Chem. Abstr.* **51**, 15682d (1957).

Babin, R., and 10 others. 1959. Some pharmacological properties of and new experimental findings on citroflavonoids. *Bull. Acad. Natl. Med. (Paris)* [3] **143**, 720-726; *Nutr. Abstr. Rev.* **31**, 108 (1961).

Bacharach, A. L., and Coates, M. E. 1942. Biological estimation of vitamin P activity. *Analyst* **67**, 313-317,

Bacharach, A. L., and Coates, M. E. 1943. The vitamin P activity of some British fruits and vegetables. *J. Soc. Chem. Ind. (London)* **62**, 85-87.

Baker, R. A., and Bruemmer, J. H. 1970. Cloud stability in the absence of various orange juice soluble components. *Citrus Ind.* **51**(1), 6-11.

Balls, A. K. 1949. Enzyme problems in the citrus industry. *Food Technol.* **3,** 96-100.

Bar Akiva, A., and Sagiv, J. 1967. Nitrate reductase in the citrus plant: Properties, assay conditions, and distribution within the plant. *Physiol. Plantarum* **20,** 500-506.

Bartholomew, E. T., and Sinclair, W. B. 1951. "The Lemon Fruit" Univ. of California Press, Berkeley, California.

Barton, D. H. R., Pradhan, S. K., Sternhell, S., and Templeton, J. F. 1961. Triterpenoids. XXV. The constitutions of limonin and related bitter principles. *J. Chem. Soc.* pp. 255-275.

Basker, H. B. 1965. Naringin crystallization in grapefruit products. *J. Assoc. Public Analysts* **3,** 83-86.

Basker, H. B. 1967. Citrus essential oils from Israel. *Am. Perfumer Cosmet.* **82,** 33-34.

Bauernfeind, J. C., and Bunnell, R. H. 1962. β-Apo-8′-carotenal–a new food color. *Food Technol.* **16** (12), 76-82.

Bauernfeind, J. C., Osadca, M., and Bunnell, R. H. 1962. β-Carotene, color and nutrient for juices and beverages. *Food Technol.* **16**(8), 101-107.

Bavetta, L. A., Nimni, M. E., Hom, D., and Jones, C. 1964. Effect of bioflavonoids and methylprednisolone on collagen synthesis. *Am. J. Physiol.* **206,** 179-182.

Bean, R. C. 1960. Carbohydrate metabolism of citrus fruits. I. Mechanisms of sucrose synthesis in oranges and lemons. *Plant Physiol.* **35**, 429-434.

Bean, R. C., and Todd, G. W. 1960. Photosynthesis and respiration in developing fruits. I. $C^{14}O_2$ uptake by young oranges in light and in dark. *Plant Physiol.* **35,** 425-429.

Bean, R. C., Porter, G. G., and Steinberg, B. M. 1961. Carbohydrate metabolism of citrus fruits. II. Oxidation of sugars by an aerodehydrogenase from young citrus fruits. *J. Biol. Chem.* **236,** 1235-1240.

Bean, R. C., Porter, G. G., and Barr, B. K. 1963. Photosynthesis and respiration in developing fruits. III. Variation in photosynthetic capacities during color changes in citrus. *Plant Physiol.* **38,** 285-290.

Beckman Instruments Inc. 1967. A very good year for grapefruit–and pH meters. *Science* **156,** 421.

Beisel, C. G., and Kitchel, R. L. 1967. Juice content–fact or fallacy. *Assoc. Food Drug Officials U.S., Quart. Bull.* **31,** 19-26.

Bellino, A., Venturella, P., and Cusmano, S. 1962. Flavonoids components of citrus. II. 5,6,7,8,4′-Pentamethoxyflavone (ponkanetin) from *Citrus deliciosa. Ann. Chim. (Rome)* **52,** 795-801; *Chem. Abstr.* **58,** 12505h (1963).

Bellomo, A. 1969. Investigations on nitrogen constituents of orange juice. *Ind. Conserve (Parma)* **44,** 3-17.

Ben-Aziz, A. 1965. Israel Patent Appl. 22754 (quoted in Ben-Aziz, 1967).

Ben-Aziz, A. 1967. Nobiletin is main fungistat in tangerines resistant to mal secco. *Science* **155,** 1026-1027.

Ben-Aziz, A., Chorin, M., Monselise, S. P., and Reichert, I. 1962. Inhibitors of *Deuterophoma tracheiphilia* in citrus varieties resistant to mal secco. *Science* **135,** 1066-1067.

Benk, E. 1961a. Clementines, mandarin-like citrus fruits. *Fruchtsaft-Ind.* **6,** 208-210.

Benk, E. 1961b. Composition of the juices of deformed oranges (monster forms). *Fruchtsaft-Ind.* **6,** 399-403.

Benk, E. 1961c. Detection of added β-carotene in orange juice products. *Deut. Lebensm.-Rundschau* **57,** 324-329; *Chem. Abstr.* **56,** 15899d (1962).

Benk, E. 1965. Content of inorganic materials, especially sodium, in natural orange juices. *Mitt. Gebiete Lebensm. Hyg.* **56,** 273-281.

Benk, E. 1968a. Detection of pulp and peel extracts in orange juices on the basis of their pentosan content. *Deut. Lebensm.-Rundschau* **64,** 146-148; *Chem. Abstr.* **69,** 18150h (1968).

Benk, E. 1968b. Natural sodium, potassium, calcium, and chloride content of orange and lemon juices. *Riechstoffe, Aromen, Koerperpflegemittel* **18,** 126-134.

Benk, E., and Bergmann, R. 1963a. Information on the sweet lime. *Fruchtsaft-Ind.* **8,** 87-90.

Benk, E., and Bergmann, R. 1963b. On tangerines and tangerine juice. *Fruchtsaft-Ind.* **8,** 290-294.

Benk, E., and Seibold, H. 1966. Detection of mandarin and tangerine juices in orange juice. *Deut. Lebensm-Rundschau* **62,** 396-398.

Benk, E., and Stein, H. 1959. Determination of juice content of fruit-juice preparations with chloramine values. *Fruchtsaft-Ind.* **4,** 154-159.

Benk, E., and Wildfeuer, I. 1961. Bergamots, bergamot juice and bergamot oil. *Fruchtsaft-Ind.* **6,** 22-25.

Benk, E., Wolff, I., and Treiber, H. 1963. Detection of added carotenoids in orange juice by thin-layer chromatography. *Deut. Lebensm.-Rundschau* **59,** 39-42.

Bernhard, R. A. 1958a. Occurrence of coumarin analogues in lemon juice. *Nature* **182,** 1171.

Bernhard, R. A. 1958b. Examination of lemon oil by gas partition chromatography. *Food Res.* **23,** 213-216.

Bernhard, R. A. 1960. Analysis and composition of oil of lemon by gas-liquid chromatography. *J. Chromatog.* **3,** 471-476.

Bernhard, R. A. 1961. Citrus flavor. Volatile constituents of the essential oil of the orange *(Citrus sinensis). J. Food Sci.* **26,** 401-411.

Berry, R. E., Wagner, C. J., and Moshonas, M. G. 1967. Flavor studies of nootkatone in grapefruit juice. *J. Food Sci.* **32,** 75-78.

Bessho, Y., Manabe, T., Koorama, M., and Kubo, S. 1964. Utilization of natsudaidai. I. Distribution of naringin in the natsudaidai fruits. *Nippon Shokuhin Kogyo Gakkaishi* **11,** 385-389; *Chem. Abstr.* **64,** 5446f (1966).

Bezanger-Beauquesne, L., Trupin, N., Guilbert, N., and Guislain, R. 1964. Choleretic effect of certain flavonoids. *Compt. Rend. Soc. Biol.* **158,** 2095-2100; *Chem. Abstr.* **63,** 1103c (1965).

Birdsall, J. J., Derse, P. H., and Teply, L. J. 1961. Nutrients in California lemons and oranges. II. Vitamin, mineral, and proximate composition. *J. Am. Dietet. Assoc.* **38,** 555-559.

Bissett, D. W. 1958. Processing freeze-damaged oranges. *Proc. Florida State Hort. Soc.* **71,** 254-259.

Blazquez, C. H. 1967. Ortanique, a new orange-tangerine cross. *Proc. Florida State Hort. Soc.* **80,** 331-337.

Boehm, K. 1967. "Flavonoids. A Review on their Physiology, Pharmacodynamics and Therapeutic Uses." Cantor, Aulendorf, Germany.

Bogin, E., and Erickson, L. C. 1965. Activity of nitochondrial preparations from Faris sweet lemon fruit. *Plant Physiol.* **40,** 566-596.

Bogin, E., and Wallace, A. 1066a. CO_2 fixation in preparations from Tunisian sweet lemons and Eureka lemon fruits. *Proc. Am. Soc. Hort. Sci.* **88,** 298-307.

Bogin, E., and Wallace, A. 1966b. Organic acid synthesis and accumulation in sweet and sour lemon fruits. *Proc. Am. Soc. Hort. Sci.* **89,** 182-194.

Böhme, H., and Pietsch, G. 1938. Stearoptene of orange peel oil. *Arch. Pharm.* **276,** 482-488; *Chem. Abstr.* **33,** 2650 (1939).

Böhme, H., and Schneider, E. 1939. Oxidative degradation and constitution of auraptene. *Chem. Ber.* **72B,** 780-784; *Chem. Abstr.* **33,** 4975 (1939).

Böhme, H., and Volcker, P. E. 1959. Knowledge of the non-volatile fraction of bitter orange oils. *Arch. Pharm.* **292,** 529-536; *Chem. Abstr.* **54,** 11386i (1960).

Bonnell, J. M. 1958. Treatment of citrus by-products with liquid anhydrous ammonia. II. Extraction of hesperidin from dehydrated, limed citrus pulp. *Proc. Florida State Hort. Soc.* **71,** 237-240.

Born, R. 1960. 3′4′,5,6,7-Penta-*O*-methoxyflavone in orange peel. *Chem. & Ind. (London)* pp. 264-265.
Bouma, D. 1959a. Growth, yield, and fruit quality in a factorial field experiment with citrus in relation to changes in phosphorus nutrition. *Australian J. Agr. Res.* **10**, 41-51.
Bouma, D. 1959b. The development of the fruit of the Washington Navel orange. *Australian J. Agr. Res.* **10**, 804-817.
Bouma, D. 1961. The development of cuttings of the Washington Navel orange to the stage of fruit set. IV. The effect of different nitrogen and phosphorus levels on fruiting cuttings. *Australian J. Agr. Res.* **12**, 1089-1099.
Bouma, D., and McIntyre, G. A. 1963. A factorial field experiment with citrus. *J. Hort. Sci.* **38**, 175-198.
Bowden, R. P. 1968. Processing quality of oranges grown in the Near North Coast area of Queensland. *Queensland J. Agr. Animal Sci.* **25**, 93-119.
Bram, B., and Solomons, G. L. 1965. Production of the enzyme naringinase by *Aspergillus niger*. *Appl. Microbiol.* **13**, 842-845.
Braverman, J. B. S. 1959. Behaviour of citrus flavonoids. *Essenze Deriv. Agrumari* **29**, 160-173.
Brewer, R. F., Garber, M. J., Guillemet, F. B., and Sutherland, F. H. 1967. Effects of accumulated fluoride on yields and fruit quality of Washington Navel oranges. *Proc. Am. Soc. Hort. Sci.* **91**, 150-156.
Bruemmer, J. H. 1969. Redox state of nicotinamide adenine dinucleotides in citrus fruit. *J. Agr. Food Chem.* **17**, 1312-1315.
Buffa, A., and Bellenot, P. 1962a. Analytical characteristics of Algerian citrus during the industrial season 1961-62. *Fruits (Paris)* **17**, 465-468.
Buffa, A., and Bellenot, P. 1962b. Variation in the naringin content of grapefruit juice in relation to technology of manufacture. *Fruits (Paris)* **17**, 479-483.
Burke, B. A., Chan, W. R., Magnus, K. E., and Taylor, D. R. 1969. Extractives of *Cedrela odorata* L. III. Structure of photogedunin. *Tetrahedron* **25**, 5009-5013.
Calapaj, R., and Sergi, G. 1967. Spectrophotometric determination of citral in lemon oil. *Ann. Fac. Econ. Com., Univ. Studi Messina* **5**(1), 1-9; *Chem. Abstr.* **69**, 21843s (1968).
Call, T. G., and Patterson, J. A. 1958. Protective activity of some flavonoid materials against the acute toxicity of adrenergic compounds to rats. *J. Am. Pharm. Assoc.* **47**, 847-848.
Calvarano, I. 1960. Analytical characteristics of juice from Italian oranges. *Essenze Deriv. Agrumari* **30**, 3-25.
Calvarano, I. 1961. Carotenoids and mineral constituents in orange juices. *Essenze Deriv. Agrumari* **31**, 5-16.
Calvarano, I. 1963a. Characteristics of Italian lemon juice. *Fruits (Paris)* **18**, 123-127.
Calvarano, I. 1963b. Amino acids in Italian orange juice. *Essenze Deriv. Agrumari* **33**, 22-33.
Calvarano, I. 1963c. The importance of some determinations in the analytical control of citrus juices. *Essenze Deriv. Agrumari* **33**, 208-227.
Calvarano, I. 1966. Chloramine and formaldehyde numbers of Italian citrus juices. *Essenze Deriv. Agrumari* **36**, 177-191.
Calvarano, I., and Calvarano, M. 1962. Mineral constituents of citrus juices. *Essenze Deriv. Agrumari* **32**, 171-186.
Calvarano, M. 1958. Vapour-phase partition chromatography of the essential oil of the mandarin. *Essenze Deriv. Agrumari* **28**, 107-118.
Calvarano, M. 1961a. Constituents of bergamot juice. II. Flavanones, sugar and minerals. *Essenze Deriv. Agrumari* **31**, 61-76.
Calvarano, M. 1961b. Coumarin of the essential oil of bergamot. *Essenze Deriv. Agrumari* **31**, 167-174.
Calvarano, M. 1962. Carotenoids and β-carotene in Italian orange juice. *Essenze Deriv. Agrumari* **32**, 92-109.

Calvarano, M. 1963. Composition of bergamot oil. I. The monoterpene hydrocarbons. *Essenze Deriv. Agrumari* **33**, 67-101.

Calvarano, M. 1965. Composition of bergamot oil. III. *Essenze Deriv. Agrumari* **35**, 197-211.

Calvarano, M., and Calvarano, I. 1964. Composition of bergamot oil. II. Use of ultraviolet spectrophotometry and gas chromatography in the analysis. *Essenze Deriv. Agrumari* **34**, 71-92.

Calvert, D. V., and Reitz, H. J. 1963. A fertilizer rate study with Valencia oranges in the Indian River area. *Proc. Florida State Hort. Soc.* **76**, 13-17.

Calvert, D. V., and Reitz, H. J. 1964. Effects of rate and frequency of fertilizer applications on yield and quality of Valencia oranges in the Indian River area. *Proc. Florida State Hort. Soc.* **77**, 36-41.

Calvert, D. V., Hunziker, R. R., and Reitz, H. J. 1962. A nitrogen source experiment with Valencia oranges on two soil types in the Indian River area. *Proc. Florida State Hort. Soc.* **75**, 77-82.

Cameron, J. W., Soost, R. K., and Olson, E. O. 1964. Chimeral basis for color in pink and red grapefruit. *J. Heredity* **55**, 23-28.

Cameron, J. W., Soost, R. K., England, A. B., and Burnett, R. H. 1966. Fruit characters of citrus hybrids in South California test plots. *Calif. Citrograph* **51**, 434, 442, 445, and 447-450.

Cananzi, V. 1958a. Chemical composition of lemon and sweet orange juices. *Conserve Deriv. Agrumari (Palermo)* **7**, 146-147.

Cananzi, V. 1958b. Adulterations of natural and concentrated citrus juices. *Riv. Ital. Essenze-Profumi, Piante Offic.-Oli. Vegetali-Saponi* **40**, 459-461.

Caporale, G. and Cingolani, E. 1958. Ultraviolet spectra of certain furocoumarins. *Rend. Ist. Super. Sanita* **21**, 943-956; *Chem. Abstr.* **53**, 9811b (1959).

Cardinali, L. R., and Seiler, F. E. E. 1958. Chemical and physical study of several varieties of citrus fruit. *Bol. Agr. Dept. Prod. Vegetal, Secretar. Agr., Ind. Com. Trabalho, Estado Minas Gerais* **7**, Nos. 7-8, 7-30; *Chem. Abstr.* **54**, 25365b (1960).

Carpena, O., and Laencina, J. 1967. Vapor phase chromatography of essential citrus oils. *Anales Real. Soc. Espan. Fis. Quim. (Madrid)* **B63**(1), 115-119; *Chem. Abstr.* **66**, 118772u (1967).

Cary, P. R. 1968. Effects of tillage, non-tillage and nitrogen on yield and fruit composition of citrus. *J. Hort. Sci.* **43**, 299-315.

Casoli, U. 1963. Free amino acids in citrus juices. *Ind. Conserve (Parma)* **38**, 113-116.

Castro, C. E., and Schmitt, R. A. 1962. Direct elemental analysis of citrus crops by instrumental neutron activation. A rapid method for total Br, Cl, Mn, Na, and K residues. *J. Agr. Food Chem.* **10**, 236-239.

Catsouras, G. C. 1965. Total carotenoids and carotene content of navel and local variety of oranges. Detection of added β-carotene to the orange juice. *Deltion Inst. Technol. Phytikon Prointon* **3**, 65-74; *Chem. Abstr.* **65**, 4545c (1966).

Cavoli, A. M. 1966. Chemical, physical and spectrophotometric characteristics of essential oils of lemon produced in the district of Palermo in 1965-6. *Conserve Deriv. Agrumari (Palermo)* **15**, 143-146.

Chaliha, B. P., Barua, A. D., Mahanta, D., and Siddappa, G. S. 1963a. Utilisation of Assam oranges. I. Recovery of peel oil and its physico-chemical composition. *Food Sci.* **12**, 240-243.

Chaliha, B. P., Barua, A. D., Mahanta, D., and Siddappa, G. S. 1963b. Utilisation of Assam oranges. II. Recovery and quality of pectin. *Food Sci.* **12**, 243-245.

Chaliha, B. P., Barua, A. D., Mahanta, D., and Siddappa, G. S. 1963c. Mandarin seed oil–physicochemical characteristics. *Indian Oil Soap J.* **29**(3), 71-74; *Chem. Abstr.* **61**, 1704b (1964).

Chaliha, B. P., Sastry, G. P., and Rao, P. R. 1964a. Chemical studies on Assam citrus fruits. I. Examination of the peels of *Citrus limonia. Indian J. Chem.* **2**, 40; *Chem. Abstr.* **60**, 12096c (1964).

Chaliha, B. P., Sastry, G. P., and Rao, P. R. 1964b. Chemical studies on Assam citrus fruits. III. Examination of Assam round lemons *(Citrus jambhiri). J. Proc. Inst. Chemists (India)* **36**, 208-210; *Chem. Abstr.* **62**, 1013g (1965).

Chaliha, B. P., Sastry, G. P., and Rao, P. R. 1965a. Chemical examination of the peel of *Citrus jambhiri* Lush. *Tetrahedron* **21**, 1441-1443.

Chaliha, B. P., Sastry, G. P., and Rao, P. R. 1965b. Flavonoids of the Assam citrus peels. *Bull. Natl. Inst. Sci., India* **31**, 63-68; *Chem. Abstr.* **66**, 55334e (1967).

Chaliha, B. P., Sastry, G. P., and Rao, P. R. 1967. Chemical investigations of *Citrus reticulata. Indian J. Chem.* **5**, 239-241; *Chem. Abstr.* **67**, 114340z (1968).

Chandler, B. V. 1958a. Anthocyanins of blood oranges. *Nature* **182**, 933.

Chandler, B. V. 1958b. Unpublished data. Food Preserv. Res. Lab., C.S.I.R.O., Ryde, N.S.W., Australia.

Chandler, B. V., and Kefford, J. F. 1953. Chemistry of bitterness in orange juice. 4. Limonexic acid. *Australian J. Sci.* **16**, 28-29.

Chandler, B. V., and Kefford, J. F. 1966. The chemical assay of limonin, the bitter principle of oranges. *J. Sci. Food Agr.* **17**, 193-197.

Chandler, B. V., and Kefford, J. F. 1969. Unpublished data. Food Preserv. Res. Lab., C.S.I.R.O., Ryde, N.S.W., Australia.

Chandler, B. V., Kefford, J. F., and Lenz, F. 1966. Absence of bitterness in navel oranges from rooted cuttings. *Nature* **210**, 868-869.

Chandler, B. V., Kefford, J. F., and Ziemelis, G. 1968. Removal of limonin from bitter orange juice. *J. Sci. Food Agr.* **19**, 83-86.

Charley, V. L. S. 1963. Some technical aspects of British comminuted drink production. *Food Technol.* **17**, 987-994.

Charley, V. L. S. 1966. Some aspects of the biochemical, physiological and nutritional significance of flavonoids of fruit origin. *Rept. Sci. Tech. Comm., Intern. Federation Juice Producers* **7**, 93-116.

Chatt, E. M. 1966. The mineral constituents of fruits. *Brit. Food Mfg. Ind. Res. Assoc. Sci. Tech. Sur.* **45**.

Chopin, J., and Dellamonica, G. 1965. Formation of aurone glycosides by spontaneous oxidation of eriocitrin, hesperidin and naringin. *Compt. Rend.* **260**, 5582-5584; *Chem. Abstr.* **64**, 3760c (1966).

Chopin, J., Dellamonica, G., and Lebreton, P. 1963. Isolation of a 6-glucoside of aureusidin from extracts of lemon peel. *Compt. Rend.* **257**, 534-536; *Chem. Abstr.* **59**, 9731f (1963).

Chopin, J., Roux, B., and Durix, A. 1964. Flavone glycosides from lemon peel. *Compt. Rend.* **259**, 3111-3113; *Chem. Abstr.* **62**, 2957d (1965).

Chopin, J., Durix, A., Bouillant, M. L., and Wallach, J. 1968. C-glucosylation of diosmetin : synthesis of the 6-(C-glycosyl) diosmetin found in lemons. *C. R. Acad. Sci., Paris, Ser. C* **267**, 1722-1725.

Chopin, J., Roux, B., Bouillant, M. L., Durix, A., D'Arcy, A., Mabry, T., and Yoshioka, H. 1969. Structure and synthesis of 6,8-bis(C-glucosyl)-apigenin of lemons. *C. R. Acad. Sci., Paris, Ser. C* **268**, 980-982.

Christiansen, K., and Boll, P. M. 1964. Structure of neohesperidin. *Planta Med.* **12**, 77-84; *Chem. Abstr.* **61**, 7090h (1964).
Chu, L-C., Shen, C-P., Chang, H-C., and Lin, Y-C. 1957. Composition of Chinese tangerines and oranges. *Ying Yang Hsueh Pao* **2**(1), 61-69.
Ciba Ltd. 1965. Carboxypeptidase C. Neth. Patent Appl. 6,407,591; *Chem. Abstr.* **63**, 675h (1965).
Cieri, U. R. 1969. Characterization of the steam non-volatile residue of bergamot oil and some other essential oil. *J. Assoc. Offic. Anal. Chemists* **52**, 719-728.
Clark, J. R., and Bernhard, R. A. 1960a. Examination of lemon oil by gas-liquid chromatography. II. The hydrocarbon fraction. *Food Res.* **25**, 389-394.
Clark, J. R., and Bernhard, R. A. 1960b. Examination of lemon oil by gas-liquid chromatography. III. The oxygenated fraction. *Food Res.* **25**, 731-738.
Clark, R. B., and Wallace, A. 1963. Dark CO_2 fixation in organic acid synthesis and accumulation in citrus fruit vesicles. *Proc. Am. Soc. Hort. Sci.* **83**, 322-332.
Clayton, E. M., Jr., Feldhaus, W. D., and Phythyon, J. M. 1967. Effect of citrus bioflavonoid, carbazochrome salicylate, and ascorbic acid on the transplacental passage of fetal erythrocytes. *Obstet. Gynecol.* **29**, 382-385; *Chem. Abstr.* **66**, 64299z (1967).
Clements, R. L. 1964a. Organic acids in citrus fruits. I. Varietal differences. *J. Food Sci.* **29**, 276-280.
Clements, R. L. 1964b. Organic acids in citrus fruits. II. Seasonal changes in the orange. *J. Food Sci.* **29**, 281-286.
Clements, R. L. 1965. Fruit proteins. Extraction and electrophoresis. *Anal. Biochem.* **13**, 390-401.
Clements, R. L. 1966. Disk electrophoresis of citrus fruit proteins. *Phytochemistry* **5**, 243-249.
Clements, R. L., and Leland, H. V. 1962a. An ion-exchange study of the free amino acids in the juices of six varieties of citrus. *J. Food Sci.* **27**, 20-25.
Clements, R. L., and Leland, H. V. 1962b. Seasonal changes in the free amino acids in Valencia orange juice. *Proc. Am. Soc. Hort. Sci.* **80**, 300-307.
Clementson, C.A.B., and Andersen, L. 1966. Plant polyphenols as antioxidants for ascorbic acid. *Ann. N.Y. Acad. Sci.* **136**, 339-376; *Chem. Abstr.* **66**, 45118y (1967).
Coffin, D. E. 1968. Correlation of the levels of several constituents of commercial orange juices. *J. Assoc. Offic. Anal. Chemists* **51**, 1199-1203.
Coffin, D. E. 1969. Gas chromatographic determination of phenolic amines. *J. Assoc. Offic. Anal. Chemists* **51**, 1199-1203.
Coggins, C. W., and Eaks, I. L. 1967. Gibberellin research on Navel oranges. *Calif. Citrograph* **52**, 475, 486, and 489-491.
Coggins, C. W., and Hield, H. Z. 1962. Navel orange fruit response to potassium gibberellate. *Proc. Am. Soc. Hort. Sci.* **81**, 227-230.
Coggins, C. W., and Lewis, L. N. 1962. Regreening of the Valencia orange as influenced by potassium gibberellate. *Plant Physiol.* **37**, 625-627.
Coggins, C. W., Hield, H. Z., and Garber, M. J. 1960a. The influence of potassium gibberellate on Valencia orange trees and fruit. *Proc. Am. Soc. Hort. Sci.* **76**, 193-198.
Coggins, C. W., Hield, H. Z., and Boswell, S. B. 1960b. The influence of potassium gibberellate on Lisbon lemon trees and fruit. *Proc. Am. Soc. Hort. Sci.* **76**, 199-207.
Coggins, C. W., Hield, H. Z., and Burns, R. M. 1962. Influence of potassium gibberellate on grapefruit trees and fruit. *Proc. Am. Soc. Hort. Sci.* **81**, 223-226.
Cohen, L., and Cohen, A. 1959. Experimental evaluation of systemic medication modifying reactions to radiotherapy. *Brit. J. Radiol.* **32**, 18-21; *Chem. Abstr.* **53**, 9459b (1959).
Cohen, M., and Reitz, H. J. 1963. Rootstocks for Valencia orange and Ruby Red grapefruit. *Proc. Florida State Hort. Soc.* **76**, 29-34.

Colburn, B., Gardner, F. E., and Horanic, G. E. 1963. A rootstock trial for Tahiti limes in Dade county, Florida. *Proc. Florida State Hort. Soc.* **76**, 24-29.

Coleman, R. L., Lund, E. D., and Moshonas, M. G. 1969. Composition of orange essence oil *J. Food Sci.* **34**, 610-611.

Connolly, J. D., Handa, K. L., McCrindle, R., and Overton, K. H. 1968a. Tetranortriterpenoids. X. Grandifolione. *J. Chem. Soc.*, pp. 2227-2230; *Chem. Commun.* No. 23, 867-868 (1966).

Connolly, J. D., Handa, K. L., and McCrindle, R. 1968b. Further constituents of nim oil: The constitution of meldenin. *Tetrahedron Letters*, pp. 437-440.

Cooper, W. C., and Henry, W. H. 1967. Effect of ascorbic acid in citrus fruit abscission. *Citrus Ind.* **48**(6), 5-7.

Cooper, W. C., Peynado, A., Furr, J. R., Hilgeman, R. H., Cahoon, G. A., and Boswell, S. B. 1963. Tree growth and fruit quality of Valencia oranges in relation to climate. *Proc. Am. Soc. Hort. Sci.* **82**, 180-192.

Cooper, W. C., Rasmussen, G. K., and Smoot, J. J. 1968. Induction of degreening of tangerines by preharvest applications of ascorbic acid, other ethylene-releasing agents. *Citrus Ind.* **49**(10), 25-27.

Coussin, B. R., and Samish, Z. 1968. Free amino acids in Israel orange juice. *J. Food Sci.* **33**, 196-199.

Cox, J. E. 1969. Quality measurements of some common orange varieties. *Agr. Gaz. N S Wales* **80**, 632-635.

Curl, A. L. 1962a. Carotenoids of Meyer lemons. *J. Food Sci.* **27**, 171-176.

Curl, A. L. 1962b. Reticulataxanthin and tangeraxanthin, two carbonyl carotenoids from tangerine peel. *J. Food Sci.* **27**, 537-543.

Curl, A. L. 1965a. Structure of the carotenoid neoxanthin. *J. Food Sci.* **30**, 426-432.

Curl, A. L. 1965b. Occurrence of β-citraurin and of β-apo-8′-carotenal in the peels of California tangerines and oranges. *J. Food Sci.* **30**, 13-18.

Curl, A. L. 1967. Apo-10′-violaxanthal, a new carotenoid from Valencia orange peels. *J. Food Sci.* **32**, 141-143.

Curl, A. L., and Bailey, G. F. 1959. Changes in the carotenoid pigments in preparation and storage of Valencia orange juice powder. *Food Technol.* **13**, 394-398.

Curl, A. L., and Bailey, G. F. 1961. The carotenoids of Navel oranges. *J. Food Sci.* **26**, 442-447.

Dalal, V. B., D'Souza, S., Subramanyam, H., and Srivastava, H. C. 1962a. Wax emulsion for extending the storage life of Sathgudi oranges. *Food Sci. (Mysore)* **11**, 232-235.

Dalal, V. B., Subramanyam, H., and Srivastava, H. C. 1962b. Studies on the effect of repeated wax coatings on the storage behaviour of Coorg oranges. *Food Sci. (Mysore)* **11** 240-244.

D'Amore, G., and Calabro, G. 1966. Constituents of essential oils. I. Gas chromatography of mandarin oil. *Ann. Fac. Econ. Com., Univ. Studi Messina* **4**, 633-661; *Chem. Abstr.* **68**, 16057p (1968).

D'Amore, G., and Calapaj, R. 1965. Fluorescent substances in the essences of lemon, bergamot, tangerine, bitter orange, and sweet orange. *Rass. Chim.* **17**, 264-269; *Chem. Abstr.* **64**, 19312h (1966).

D'Amore, G., and Corigliano, F. 1966. Spectrophotometric characterization of oils of mandarin, lemon, and bergamot. *Ann. Fac. Econ. Com., Univ. Studi Messina* **4**, 413-433; *Chem. Abstr.* **68**, 16063n (1968).

Danziger, M. T., and Mannheim, C. H. 1967. Constituents of Israeli orange juice as affected by extraction conditions. *Fruchtsaft-Ind.* **12**, 124-129.

Dastoli, F. R., Lopiekes, D. V., and Doig, A. R. 1968. Bitter-sensitive protein from porcine taste buds. *Nature* **218**, 884-885.

Datta, S., and Nicholas, H. J. 1968. Incorporation of mevalonic acid-2-^{14}C into the triterpene limonin. *Phytochemistry* **7**, 955-956.

Davidek, J. 1963. Flavonoids, their determination, isolation and utilization in food industry. *Veda Vyzkum Prumyslu Potravinarskem* **12**, 179-209; *Chem. Abstr.* **62**, 7027d (1965).

Davis, P. L. 1966. A rapid procedure for extraction of naringin from grapefruit rind. *Proc. Florida State Hort. Soc.* **79**, 325-326.

Davis, W. B. 1947. Determination of flavanones in citrus fruits. *Anal. Chem.* **19**, 476-478.

Dawes, S. N. 1969. Composition of New Zealand fruit juices. I. Lemon Juice. *New Zealand J. Sci.* **12**, 129-138.

Dean, H. A., and Hoelscher, C. E. 1967. Responses of Pineapple orange trees to selected petroleum oil fractions. *J. Econ. Entomol.* **60**, 1668-1672.

De Eds, F. 1959. Physiological effects and metabolic fate of flavonoids. *Pharmacol. Plant Phenolics, Proc. Symp. Oxford, 1958,* pp. 91-102. Academic Press, New York.

De Fossard, R. A., and Lenz, F. H. 1967. Influence of nitrogen fertilization on quality, respiration, and storage life of Washington Navel oranges. *Qualitas Plant Mater. Vegetabiles* **14**, 289-304.

Deszyck, E. J., and Ting, S. V. 1958. Seasonal changes in acid content of Ruby Red grapefruit as affected by lead arsenate sprays. *Proc. Am. Soc. Hort. Sci.* **72**, 304-308.

Deszyck, E. J., and Ting, S. V. 1960a. Processed products from Murcott orange. I. Availability and characteristics of fruit. *Proc. Florida State Hort. Soc.* **73**, 276-279.

Deszyck, E. J., and Ting, S. V. 1960b. Sugar composition, bioflavonoid content, and pH of grapefruit as affected by lead arsenate sprays. *Proc. Am. Soc. Hort. Sci.* **75**, 266-270.

Deyoe, C. W., Deacon, L. E., and Couch, J. R. 1962. Citrus bioflavonoids in broiler diets. *Poultry Sci.* **41**, 1088-1090.

Diemair, W., and Pfeifer, K. 1962. Free phosphoric acid in fruit juices. *Z. Lebensm.-Untersuch. Forsch.* **117**, 209-215.

Di Giacomo, A., and Lo Presti, V. 1959. Extraction of hesperidin from tangerines. *Essenze Deriv. Agrumari* **29**, 57-59; *Chem. Abstr.* **54**, 3853d (1960).

Di Giacomo, A., and Rispoli, G. 1962a. Characterization of the lemon essence extracted by the "sponge method" in comparison with those mechanically obtained, by vapor-phase chromatography. *Essenze Deriv. Agrumari* **32**, 260-273.

Di Giacomo, A., and Rispoli, G. 1962b. The oxidation of essential oils in citrus beverages in relation to the presence or the absence of bioflavonoid substances. *Ind. Conserve (Parma)* **37**, 107.

Di Giacomo, A., and Rispoli, G. 1964. Argentinian lemon essential oils. I. General notices and analysis of terpene constituents. *Riv. Ital. Essenze-Profumi, Piante Offic.-Aromi-Saponi-Cosmet.* **46**, 381-386; *Chem. Abstr.* **62**, 5135h (1965).

Di Giacomo, A., and Rispoli, G. 1966. Ratio between chloramine and formaldehyde number. *Riv. Ital. Essenze-Profumi, Piante Offic., Aromi-Saponi, Cosmet.-Aerosol.* **48**, 723-725; *Chem. Abstr.* **66**, 104169r (1967).

Di Giacomo, A., Rispoli, G., and Crupi, F. 1962. Gas chromatography of the terpene fraction from Sicilian lemon essential oils. *Essenze Deriv. Agrumari* **32**, 126-134.

Di Giacomo, A., Rispoli, G., and Tracuzzi, M. L. 1963. Analysis of terpene constituents of citrus essential oils. *Riv. Ital. Essenze-Profumi -Piante Offic.-Aromi-Saponi-Cosmet.* **45**, 269-281; *Chem. Abstr.* **59**, 11183g (1963).

Di Giacomo, A., Rispoli, G., and Tracuzzi, M. L. 1964. The composition of bitter orange oil and methods useful to determine its purity. *Essenze Deriv. Agrumari* **34**, 3-16.

Di Giacomo, A., Pennisi, L., and Rispoli, G. 1965. Oxygen-containing substances from Sicilian lemon essence studied by new experimental techniques. *Essenze Deriv. Agrumari* **35**, 9-22.

Di Giacomo, A., Rispoli, G., and Pennisi, L. 1968a. Amino acid composition of Italian lemon juice. *Ind. Conserve (Parma)* **43**, 23-28.

Di Giacomo, A., Rispoli, G., and Aversa, M. C. 1968b. Detection of the addition of mandarin juice to orange juice through the analysis of the carotenoids present. *Ind. Conserve (Parma)* **43**, 123-128.

Di Giacomo, A., Rispoli, G., and Tita, S. 1968c. Carotenoids of Italian tangerine juice. *Riv. Ital. Essenze-Profumi, Piante Offic., Aromi-Saponi, Cosmet.-Aerosol.* **50**, 64-67; *Chem. Abstr.* **69**, 34786a (1968).

Di Giacomo, A., Rispoli, G., and Tita, S. 1968d. Mineral composition of Italian mandarin juice. *Riv. Ital. Essenze-Profumi, Piante Offic., Aromi-Saponi, Cosmet.-Aerosol.* **50**, 241-245; *Chem. Abstr.* **69**, 85558k (1969).

Di Giacomo, A., Rispoli, G., and Tita, S. 1968e. Amino acids in Italian mandarin juice. *Riv. Ital. Essenze-Profumi, Piante Offic., Aromi-Saponi, Cosmet.-Aerosol.* **50**, 297-302; *Chem. Abstr.* **69**, 105217d (1969).

Di Giacomo, A., Pennisi, L., and Raciti, G. 1969. Yield and quality of the essential oil from "Femminello Comune" lemons in relation to the growing environment and the nutritional level of the plants. *Ind. Conserve (Parma)* **44**, 110-116.

Dodge, F. D. 1938. Lactones of the citrus oils. *Am. Perfumer* **37**(5), 34-36.

Dougherty, M. H. 1968. A method for measuring the water-soluble volatile constituents of citrus juices and products. *Food Technol.* **22**, 1455-1456.

Drawert, F., Heimann, W., and Ziegler, A. 1966. Thin-layer chromatographic separation and spectrophotometric determination of the vitamin P factors, rutin, hesperidin, and naringin. *Z. Anal. Chem.* **217**, 22-31.

Dreyer, D. L. 1964. A biogenetic proposal for the simaroubaceous bitter principles. *Experientia* **20**, 297-299.

Dreyer, D. L. 1965a. Citrus bitter principles. II. Application of NMR to structural and stereochemical problems. *Tetrahedron* **21**, 75-87.

Dreyer, D. L. 1965b. Citrus bitter principles. III. Isolation of deacetylnomilin and deoxylimonin. *J. Org. Chem.* **30**, 749-751.

Dreyer, D. L. 1966a. Citrus bitter principles. V. Botanical distribution and chemotaxonomy in the Rutaceae. *Phytochemistry* **5**, 367-378.

Dreyer, D. L. 1966b. Citrus bitter principles. VI. Ichangin. *J. Org. Chem.* **31**, 2279-2281.

Dreyer, D. L. 1967. Citrus bitter principles. VII. Rutaevin. *J. Org. Chem.* **32**, 3442-3445.

Dreyer, D. L. 1968a. Citrus bitter principles. VIII. Application of optical rotatory dispersion and circular dichroism to stereochemical problems. *Tetrahedron* **24**, 3273-3283.

Dreyer, D. L. 1968b. Citrus bitter principles. IX. Extractives of *Casimiroa edulis* Llave et Lex. The structure of zapoterin. *J. Org. Chem.* **33**, 3577-3582.

Dreyer, D. L. 1968c. Limonoid bitter principles. *Fortschr. Chem. Org. Naturstoffe* **26**, 190-244.

Dreyer, D. L., Tabata, S., and Horowitz, R. M. 1964. Flavonoids of citrus. VIII. Synthesis of limocitrol, limocitrin, and spinacetin. *Tetrahedron* **20**, 2977-2983; *Chem. Abstr.* **62**, 2757h (1965).

Drummond, F. E. 1960. Citrus fluorescent dyes. British Patent 840,887.

Dunlap, W. J., and Wender, S. H. 1960. Purification and identification of flavanone glycosides in the peel of the sweet orange. *Arch. Biochem. Biophys.* **87**, 228-231.

Dunlap, W. J., and Wender, S. H. 1962. Identification studies on some minor flavonoid constituents of the grapefruit. *Anal. Biochem.* **4**, 110-115.

Dunlap, W. J., Hagen, R. E., and Wender, S. H. 1962a. Preparation and properties of rhamnosidase and glucosidase fractions from a fungal flavonoid glycosidase preparation, "Naringinase C-100." *J. Food Sci.* **27**, 597-601.

Dunlap, W. J., Nakagawa, Y., and Wender, S. H. 1962b. Preparation of highly purified flavonoid aglycones for biological studies. *Anal. Biochem.* **3**, 353-354.

Eaks, I. L. 1961. Effect of temperature and holding period on some physical and chemical characteristics of lemon fruits. *J. Food Sci.* **26**, 593-599.

Eaks, I. L. 1964. Ascorbic acid content of citrus during growth and development. *Botan. Gaz.* **125**, 186-191.

Eaks, I. L., and Masias, E. 1965. Chemical and physiological changes in lime fruits during and after storage. *J. Food Sci.* **30**, 509-515.

El Ashiry, G. M., Ringsdorf, W. M., and Cheraskin, E. 1964. Local and systemic influences in periodontal disease. IV. Effect of prophylaxis and natural versus synthetic Vitamin C upon clinical tooth mobility. *Intern. J. Vitamin Res.* **34**, 202-218.

El Tobshy, Z., and Sinclair, J. B. 1964. Inhibition *in vitro* of *Phomopsis citri* by extracts from orange peel and by fungicides. *Plant Disease Reptr.* **48**, 925-928.

El Zhorkani, A. S. 1968. Seasonal changes in physical character and juice composition of some citrus varieties as affected by soil type. *Agr. Res. Rev. (Cairo)* **46**, 72-83; *Chem. Abstr.* **71**, 90405d (1969).

Embleton, T. W., Jones, W. W., and Page, A. L. 1967. Potassium and phosphorus effects on deficient Eureka lemon trees and some salinity problems. *Proc. Am. Soc. Hort. Sci.* **91**, 120-127.

Emerson, O. H. 1948. Bitter principles in citrus fruit. I. Isolation of nomilin, a new bitter principle from the seeds of oranges and lemons. *J. Am. Chem. Soc.* **70**, 545-549.

Emerson, O. H. 1949. The bitter principle in Navel oranges. *Food Technol.* **3**, 248-250.

Erickson, L. C. 1957. Citrus fruit grafting. *Science* **125**, 994.

Ershoff, B. H., and Steers, C. W., Jr. 1960. Bioflavonoids and survival time of mice exposed to multiple sublethal doses of X-irradiation. *Proc. Soc. Exptl. Biol. Med.* **105**, 283-286.

Exarchos, C. D., and Aspridis, J. A. 1962. Seasonal changes in Greek oranges. I. Region of Sparta. Season 1960-1961. *Bull. Technol. Inst. Plant Prod. (Athens)* **2**, 4-47.

Exarchos, C. D., and Aspridis, J. A. 1965. Seasonal changes in Greek oranges. II. Region of Sparta. 1960-1963. *Bull. Technol. Inst. Plant Prod. (Athens)* **3**, 9-42.

Exarchos, C. D., and Aspridis, J. A. 1968. Seasonal changes in physical characters and chemical constituents of oranges, lemons, and mandarins in Greece. *Bull. Technol. Inst. Plant Prod. (Athens)* **4**, 5-42.

Fabianek, J. 1961. Scurvy, Vitamin C, and bioflavanoids. *Ann. Nutr. Aliment.* **15**, 67-102.

Farid, S. 1968. A new flavone from bergamot oil. *Tetrahedron* **24**, 2121-2123.

Farkas, L., Nogradi, M., and Zubovics, Z. 1967. New synthesis of sudachitin and demethoxysudachitin. *Acta Chim. Acad. Sci. Hung.* **52**, 301-304; *Chem. Abstr.* **67**, 64179f (1967).

Fazl-i-Rubbi, S., Mukherjee, B. D., Moslem Ali, A. K. M., Rahman, H., and Khan, N. A. 1959. Fruits and fruit products. I. Some valuable constituents of citrus fruits of East Pakistan. *Pakistan J. Biol. Agr. Sci.* **2**, 73-83; *Chem. Abstr.* **54**, 3779i (1960).

Feldman, A. W. 1968. Chemical changes induced in citrus plants by viruses. *Intern. Citrus Symp., 1st, Riverside,* Vol. 3, 1495-1503.

Feldman, J. R., Hamell, M., and Ward, W. W. 1968. Dihydrochalcone derivatives and method for their production. U.S. Patent 3,364,196.

Feliu, A. R., and Bernal, A. A. 1966. Tests on the possible industrial utilization of the Navelate orange. *Inform. Conservera* **14**, 311-314.

Fernandez, B., Fagerson, I. S., and Nawar, W. W. 1963. Applicability of gas chromatography to detection of changes in orange oil. *J. Gas Chromatog.* **1**(9), 21-22.

Fernandez-Flores, E., Johnson, A. R., and Blomquist, V. H. 1968. Collaborative study of a polarimetric method for l-malic acid. *J. Assoc. Offic. Anal. Chemists* **51**, 934-936.

Fischer, I. R., and Still, F. 1967. Genuine essential oils. *Planta Med.* **15**, 6-16.

Fisher, J. F. 1968. A procedure for obtaining radioactive naringin from grapefruit leaves fed L-phenylalanine-^{14}C. *Phytochemistry* **7**, 769-771.

Fisher, J. F., and Nordby, H. E. 1965. Isolation and spectral characterization of coumarins in Florida grapefruit peel oil. *J. Food Sci.* **30**, 869-873.

Fisher, J. F., and Nordby, H. E. 1966. Two new coumarins from grapefruit peel oil. *Tetrahedron* **22**, 1489-1493.

Fisher, J. F., Nordby, H. E., and Kew, T. J. 1966. Thin-layer chromatographic-colorimetric method for determining naringin in grapefruit. *J. Food Sci.* **31**, 947-950.

Fisher, J. F., Nordby, H. E., Waiss, A. C., and Stanley, W. L. 1967. A new coumarin from grapefruit peel oil. *Tetrahedron* **23**, 2523-2528.

Fishman, G. M. 1969. Biochemical characteristics for food and medical purposes of grapefruit and shaddock. *Subtrop. Kul't.* **1969**, 52-58; *Chem. Abstr.* **72**, 2297n (1970).

Fishman, G. M., and Gumanitskaya, M. N. 1966. Objective method of quality control of citrus fruit products. *Konserv. i Ovoshchesushil'. Prom.* **21**(11), 22-24; *Chem. Abstr.* **66**, 27832x (1967).

Fishman, G. M., and Gumanitskaya, M. N. 1967. Use of enzyme systems for eliminating the bitter taste from grapefruit juices. *Tr., Gruzinsk. Nauch.-Issled. Inst. Pishch. Prom.* **3**, 70-79; *Chem. Abstr.* **69**, 18146m (1968).

Flath, R. A., Lundin, R. E., and Teranishi, R. 1966. Structure of β-sinensal. *Tetrahedron Letters* pp. 295-302.

Flavian, S., and Levy, A. 1970. A study of the natural disappearance of limonin monolactone in the peel of Shamouti oranges. *J. Food Tech.* **5** (2), in press.

Fletcher, W. A., and Hollies, M. 1965. Maturity and quality in New Zealand oranges evaluated. *New Zealand J. Agr.* **110**, 153, 155, and 159-160.

Floyd, K. M., and Rogers, G. R. 1969. Chemical composition of Florida juices and concentrates. *J. Agr. Food Chem.* **17**, 1119-1122.

Floyd, K. M., Rogers, G. R. Harrell, J. E., and Wilkes, P. S. 1969. Chemical composition of Florida orange juice. *J. Assoc. Offic. Anal. Chemists* **52**, 1150-1152.

Fontanelli, L., and Silvestri, S. 1968. Pharmaceutical product analyses. VI. Determination of hesperidin in the presence of other bioflavonoids. *Farmaco (Pavia), Ed. Prat.* **23**, 139-143 (1968); *Chem. Abstr.* **68**, 89888v (1968).

Foote, C. S., Wuesthoff, M. T., Wexler, S., Burstain, I. G., Denny, R., Schenck, G. O., and Schulte-Elte, K-H. 1967. Photosensitized oxygenation of alkyl-substituted furans. *Tetrahedron* **23**, 2583-2599.

Franguelli, L., and Mariani, E. 1959. Oil of lemon seeds. *Olii Minerali, Grassi Saponi, Colori Vernici* **36**, 407-409; *Chem. Abstr.* **54**, 7186f (1960).

Frank, N. A., Pounden, W. D., and Kesterson, J. W. 1963. Hesperidin complex lacks estrogenic activity. *Citrus Ind.* **44**(2), 9.

Freedman, L., and Merritt, A. J. 1963. Citrus flavonoid complex: Chemical fractionation and biological activity. *Science* **139**, 344-345.

Freedman, L., Merritt, A., and Sadin, S. 1959. Citrus bioflavonoids solution. U.S. Patent 2,888,381.

Freedman, L., Merritt, A., and Sadin, S. 1960. Selective extraction of water soluble complex mixtures of biologically active flavonoids. U.S. Patent 2,952,674.

Freedman, S. O., Siddiqi, A. I., Krupey, J., and Sehon, A. H. 1962. Identification of a simple chemical compound (chlorogenic acid) as an allergen in plant materials causing human atopic disease. *Am. J. Med. Sci.* **244**, 548-555.

Freedman, S. O., Schulman, R., and Krupey, J. 1964. Loss of allergenic action of chlorogenic acid in gastrointestinal tract. *J. Allergy* **35**, 108-116; *Chem. Abstr.* **63**, 18775e (1965).

Friz, M. 1959. Animal experiments on the problem of fertility inhibition with hyaluronidase inhibitors. *Zentr. Gynaekol.* **81**, 1635-1642; *Chem. Abstr.* **54**, 4901b (1960).

Fukumoto, J., and Okada, S. 1963. Bitter taste of *Citrus natsudaidai. Kagaku (Kyoto)* **18**, 614-617; *Chem. Abstr.* **60**, 9095c (1964).

Gamkrelidze, I. D. 1965. Role of phosphorus fertilizers in citrus crops. *Subtropich. Kul'tury, Min. Sel'sk. Khoz. SSSR* **1965** (2), 43-76; *Chem. Abstr.* **66**, 28093u (1967).

Gardner, F. E., and Horanic, G. E. 1961. A comparative evaluation of rootstocks for Valencia and Parson Brown oranges on Lakeland fine sand. *Proc. Florida State Hort. Soc.* **74**, 123-127.

Gardner, F. E., and Horanic, G. E. 1966. Growth, yield, and fruit quality of Marsh grapefruit on various rootstocks on the Florida east coast–a preliminary report. *Proc. Florida State Hort. Soc.* **79**, 109-114.

Gardner, F. E., and Horanic, G. E. 1967. *Poncirus trifoliata* and some of its hybrids as rootstocks for Valencia sweet orange. *Proc. Florida State Hort. Soc.* **80**, 85-88.

Gardner, F. E., and Reece, P. C. 1960. Evaluation of 28 Navel orange varieties in Florida. *Proc. Florida State Hort. Soc.* **73**, 23-28.

Gardner, F. E., Hutchinson, D. J., Horanic, G. E., and Hutchins, P. C. 1967. Growth and productivity of virus-infected Valencia orange trees on 25 rootstocks. *Proc. Florida State Hort. Soc.* **80**, 89-92.

Gentili, B., and Horowitz, R. M. 1964. Flavonoids of citrus. VII. Limocitrol and isolimocitrol. *Tetrahedron* **20**, 2313-2318.

Gentili, B., and Horowitz, R. M. 1965. Isolation and characterization of some new flavanone rutinosides. *Bull. Natl. Inst. Sci. India* **31**, 78-82; *Chem. Abstr.* **66**, 46556b (1967).

Gentili, B., and Horowitz, R. M. 1968. Flavonoids of citrus. IX. Some new *C*-glycosylflavones and a nuclear magnetic resonance method for differentiating 6- and 8-*C*-glycosyl isomers. *J. Org. Chem.* **33**, 1571-1577.

Gerngross, O., and Renda, N. 1966. Occurrence and quantitative estimation of naringin in citrus species. *Ann. Chem.* **691**, 186-189.

Gerritz, H. W. 1969. Report on fruit products. *J. Assoc. Offic. Anal. Chemists* **52**, 260-261.

Gershtein, L. A. 1962. Changes in the chemical composition of oranges during ripening and storage. *Biokhim. Plodov i Ovoshchei, Akad. Nauk. SSSR* **1962**(7), 160-180; *Chem. Abstr.* **57**, 12970d (1962).

Ghosh, B. P., Mukherjee, A. K., and Banerjee, S. 1955. New pyrone derivative, cirantin, an oral contraceptive. *Naturwissenschaften* **42**, 77.

Gierschner, K., and Baumann, G. 1966. Evaluation of citrus juices on the basis of specific analytical data with particular reference to the formol value. *Rept. Sci. Tech. Com., Intern. Fed. Fruit Juice Producers* **7**, 177-200.

Gjessing, L., and Armstrong, M. D. 1963. Occurrence of (–)-sympatol in oranges. *Proc. Soc. Exptl. Biol. Med.* **114**, 226-229.

Goad, L. J., Williams, B. L., and Goodwin, T. W. 1967. Phytosterol biosynthesis. The presence of 4α,14α-dimethyl-$\Delta^{8,24(28)}$-ergostadien-3β-ol in grapefruit peel and its co-occurrence with cycloeucalenol in higher plant tissues. *European J. Biochem.* **3**, 232-236.

Goren, R. 1965. Hesperidin content in the Shamouti orange fruit. *Proc. Am. Soc. Hort. Sci.* **86**, 280-287.

Goren, R., and Goldschmidt, E. E. 1966. Peroxidase activity in citrus tissues. *Phytochemistry* **5**, 153-159.

Goren, R., and Monselise, S. P. 1964a. Morphological features and changes in nitrogen content in developing Shamouti orange fruits. *Israel J. Agr. Res.* **14**, 65-74; *Chem. Abstr.* **62**, 830b (1965).

Goren, R., and Monselise, S. P. 1964b. Determination of hesperidin in dry matter of citrus tissues by ultraviolet spectrophotometry method. *J. Assoc. Offic. Agr. Chemists* **47**, 677-681.

Goren, R., and Monselise, S. P. 1965a. Inter-relations of hesperidin, some other natural components and certain enzyme systems in developing Shamouti orange fruits. *J. Hort. Sci.* **40**, 83-99.

Goren, R., and Monselise, S. P. 1965b. Fluctuations in peel tissue components and enzymic activities of Shamouti orange fruits during a 24-hour cycle. *Botan. Gaz.* **126**, 31-35.

Goretti, G., Laencina, J., and Liberti, A. 1967a. Gas chromatographic study of essential lemon oils using a glass capillary column. *Riv. Ital. Essenze-Profumi, Piante Offic., Aromi-Saponi, Cosmet.-Aerosol.* **49**, 145-148; *Chem. Abstr.* **68**, 33093q (1968).

Goretti, G., Nota, G., and Zocolillo, L. 1967b. Use of high-resolution gas chromatography on the analysis of citrus essential oils. *Essenze Deriv. Agrumari* **37**, 209-220.

Gori, C. 1958. Preliminary observations on the cytological action of some furano-coumarin compounds of bergamot oil. *Caryologia* **11**, 68-71; *Chem. Abstr.* **53**, 9391b (1959).

Gorin, P.A.J., and Perlin, A. S. 1959. Configuration of glycosidic linkages in oligosaccharides. VIII. Synthesis of 2-D-mannopyranosyl-and 2-L-rhamnopyranosyl-disaccharides by the Koenigs-Knoor reaction. *Can. J. Chem.* **37**, 1930-1933.

Gounelle, H., Boudene, C., and Marnay, C. 1965. Effects of preservation treatments with boron-containing products, on levels of boron and vitamin C in oranges. *Ann. Nutr. Aliment.* **19**, 71-85; *Chem. Abstr.* **64**, 14860g (1966).

Grierson, W. 1968. Harvesting and market preparation techniques for Florida lemons. *Proc. Am. Soc. Hort. Sci.* **92**, 797-806.

Grierson, W., and Ting, S. V. 1960. Florida lemons for fresh and cannery use: Variety selection and degreening methods. *Proc. Florida State Hort. Soc.* **73**, 284-289.

Griffiths, F. P., and Lime, B. J. 1959. Debittering grapefruit products with naringinase. *Food Technol.* **13**, 430-433.

Guadagni, D. G., Okano, S., Buttery, G., and Burr, H. K. 1966. Correlation of sensory and gas-liquid chromatographic measurements of apple volatiles. *Food Technol.* **20**, 518-521.

Guenther, F., Burckhart, O., and Oostinga, I. 1968. Composition of lemon juices. *Ind. Obst. Gemueseverwert.* **53**, 422-427; *Chem. Abstr.* **69**, 95231n (1968).

Guenther, H. 1968. Gas-chromatographic and infrared-spectroscopic studies of lemon oils. *Deut. Lebensm.-Rundschau,* **64**, 104-111.

Gutfinger, T., and Zimmermann, G. 1962. Characterisation of oxidative changes in vegetable oils with and without citrus flavonoids as possible antioxidants. *Abstr. 1st Intern. Congr. Food Sci. Technol., London, 1962,* p. 18.

Hagen, R. E., Dunlap, W. J., Mizelle, J. W., Wender, S. H., Lime, B. J., Albach, R. E., and Griffiths, F. P. 1965. A chromatographic-fluorometric method for determination of naringin, naringenin rutinoside, and related flavanone glycosides in grapefruit juice and juice sacs. *Anal. Biochem.* **12**, 472-482.

Hagen, R. E., Dunlap, W. J., and Wender, S. H. 1966. Seasonal variation of naringin and certain other flavanone glycosides in juice sacs of Texas Ruby Red grapefruit. *J. Food Sci.* **31**, 542-547.

Hänsel, R. 1965. Flavonoids: Chemical evaluation and therapeutic effect. *Pharm. Weekblad* **100**, 1425-1438; *Chem. Abstr.* **65**, 15925h (1966).

Hanson, J. 1968. Navel orange juice and process for its preparation. British Patents 1,134,279 and 1,134,280.

Hara, T., and Koaze, Y. 1965. Naringinase. Japanese Patent 20,229; *Chem. Abstr.* **65**, 12827a (1966).

Harborne, J. B. 1967. "The Comparative Biochemistry of Bioflavonoids." Academic Press, New York.

Hardegger, E., and Braunshweker, H. 1961. Absolute configuration of the aglycone from neohesperidin. *Helv. Chim. Acta* **44**, 1413-1417.

Harding, P. L., and Sunday, M. B. 1964. Forecasting quality and pounds-solids in Florida oranges. *Citrus Ind.* **45**(10), 7-12.

Harding, P. L., Sunday, M. B., and Davis, P. L. 1959. Seasonal changes in Florida tangelos. *U.S. Dept. Agr., Tech. Bull.* **1205.**

Hart, B. F. 1960. Flavanone compounds and preparation thereof. U.S. Patent 2,926,162.

Hart, S. H. 1961. Authentic data for Texas oranges. *J. Assoc. Offic. Agr. Chemists* **44**, 633-639.

Haskell, G. 1965. Biochemical differences in color sectors of a chimeral orange fruit. *J. Heredity* **56**, 35-37; *Chem. Abstr.* **63**, 923d (1965).

Hatanaka, C., and Ozawa, J. 1968. Enzyme degradation of pectic acid. VIII. Neutral sugar components of the pectic substances of Citrus unshiu. *Nippon Nogei Kagaku Kaishi* **42**, 698-702; *Chem. Abstr.* **70**, 64431u (1969).

Hattori, S., and Konishi, T. 1963. Darkening of natsudaidai oil. *Nippon Shokuhin Kogyo Gakkaishi* **10**, 8-15; *Chem. Abstr.* **63**, 813d (1965).

Hattori, S.· Okamoto, Y., and Shiizaki, T. 1959. Distribution of components in orange. *Nosan Kako Gijutsu Kenkyu Kaishi* **6**, 37-40; *Chem. Abstr.* **53**, 22594g (1959).

Heintze, K. 1965. Vitamin P flavonoids and their significance in food plants. *Deut. Lebensm.-Rundschau* **61**, 309-311.

Hendershott, C. H., and Walker, D. R. 1959. Identification of a growth inhibitor from extracts of dormant peach flowers. *Science* **130**, 798-799.

Hendrickson, R., and Kesterson, J. W. 1956. Purification of naringin. *Proc. Florida State Hort. Soc.* **69**, 149-152.

Hendrickson, R., and Kesterson, J. W. 1961. Grapefruit seed oil. *Proc. Florida State Hort. Soc.* **74**, 219-223.

Hendrickson, R., and Kesterson, J. W. 1962. Three year study on hesperidin recovery as related to Valencia orange maturity. *Proc. Florida State Hort. Soc.* **75**, 289-291.

Hendrickson, R., and Kesterson, J. W. 1963a. Seed oils from *Citrus sinensis* (oranges). *J. Am. Oil Chemists' Soc.* **40**, 746-747.

Hendrickson, R., and Kesterson, J. W. 1963b. Florida lemon seed oil. *Proc. Florida State Hort. Soc.* **76**, 249-253.

Hendrickson, R., and Kesterson, J. W. 1964a. Seed oils from Florida mandarins and related varieties. *Proc. Florida State Hort. Soc.* **77**, 347-351.

Hendrickson, R., and Kesterson, J. W. 1964b. Hesperidin in Florida oranges. *Florida, Univ., Agr. Expt. Sta. (Gainesville), Bull.* **684.**

Hendrickson, R., and Kesterson, J. W. 1965. By-products of Florida citrus. *Florida, Univ., Agr. Expt. Sta. (Gainesville), Bull.* **698.**

Hendrickson, R., Kesterson, J. W., and Edwards, G. J. 1958a. Effect of processing variables on UV absorption of grapefruit juice. *Proc. Florida State Hort. Soc.* **71**, 190-194.

Hendrickson, R., Kesterson, J. W., and Edwards, G. J. 1958b. Ultra-violet absorption technique to determine the naringin content of grapefruit juice. *Proc. Florida State Hort. Soc.* **71**, 194-198.

Hendrickson, R., Kesterson, J. W., and Edwards, G. J. 1959. Hesperidin in orange juice and peel extracts determined by UV absorption. *Proc. Florida State Hort. Soc.* **72**, 258-263.

Herrman, K. 1962. Bioflavonoids and carotenoids of cultivated citrus fruits. *Fruchtsaft-Ind.* **7**, 320-331.

Herzog, P., and Monselise, S. P. 1968. Growth and development of grapefruit in two different climatic districts of Israel. *Israel J. Agr. Res.* **18**, 181-186.

Hield, H. Z., Burns, R. M., and Coggins, C. W. 1962. Some fruit thinning effects of naphthalene acetic acid on Wilking mandarin. *Proc. Am. Soc. Hort. Sci.* **81**, 218-222.
Higby, R. H. 1941. Canning Navel orange juice. *Calif. Citrograph* **26**, 360, 380-382.
Higby, W. K. 1962. A simplified method for determination of some aspects of the carotenoid distribution in natural and carotene-fortified orange juice. *J. Food Sci.* **27**, 42-49.
Higby, W. K. 1963. Analysis of orange juice for total carotenoids, carotenes, and added β-carotene. *Food Technol.* **17**, 331-334.
Hilgeman, R. H. 1966. Effect of climate of Florida and Arizona on grapefruit enlargement and quality; apparent transpiration and internal water stress. *Proc. Florida State Hort. Soc.* **79**, 99-106.
Hilgeman, R. H., Rodney, D. R., Dunlap, J. A., and Hales, T. A. 1966. Rootstock evaluation for lemons on two soil types in Arizona. *Proc. Am. Soc. Hort. Sci.* **88**, 280-290.
Hilgeman, R. H., Dunlap, J. A., and Sharp, F. O. 1967a. Effect of time of harvest of Valencia oranges in Arizona on fruit grade and size and yield the following year. *Proc. Am. Soc. Hort. Sci.* **90**, 103-109.
Hilgeman, R. H., Dunlap, J. A., and Sharples, G. C. 1967b. Effect of time of harvest of Valencia oranges on leaf carbohydrate content and subsequent set of fruit. *Proc. Am. Soc. Hort. Sci.* **90**, 110-116.
Hill, E. C., Wenzel, F. W., and Huggart, R. L. 1960. Effect of variety and maturity of fruit on acetylmethylcarbinol and diacetyl content of fresh citrus juices. *Food Technol.* **14**, 268-270.
Hirose, Y. 1963. Structure of evodol, a principle of *Evodia rutaecarpa. Chem. & Pharm. Bull. (Tokyo)* **11**, 535-536; *Chem. Abstr.* **59**, 5143e (1963).
Hirose, Y., Kondo, K., Arita, H., and Fujita, A. 1967. Components of evodia fruit. *Shoyakugaku Zasshi* **21**, 126-127; *Chem. Abstr.* **69**, 65151u (1969).
Hirota, I. 1962a. Nongalacturonide constituents of *Citrus unshiu* pectin. *Nippon Nogeikagaku Kaishi* **36**, 774-777; *Chem. Abstr.* **59**, 6649c (1963).
Hirota, I. 1962b. Preparation of pectin and hesperidin and utilization of sugars from the rind of mandarin orange. *Nippon Shokuhin Kogyo Gakkaishi* **9**, 205-211; *Chem. Abstr.* **59**, 13273g (1963).
Hobson, G. F. 1962. Determination of polygalacturonase in fruits. *Nature* **195**, 804-805.
Hodgson, R. W. 1967. Horticultural varieties of Citrus. *In* "The Citrus Industry" (W. Reuther, H. J. Webber, and L. D. Batchelor, eds.), rev. ed., Vol. I, pp. 431-591. Div. Agr. Sci., Univ. of California, Berkeley, California.
Holeman, E. H. (chairman). 1963. Statement of policy concerning definitions and standards of identity for fruit juice beverages. *Assoc. Food Drug Officials U. S. Quart. Bull.* **27**, Proc. Issue, 37-52.
Hopkins, G. A., and Walkley, V. T. 1967. A note on the potassium and phosphorus content of orange juice. *J. Assoc. Public Analysts* **5**, 39-40.
Hörhammer, L., and Wagner, H. 1962. Citrus bioflavonoids. *Deut. Apotheker-Ztg.* **102**, 759-765; *Chem. Abstr.* **57**, 12631d (1962).
Horie, T., Masumura, M., and Shigeo, F. 1961. Sudachitin, a new flavone pigment of *Citrus sudachi. Bull. Chem. Soc. Japan* **34**, 1547-1548; *Chem. Abstr.* **56**, 11989e (1962).
Horie, T., Shimoo, H., Masumura, M., and Okumara, F. S. 1962. Demethoxysudachitin. *Nippon Kagaku Zasshi* **83**, 602-604; *Chem. Abstr.* **59**, 6346b (1963).
Horowitz, R. M. 1956. Flavonoids of citrus. I. Isolation of diosmin from lemons *(Citrus limon). J. Org. Chem.* **21**, 1184-1185.
Horowitz, R. M. 1957. Flavonoids of citrus. II. Isolation of a new flavonol from lemons. *J. Am. Chem. Soc.* **79**, 6561-6562.

Horowitz, R. M. 1958. Eriodictyol. U.S. Patent 2,857,318.

Horowitz, R. M. 1961a. The citrus flavonoids. *In* "The Orange" (W. B. Sinclair, ed.), pp. 334-372. Univ. of California, Berkeley, California.

Horowitz, R. M. 1961b. Structure and bitterness of the flavonoid glycosides of citrus. *Symp. Biochem. Plant Phenolic Substances, Fort Collins, Colo., 1961* pp. 1-8.

Horowitz, R. M. 1964. Relations between the taste and structure of some phenolic glycosides. *In* "Biochemistry of Phenolic Compounds" (J. B. Harbone, ed.), pp. 545-571. Academic Press, New York.

Horowitz, R. M., and Gentili, B. 1959. Use of the Davis method to estimate flavanones. *Food Res.* **24,** 757-759.

Horowitz, R. M., and Gentili, B. 1960a. Flavonoid compounds of citrus. III. Isolation and structure of eriodictyol glycoside. *J. Am. Chem. Soc.* **82,** 2803-2806.

Horowitz, R. M., and Gentili, B. 1960b. Flavonoid compounds of citrus. IV. Isolation of some aglycones from the lemon *(Citrus limon). J. Org. Chem.* **25,** 2183-2187.

Horowitz, R. M., and Gentili, B. 1960c. Flavonoids of the Ponderosa lemon. *Nature* **185,** 319.

Horowitz, R. M., and Gentili, B. 1961a. Phenolic glycosides of grapefruit: A relation between bitterness and structure. *Arch. Biochem. Biophys.* **92,** 191-193.

Horowitz, R. M., and Gentili, B. 1961b. Flavonoids of citrus. V. Structure of limocitrin. *J. Org. Chem.* **26,** 2899-2902.

Horowitz, R. M., and Gentili, B. 1963a. Flavonoids of citrus. VI. Structure of neohesperidose. *Tetrahedron* **19,** 773-782.

Horowitz, R. M., and Gentili, B. 1963b. Dihydrochalcone derivatives and their use as sweetening agents. U.S. Patent 3,087,821.

Horowitz, R. M., and Gentili, B. 1964. Structure of vitexin and isovitexin. *Chem. & Ind. (London)* pp. 498-499.

Horowitz, R. M., and Gentili, B. 1966. Long range proton shielding in *C*-glycosyl compounds: Structure of some new C-glycosyl flavones. *Chem. & Ind. (London)* pp. 625-627.

Horowitz, R. M., and Gentili, B. 1968. Neohesperidin and neohesperidin chalcone. U. S. Patent 3,375,242.

Horowitz, R. M., and Gentili, B. 1969. Taste and structure in phenolic glycosides. *J. Agr. Food Chem.* **17,** 696-700.

Huet, R. 1961a. Bitterness in Navel orange juice. *Fruits (Paris)* **16,** 61-65.

Huet, R. 1961b. The influence of the method of extraction on the bitterness of grapefruit juice. *Fruits (Paris)* **16,** 327-332.

Huet, R. 1962a. The flavonoids of citrus. *Fruits (Paris)* **17,** 251-256.

Huet, R. 1962b. Chemical characteristics of different varieties of oranges and grapefruit of Morocco. *Fruits (Paris)* **17,** 469-475.

Huet, R. 1967. Identification of essential oils from citrus fruits by gas chromatography. *Fruits (Paris)* **22,** 177-181.

Huet, R. 1968. The aromas of citrus juices. *Fruits (Paris)* **23,** 453-472.

Huet, R. 1969. The essential oils contained in citrus juices. *Fruits (Paris)* **24,** 129-136.

Huet, R., and Dupuis, C. 1968. Essential oil of bergamot in Africa and in Corsica. *Fruits (Paris)* **23,** 301-311.

Huffaker, R. C., and Wallace, A. 1959. Dark fixation of CO_2 in homogenates from citrus leaves, fruits, and roots. *Proc. Am. Soc. Hort. Sci.* **74,** 348-356.

Hulme, B., Morries, P., and Stainsby, W. J. 1965. Analysis of citrus fruit 1962-1964. *J. Assoc. Public Analysts* **3,** 113-117.

Hunter, G.L.K., and Brogden, W. B. 1964a. 2,4-*p*-Menthadiene. A new monoterpene from Valencia orange oil. *J. Org. Chem.* **29,** 498-499.

Hunter, G.L.K., and Brogden, W. B. 1964b. Structure of ylangene. *J. Org. Chem.* **29**, 982-983.

Hunter, G.L.K., and Brogden, W. B. 1964c. β-Ylangene, a new sesquiterpene hydrocarbon from orange oil. *J. Org. Chem.* **29**, 2100.

Hunter, G.L.K., and Brogden, W. B. 1964d. A rapid method for isolation and identification of sesquiterpene hydrocarbons in cold-pressed grapefruit oil. *Anal. Chem.* **36**, 1122-1123.

Hunter, G.L.K., and Brogden, W. B. 1965a. Terpenes and sesquiterpenes in cold-pressed orange oil. *J. Food Sci.* **30**, 1-4.

Hunter, G.L.K., and Brogden, W. B. 1965b. Analysis of the terpene and sesquiterpene hydrocarbons in some citrus oils. *J. Food Sci.* **30**, 383-387.

Hunter, G.L.K., and Brogden, W. B. 1965c. Conversion of valencene to nootkatone. *J. Food Sci.* **30**, 876-878.

Hunter, G.L.K., and Moshonas, M. G. 1965. Isolation and identification of alcohols in cold-pressed Valencia orange oil by liquid-liquid extraction and gas chromatography. *Anal. Chem.* **37**, 378-380.

Hunter, G.L.K., and Moshonas, M. G. 1966. Analysis of alcohols in essential oils of grapefruit, lemon, lime, and tangerine. *J. Food Sci.* **31**, 167-171.

Hunter, G.L.K., and Parks, G. L. 1964. Isolation of beta-elemene from orange oil. *J. Food Sci.* **29**, 25-26.

Ichikawa, N., and Yamashita, T. 1941. The constitution of ponkanetin, a new flavanone derivative from the peel of *Citrus poonensis*. *J. Chem. Soc. Japan* **62**, 1006-1010; *Chem. Abstr.* **41**, 2775h (1947).

Iizuka, H., Ito, T., and Takiguchi, H. 1964. Fungal naringinase. Japanese Patent 29,807; *Chem. Abstr.* **63**, 9028c (1965).

Ikeda, R. M., and Spitler, E.M. 1964. Isolation, identification, and gas-chromatographic estimation of some esters and alcohols of lemon oil. *J. Agr. Food Chem.* **12**, 114-117.

Ikeda, R. M., Stanley, W. L., Vannier, S. H., and Rolle, L. A. 1961. Deterioration of lemon oil. Formation of *p*-cymene from gamma-terpinene. *Food Technol.* **15**, 379-380.

Ikeda, R. M., Rolle, L. A., Vannier, S. H., and Stanley, W. L. 1962a. Isolation and identification of aldehydes in cold-pressed lemon oil. *J. Agr. Food Chem.* **10**, 99-102.

Ikeda, R. M., Stanley, W. L., Rolle, L. A., and Vannier, S. H. 1962b. Monoterpene hydrocarbon composition of citrus oils. *J. Food Sci.* **27**, 593-596.

Inagaki, C., Igarashi, O., Arakawa, N., and Ohta, T. 1968. Antioxidative components in the flavedo-oil from Unshiu-orange. *Nippon Nogeikagaku Kaishi* **42**, 731-734.

Inoue, H., and Yamamoto, H. 1968. Physiological studies on the growth of *Citrus natsudaidai*. II. Comparison of nutrient contents in mature fruits of different sizes. *Kagawa Daigaku Nogakubu Gakuzyutu Hokoku* **19**, 122-129; *Chem. Abstr.* **69**, 95242s (1968).

Inoue, H., Yamamoto, H., and Fukuda, M. 1968. Physiological studies on the growth of *Citrus natsudaidai*. I. Comparison of the morphological and physiological characters of mature fruit of different sizes. *Kagawa Daigaku Nogakubu Gakuzyutu Hokoku* **19**, 115-121; *Chem. Abstr.* **69**, 95241 (1968).

Ismail, M. A. 1966. Polymeric nucleic acids from citrus fruits and leaves in various stages of growth and development, including chemically modified senescence. Ph.D. Thesis, University of Florida; *Dissertation Abstr.* **28**, 1296 (1968).

Ismail, M. A., and Wolford, R. W. 1967. Changes in organic nitrogen including free amino acids during processing of Florida orange concentrate. *Proc. Florida State Hort. Soc.* **80**, 261-267.

Ismail, M. A., Biggs, R. H., and Oberbacher, M. F. 1967. Effect of gibberellic acid on color changes in the rind of three sweet orange cultivars. *Proc. Am. Soc. Hort. Sci.* **91**, 143-149.

Ito, S., and Izumi, Y. 1966. Remaining arsenic on summer orange fruits sprayed with lead arsenate. *Nippon Shokuhin Kogyo Gakkaishi* **13**, 486-488; *Chem. Abstr.* **66**, 104060y (1967).

Jahn, O. L., and Sunday, M. B. 1965. Color changes in citrus fruit as measured by light transmittance techniques. *Proc. Florida State Hort. Soc.* **78**, 229-332.

Janku, I. 1957. Apigenin–a spasmolytic substance from *Matricaria chamomilla. Compt. Rend. Soc. Biol.* **151**, 241; *Chem. Abstr.* **52**, 2254c (1958).

Jansen, E. F., Jang, R., and Bonner, J. 1960. Orange pectinesterase binding and its activity. *Food Res.* **25**, 64-72.

Jennings, W. G. 1962. The chemical characterization of flavor volatiles from Bartlett pears, with some attention to the role of the individual fractions in pear flavor. *Intern. Fruchtsaft-Union, Ber. Wiss.-Tech. Komm.* **4**, 337-347; *Chem. Abstr.* **58**, 13057d (1963).

Jones, R. W., Stout, M. G., Reich, H., and Huffmann, M. N. 1964. Cytostatic activities of certain flavonoids against zebra-fish embryos. *Cancer Chemotherapy Rept.* **34**, 19-20.

Jones, W. W., and Embleton, T. W. 1967. Yield and fruit quality of Washington Navel orange trees as related to leaf nitrogen and nitrogen fertilization. *Proc. Am. Soc. Hort. Sci.* **91**, 138-142.

Jones, W. W., Embleton, T. W., and Cree, C. B. 1962. Temperature effects on acid, Brix in Washington Navel oranges. *Calif. Citrograph* **47**, 132-134.

Jones, W. W., Embleton, T. W., and Platt, R. G. 1968. Leaf analysis and nitrogen fertilization of oranges. *Calif. Citrograph* **53**, 367.

Joseph, G. H., Stevens, J. W., and MacRill, J. R. 1961. Nutrients in California lemons and oranges. I. Source and treatment of samples. *J. Am. Dietet. Assoc.* **38**, 552-554.

Kadota, R., and Nakamura, T. 1963. Hyuga-natsu, a variety of summer orange. *Nippon Shokuhin Kogyo Gakkaishi* **10**, 288-292; *Chem. Abstr.* **63**, 2317h (1965).

Kadota, R., and Nakamura, T. 1967. Summer orange. II. Detection of flavor substances by thin-layer chromatography. *Nippon Shokuhin Kogyo Gakkaishi* **14**, 7-10; *Chem. Abstr.* **68**, 6098p (1968).

Kamikawa, T. 1962. Constitution of obacunone. *Nippon Kagaku Zasshi* **83**, 625-630; *Chem. Abstr.* **59**, 534b (1963).

Kamikawa, T., and Kubota, T. 1961. Natural furan derivatives. VI. Presence of a seven-membered α,β-unsaturated lactone in obacunone. *Tetrahedron* **12**, 262-268.

Kamiya, S., Esaki, S., and Hama, M. 1967. Glycosides and oligosaccharides in the L-rhamnose series. I. Enzymatic partial hydrolysis of flavonoid glycosides. *Agr. Biol. Chem. (Tokyo)* **31**, 133-136.

Kariyone, T., and Matsuno, T. 1953. Constituents of orange oil. I. Structure of auraptene. *Pharm. Bull.* **1**, 119-122; *Chem. Abstr.* **48**, 6080g (1954).

Kariyone, T., and Matsuno, T. 1954. Components of citrus species. II. *J. Pharm. Soc. Japan* **74**, 363-365; *Chem. Abstr.* **48**, 9020g (1954).

Karrer, W. 1949. Presence of hesperidin and neohesperidin in unripe oranges and in the seed vessels and stigmas of orange flowers. *Helv. Chim. Acta* **32**, 714-717.

Kefford, J. F. 1959. The chemical constituents of citrus fruits. *Advan. Food Res.* **9**, 285-372.

Kefford, J. F. 1966. Citrus fruits and processed citrus products in human nutrition. *World Rev. Nutr. Dietet.* **6**, 197-249.

Kefford, J. F., and Chandler, B. V. 1961. Influence of rootstocks on the composition of oranges, with special reference to bitter principles. *Australian J. Agr. Res.* **12**, 56-68.

Kennedy, B. M., and Schelstraete, M. 1965. Ascorbic acid, acidity, and sugar in Meyer lemons. *J. Food Sci.* **30**, 77-79.

Kertesz, Z. I. 1955. Pectic enzymes. *In* "Methods in Enzymology" (S. P. Colowick and N. O. Kaplan, eds.), Vol. I, pp. 158-166. Academic Press, New York.

Kesterson, J. W., and Hendrickson, R. 1957. Naringin, a bitter principle of grapefruit. *Florida, Univ., Agr. Expt. Sta. (Gainesville), Bull.* **511A.**

Kesterson, J. W., and Hendrickson, R. 1958. Evaluation of coldpressed Florida lemon oil and lemon bioflavonoids. *Proc. Florida State Hort. Soc.* **71**, 132-140.

Kesterson, J. W., and Hendrickson, R. 1960. Florida coldpressed Murcott oil. *Am. Perfumer Aromat.* **75**(11), 35-37.

Kesterson, J. W., and Hendrickson, R. 1962. Composition of Valencia orange oil as related to fruit maturity. *Am. Perfumer Cosmet.* **77**(12), 21-24.

Kesterson, J. W., and Hendrickson, R. 1963. Evaluation of cold-pressed Marsh grapefruit oil. *Am. Perfumer Cosmet.* **78**(5), 32-35.

Kesterson, J. W., and Hendrickson, R. 1964. Comparison of red and white grapefruit oils. *Am. Perfumer Cosmet.* **79**(1), 34-36.

Kesterson, J. W., and Hendrickson, R. 1966. Aldehyde content of Valencia orange oil as related to total rainfall. *Am. Perfumer Cosmet.* **81**(2), 39-40.

Kesterson, J. W., and Hendrickson, R. 1967. Curing Florida grapefruit oils. *Am. Perfumer Cosmet.* **82**(1), 37-40.

Kesterson, J. W., and Hendrickson, R. 1969. Florida expressed tangelo oil. *Am. Perfumer Cosmet.* **84**, 51-54.

Kesterson, J. W., Hendrickson, R., and Edwards, G. J. 1959. Cold-pressed orange oil. *Am. Perfumer Aromat.* **74**(4), 33-34.

Kesterson, J. W., Hendrickson, R., Seiler, R. R., Huffman, C. E., Brent, J. A., and Griffiths, J. T. 1965a. Nootkatone content of expressed Duncan grapefruit oil as related to fruit maturity. *Am. Perfumer Cosmet.* **80**(12), 29-31.

Kesterson, J. W., Hendrickson, R., Seiler, R. R., Huffman, C. E., Brent, J. A., and Griffiths, J. T. 1965b. Flavor of expressed Duncan grapefruit oil as related to fruit maturity. *Proc. Florida State Hort. Soc.* **78**, 207-210.

Khadr, A., and Wallace, A. 1964. Uptake and translocation of radioactive iron and zinc by trifoliate orange and rough lemon. *Proc. Am. Soc. Hort. Sci.* **85**, 189-200.

Khalifah, R. A., and Kuykendall, J. R. 1965. Effect of maturity, storage temperature, and prestorage treatment on storage quality of Valencia oranges. *Proc. Am. Soc. Hort. Sci.* **86**, 288-296.

Khalifah, R. A., and Lewis, L. N. 1966. Cytokinins in citrus: Isolation of a cell division factor from lemon seeds. *Nature* **212**, 1472-1473.

Khalifah, R. A., Lewis, L. N., and Coggins, C. W. 1963. New natural growth promoting substance in young citrus fruit. *Science* **142**, 399-400.

Khalifah, R. A., Lewis, L. N., and Coggins, C. W. 1965. Isolation and properties of gibberellin-like substances from citrus fruits. *Plant Physiol.* **40**, 441-445.

Khalifah, R. A., Lewis, L. N., and Coggins, C. W. 1966. Differentiation between indoleacetic acid and the citrus auxin by column chromatography. *Plant Physiol.* **41**, 208-210.

Khan, N. A., Yunus, M., Rahman, H., and Khuda, M. Q. 1959. Pectin and pectin-like products—pectin, gelatin, their resources and utilisation. *Pakistan J. Sci. Res.* **11**, 5-8; *Chem. Abstr.* **53**, 22593a (1959).

Kilburn, R. W. 1958. The taste of citrus juice. I. Relationship between Brix, acid and pH. *Proc. Florida State Hort. Soc.* **71**, 251-254.

Kilburn, R. W., and Davis, T. T. 1959. The taste of citrus juice. II. Citrate salts and pH. *Proc. Florida State Hort. Soc.* **72**, 271-276.

Kirchner, J. G., and Miller, J. M. 1953. Volatile oil constituents of grapefruit juice. *J. Agr. Food Chem.* **1**, 512-518.

Kishi, K. 1955. Naringinase. I. Selection of naringinase secreting strains. *Kagaku To Kogyo (Osaka)* **29**, 140-145; *Chem. Abstr.* **49**, 14106i (1955).

Kishi, K. 1957. Naringinase. II. Culture conditions and naringinase formation. *Kagaku To Kogyo (Osaka)* **31**, 328-333; *Chem. Abstr.* **52**, 3923b (1958).

Kishi, K. 1958. III. Soybean constituent effective for the formation of naringinase. *Kagaku To Kogyo (Osaka)* **32**, 389-392; *Chem. Abstr.* **53**, 4430i (1959).

Kishi, K. 1959. Naringinase. IV. End products of enzyme hydrolysis. *Kagaku To Kogyo (Osaka)* **33**, 185-187; *Chem. Abstr.* **54**, 8928g (1960).

Kita, Y., Nakatani, Y., Kobayashi, A., and Yamanishi, T. 1969. Composition of peel oil from Citrus Unshiu. *Agr. Biol. Chem. (Tokyo)* **33**, 1559-1565.

Koch, J., and Haase-Sajak, E. 1964. Chemical composition of orange products. I. Freshly pressed orange juice. *Fruchtsaft-Ind.* **9**, 26-35.

Koch, J., and Haase-Sajak, E. 1965a. Chemical composition of orange products. II. Whole-orange drinks and comminuted bases. *Deut. Lebensm.-Rundschau* **61**, 199-209.

Koch, J., and Haase-Sajak, E. 1965b. Natural coloring material of citrus fruits. I. Orange and mandarin orange carotenoids. *Z. Lebensm.-Untersuch.- Forsch.* **126**, 260-271.

Koch, J., and Haase-Sajak, E. 1965c. Natural pigments of citrus fruits. II. Anthocyanins in blood oranges. *Z. Lebensm.-Untersuch.- Forsch.* **127**, 1-4.

Koch, J., and Hess, D. 1964. Pectin determination in fruit juices. *Z. Lebensm.-Untersuch.-Forsch.* **126**, 25-38.

Kodama, M., Manabe, T., Bessho, Y., and Kubo, S. 1964. Utilization of *Citrus natsudaidai.* III. Properties of juice sac of *Citrus natsudaidai. Nippon Shokuhin Kogyo Gakkaishi* **11**, 419-423; *Chem. Abstr.* **64**, 20199h (1966).

Koeppen, B. H. 1968. Synthesis of neohesperidose. *Tetrahedron* **24**, 4963-4966.

Koeppen, B. H., and Smit, C.J.B. 1960. Paper-chromatographic identification of naringin deposits in canned grapefruit. *S. African J. Agr. Sci.* **3**, 649-650.

Kohli, R. R., and Bhambota, J. R. 1965. Storage of oranges. *Indian J. Hort.* **22**, 167-174; *Chem. Abstr.* **68**, 38244z (1968).

Kolle, F., and Gloppe, K. 1936. A new hesperidin. *Pharm. Zentralhalle* **77**, 421-425; *Chem. Abstr.* **30**, 6508 (1936).

Komatsu, S., Tanaka, S., Ozawa, S., Kubo, R., Ono, Y., and Matsuda, Z. 1930. Biochemical studies on grapefruits, *Citrus aurantium* L. *J. Chem. Soc. Japan* **51**, 478-498; *Chem. Abstr.* **26**, 717 (1932).

Koo, R.C.J. 1962. Use of leaf, fruit, and soil analyses in estimating potassium status of orange trees. *Proc. Florida State Hort. Soc.* **75**, 67-72.

Koo, R.C.J. 1963. Effects of frequency of irrigations on yield of orange and grapefruit. *Proc. Florida State Hort. Soc.* **76**, 1-5.

Koo, R.C.J., and McCornack, A. A. 1965. Effects of irrigation and fertilization on production and quality of "Dancy" tangerine. *Proc. Florida State Hort. Soc.* **78**, 10-15.

Kordan, H. A. 1965. Phenolic or phenolic-like substances produced by lemon fruit tissue growing *in vitro. Advan. Frontiers Plant Sci.* **10**, 59-62; *Chem. Abstr.* **63**, 3319a (1965).

Kordan, H. A., and Morgenstern, L. 1962. Flavonoid production by mature citrus fruit tissue proliferating *in vitro. Nature* **195**, 163-164.

Kovats, E. 1963. Essential oils. IV. The so-called "distilled" oil of lime. *Helv. Chim. Acta* **46**, 2705-2731.

Kretchman, D. W., and Jutras, P. J. 1962. Influence of pruning on size and quality of Florida grapefruit. *Proc. Florida State Hort. Soc.* **75**, 35-42.

Krewson, C. F., and Couch, J. F. 1948. Isolation of rutin from a citrus hybrid. *J. Am. Chem. Soc.* **70**, 257-258.

Krummel, G. 1963. Composition of juice from Spanish and Moroccan oranges. *Deut. Lebensm.-Rundschau* **59**, 50-52.

Kubo, S., Bessho, Y., Manabe, T., and Kodama, M. 1966a. Utilization of *Citrus natsudaidai.* VI. Removal of the bitter taste of natsudaidai fruits by naringinase. *Nippon Shokuhin Kogyo Gakkaishi* **13**, 511-517; *Chem. Abstr.* **67**, 2250p (1967).

Kubo, S., Bessho, Y., Manabe, T., and Kodama, M. 1966b. Utilization of *Citrus natsudaidai.* V. Changes of temperature and sugar composition during pasteurization of canned *Citrus natsudaidai. Nippon Shokuhin Kogyo Gakkaishi* **13**, 230-236; *Chem. Abstr.* **65**, 19225h (1966).

Kubota, T., Kamikawa, T., Tokorayama, T., and Matsuura, T. 1960. Chemical constitution of obacunone. *Tetrahedron Letters* No. 8, 1-10.

Kubota, T., Matsuura, T., Tokoroyama, T., Kamikawa, T., and Matsumoto, T. 1961. Correlation of obacunone and limonin. *Tetrahedron Letters* No. 10, 325-332.

Kugler, E., and Kovats, E. 1963. Essential oils. I. Mandarin peel oil *(Citrus reticulata). Helv. Chim. Acta* **46**, 1480-1513.

Kunjukutty, N., Sankunny, T. R., and Menarchery, M. 1966. Chemical composition and feeding values of lime *(Citrus aurantrifolia)* and lemon *(C. limon). Indian Vet. J.* **43**, 453; *Chem. Abstr.* **65**, 11246c (1966).

Kunkar, A. 1964a. Bitter orange. *Riv. Ital. Essenze-Profumi-Piante Offic.-Aromi-Saponi-Cosmet.* **46**, 17-19; *Chem. Abstr.* **61**, 4879a (1964).

Kunkar, A. 1964b. *Citrus aurantium.* III. Essential oil of bitter orange. *Riv. Ital. Essenze-Profumi-Piante Offic.-Aromi-Saponi-Cosmet.* **46**, 228-233; *Chem. Abstr.* **61**, 11844b (1964).

Kunkar, A. 1965. Bitter orange *(Citrus aurantium).* IV. The residue on evaporation and the coumarins in the essential oil. *Riv. Ital. Essenze-Profumi-Piante Offic.-Aromi-Saponi-Cosmet.* **47**, 239-242; *Chem. Abstr.* **63**, 14631c (1965).

Kunkar, A. 1968. Anthocyanins in Calabrian orange juice. *Riv. Ital. Essenze-Profumi, Piante Offic., Aromi-Saponi, Cosmet.-Aerosol.* **50**, 180-184; *Chem. Abstr.* **69**, 75738e (1968).

Kwietny, A., and Braverman, J.B.S. 1959. Critical evaluation of the cyanidin reaction for flavonoid compounds. *Bull. Res. Council Israel* **C7**, 187-196.

Kwietny, A., and Zimmermann, G. 1960. The gravimetric determination of hesperidin. *J. Assoc. Offic. Agr. Chemists* **43**, 235-237.

Labanauskas, C. K., Jones, W. W., and Embleton, T. W. 1963. Effects of foliar applications of manganese, zinc, and urea on yield and fruit quality of Valencia oranges, and nutrient concentrations in the leaves, peel and juice. *Proc. Am. Soc. Hort. Sci.* **82**, 142-152.

Labruto, G., and Di Giacomo, A. 1963. Some components of exhausted solid residues after mechanical extraction of bergamot oil. *Atti Soc. Peloritana Sci. Fis. Mat. Nat.* **9**, 163-175; *Chem. Abstr.* **64**, 3277d (1966).

La Face, F. 1960. The properties of Italian orange juices. *Essenze Deriv. Agrumari* **30**, 161-173 (1960).

La Face, F. 1961. Bergamot oil and its nonvolatile constituents. *Fruchtsaft-Ind.* **6**, 210-211.

Langendorf, H., and Lang, K. 1961. Occurrence of free phosphoric acid in foods. *Z. Lebensm.-Untersuch.-Forsch.* **115**, 400-402.

Latz, H. W., and Madsen, B. C. 1969. Total luminescence of coumarin derivatives isolated from expressed lime oil. *Anal. Chem.* **41**, 1180-1185.

Laughton, P. M., Skakum, W., and Levi, L. 1962. Determination of citral in food and drug products by the barbituric acid condensation method. *J. Agr. Food Chem.* **10**, 49-51.

Lavollay, J., and Neumann, J. 1959. Problems posed by activity of certain flavonoids on vascular resistance. *Pharmacol. Plant Phenolics, Proc. Symp. Oxford, 1958* pp. 103-122. Academic Press, New York.

Leach, E. H., and Lloyd, J.P.F. 1956. Citral poisoning. *Proc. Nutr. Soc. (Engl. Scot.)* **15**, xv-xvi.

Lecomte, J. 1959. Rutosides do not modify anaphylactic shock. *Compt. Rend. Soc. Biol.* **153**, 1086-1088; *Chem. Abstr.* **54**, 3715a (1960).

Lee, R. E. 1960. Unique role of ascorbic acid in peripheral vascular physiology as compared with rutin and hesperidin. *J. Nutr.* **72**, 203-209.

Leger, H., Babin, R., Beavieux, J., and Coustun, F. 1960. Effect of certain flavones and of their chelates upon the elastic frame of arteries during experimental atheromatosis in chickens and rabbits. *Therapie* **15**, 1085-1095; *Chem. Abstr.* **59**, 4387c (1963).

Leonard, C. D., and Graves, H. B. 1966. Effect of air-borne fluorides on Valencia orange yields. *Proc. Florida State Hort. Soc.* **79**, 79-86.

Leonard, C. D., Stewart, I., and Wander, I. W. 1961. A comparison of ten nitrogen sources for Valencia oranges. *Proc. Florida State Hort. Soc.* **74**, 79-86.

Levi, L., and Laughton, P. M. 1959. Determination of citral in lemongrass and citrus oils by condensation with barbituric acid. *J. Agr. Food Chem.* **7**, 850-859.

Lewis, L. N., and Coggins, C. W. 1964. Inhibition of carotenoid accumulation in Navel oranges by gibberellin A3, as measured by thin-layer chromatography. *Plant Cell Physiol.* **5**, 457-463.

Lewis, L. N., Coggins, C. W., and Garber, M. J. 1964. Chlorophyll concentration in the Navel orange rind as related to potassium gibberellate, light intensity, and time. *Proc. Am. Soc. Hort. Sci.* **84**, 177-180.

Lewis, L. N., Khalifah, R. A., and Coggins, C. W. 1965a. Existence of the non-indolic citrus auxin in several plant families. *Phytochemistry* **4**, 203-205.

Lewis, L. N., Khalifah, R. A., and Coggins, C. W. 1965b. Seasonal changes in citrus auxin and two auxin antagonists as related to fruit development. *Plant Physiol.* **40**, 500-505.

Lewis, L. N., Coggins, C. W., Labanauskas, C. K., and Dugger, W. M. 1967. Biochemical changes associated with natural and gibberellin A3 delayed senescence in the Navel orange rind. *Plant Cell Physiol. (Tokyo)* **8**, 151-160.

Lewis, W. M. 1966. Chemical evaluation of orange juice in compounded soft drinks. *J. Sci. Food Agr.* **17**, 316-320.

Lezhava, V. V. 1967. Effect of glauconite on yield and quality of mandarins. *Subtropich. Kul'tury Min. Sel'sk. Khoz. SSSR* **1967**(1), 85-87; *Chem. Abstr.* **67**, 90132e (1967).

Lifshitz, A., Stepak, Y., and Basker, H. B. 1969. Characterisation of Israel lemon oil and detection of its adulteration. *J. Food Sci.* **34**, 254-257.

Lime, B. J., and Tucker, D. M. 1961. Seasonal variation of Texas Valencia orange juice. *J. Rio Grande Valley Hort. Soc.* **15**, 29-31.

Lisle, D. B. 1965. Analysis of orange juice. *Proc. Soc. Anal. Chem.* **2**, 123-125.

Lockett, M. F. 1959. Capillary structure and action of flavonoids. *Pharmacol. Plant Phenolics, Proc. Symp. Oxford, 1958* pp. 81-90. Academic Press, New York.

Lodh, S. B., De, S., Mukherjee, S. K., and Bose, A. N. 1963. Storage of mandarin oranges. II. Effects of hormones and wax coatings. *J. Food Sci.* **28**, 519-524.

Lombard, P. B. 1963. Maturity study of mandarins, tangelos, and tangors. *Calif. Citrograph* **48**, 171-176.

Lombard, P. B., and Brunk, H. D. 1963. Evaluating the relation of juice composition of mandarin oranges to percent unacceptance of a taste panel. *Food Technol.* **17**, 1325-1327.

Long, S. K., Hill, E. C., and Wheaton, T. A. 1967. Florida citrus molasses as a fermentation substrate. I. Free amino acids in molasses from early and midseason varieties of citrus fruits during 2 years of production. *Appl. Microbiol.* **15**, 1091-1094.

Long, W. G. 1962. Relationships among several physical and chemical measurements made on oranges. *Proc. Florida State Hort. Soc.* **75**, 292-294.

Long, W. G., Harding, P. L., Soule, M. J., and Sunday, M. B. 1959. Variations in quality of Marsh grapefruit. *U.S. Dept. Agr., AMS* **336.**

Long, W. G., Harding, P. L., Soule, M. J., and Sunday, M. B. 1961. Variations in quality of Florida grown Duncan grapefruit. *U.S. Dept. Agr., AMS* **420.**

Long, W. G., Sunday, M. B., and Harding, P. H. 1962. Seasonal changes in Florida Murcott Honey oranges. *U.S. Dept. Agr., Tech. Bull.* **1271.**

Loori, J. J., and Cover, A. R. 1964. The mechanism of formation of *p*,α-dimethylstyrene in the essential oil of distilled lime *(Citrus aurantifolia). J. Food Sci.* **29,** 576-582.

Ludin, A., and Samish, Z. 1962. Preliminary data on the composition of Israel citrus fruits. *Fruits (Paris)* **17,** 497-500.

McCarty, C. D., Kemper, W. C., and Hield, H. Z. 1965. Mandarin, tangelo, and tangor maturity studies–1963-64 season. *Calif. Citrograph* **50,** 122, 134-136, and 138-139.

McConnell, B., and Sokoloff, B. 1959. Little stroke: possible use of citrus fruit in its prevention. *Citrus Ind.* **40**(3), 3, 7, and 41-42.

McCready, R. M., and Gee, M. 1960. Determination of pectic substances by paper chromatography. *J. Agr. Food Chem.* **8,** 510-513.

McCready, R. M., and Seegmiller, C. G. 1954. Action of pectic enzymes on oligogalacturonic acids and some of their derivatives. *Arch. Biochem. Biophys.* **50,** 440-450.

McDuff, O. R. 1967. Florida Canners Association taste panel for frozen concentrated orange juice. *Citrus Ind.* **48**(4), 29-31.

McFadden, W. H., Teranishi, R., Black, D. R., and Day, J. C. 1963. Use of capillary gas chromatography with a time-of-flight mass spectrometer. *J. Food Sci.* **28,** 316-319.

McGraw, J. Y. 1960. Influence of vitamin P. *Compt. Rend. Soc. Biol.* **154,** 1922-1927; *Chem. Abstr.* **55,** 18911h (1961).

MacLeod, W. D. 1965. Constitution of nootkatone, nootkatene, and valencene. *Tetrahedron Letters* pp. 4779-4783.

MacLeod, W. D. 1966. Nootkatone, grapefruit flavor and the citrus industry. *Calif. Citrograph* **51,** 120-123.

MacLeod, W. D. 1968a. Lemon oil analyses. III. Rapid capillary gas chromatography with combined flow and temperature programming. *J. Food Sci.* **33,** 436-437.

MacLeod, W. D. 1968b. Rapid analysis of natural essences by combined flow and temperature-programmed capillary gas chromatography. *J. Agr. Food Chem.* **16,** 884-886.

MacLeod, W. D., and Buigues, N. M. 1964. Sesquiterpenes. I. Nootkatone, a new grapefruit flavor constituent. *J. Food Sci.* **29,** 565-568.

MacLeod, W. D., and Buigues, N. M. 1966. Lemon oil analysis. I. Two-dimensional thin-layer chromatography. *J. Food Sci.* **31,** 588-590.

MacLeod, W. D., McFadden, W. H., and Buigues, N. M. 1966. Lemon oil analysis. II. Gas-liquid chromatography on a temperature-programmed long, open tubular column. *J. Food Sci.* **31,** 591-594.

Maekawa, K., Kodama, M., Kushii, M., and Mitamura, M. 1967. Essential oils of some orange peels. *Agr. Biol. Chem. (Tokyo)* **31,** 373-377.

Maier, V. P., and Beverly, G. O. 1968. Limonin monolactone, the non-bitter precursor responsible for delayed bitterness in certain citrus juices. *J. Food Sci.* **33,** 488-492.

Maier, V. P., and Dreyer, D. L. 1965. Citrus bitter principles. IV. Occurrence of limonin in grapefruit juice. *J. Food Sci.* **30,** 874-875.

Maier, V. P., and Grant, E. R. 1970. Specific thin-layer chromatography assay of limonin, a citrus bitter principle. *J. Agr. Food Chem.* **18,** 250-252.

Maier, V. P., and Margileth, D. A. 1969. Limonoic acid A-ring lactone, a new limonin derivative in citrus. *Phytochemistry* **8,** 243-248.

Maier, V. M., and Metzler, D. M. 1967a. Grapefruit phenolics. I. Identification of dihydrokaempferol and its cooccurrence with naringenin and kaempferol. *Phytochemistry* **6,** 763-765.

Maier, V. M., and Metzler, D. M. 1967b. Grapefruit phenolics. II. Principal aglycones of endocarp and peel and their possible biosynthetic relation. *Phytochemistry* **6,** 1127-1135.

Maier, V. P., Hasegawa, S., and Hera, E. 1969. Limonin D-ring-lactone hydrolase. A new enzyme from citrus seeds. *Phytochemistry* **8**, 405-407.

Maleki, M., and Sarkissian, S. 1967. Effect of canning and storage on the chemical composition and organoleptic quality of juices of different varieties of oranges grown in Lebanon. *J. Sci. Food Agr.* **18**, 501-504.

Manabe, T., Bessho, Y., Kodama, M., and Kubo, S. 1964. Utilization of natsudaidai. II. Behaviour of naringin in natsudaidai fruits. *Nippon Shokuhin Kogyo Gakkaishi* **11**, 389-394; *Chem. Abstr.* **64**, 5446g (1966).

Mannheim, C. H., and Siv, S. 1969. Occurrence of polygalacturonase in citrus fruit. *Fruchtsaft-Ind.* **14**, 98-103.

Manwaring, D. G., and Rickards, R. W. 1968. Identity of cirantin, a reported anti-fertility agent, with hesperidin. *Phytochemistry* **7**, 1881-1882.

Maraulja, M. D., Barron, R. W., Huggart, R. L., and Hill, E. C. 1963. Characteristics of commercial frozen orange juice concentrate packed during seasons when freezes occurred in Florida. *Proc. Florida State Hort. Soc.* **76**, 285-290.

Markh, A. T. 1953. Biochemical alterations of citrus juices. *Tr. Odessk. Tekhnol. Inst. Pishchevoi i Kholodil'n. Prom.* **5**(2), 80-89; *Chem. Abstr.* **49**, 6499i (1955).

Markh, A. T., and Fel'dman, A. L. 1949. Removing the bitter taste from citrus products. U.S.S.R. Patent 77,160; *Chem. Abstr.* **47**, 10150i (1953).

Markh, A. T., and Fel'dman, A. L. 1950. Biochemical transformations of flavanone glucosides of citrus fruits. *Biokhimiya* **15**, 230-235; *Chem. Abstr.* **44**, 9582i (1950).

Marloth, R. H. 1959. Citrus rootstock research in South Africa. *5th Intern. Congr. Mediterranean Citrus Growers, Catania, 1959* preprint.

Marloth, R. H., and Basson, W. J. 1959. Relative performance of Washington Navel orange selections and other Navel varieties. *J. Hort. Sci.* **34**, 133-141.

Marloth, R. H., and Basson, W. J. 1960. Empress mandarin as a rootstock for citrus. *J. Hort. Sci.* **35**, 282-292.

Masri, M. S., and De Eds, F. 1958. Effect of certain flavonoids on the pituitary-adrenal axis. *Proc. Soc. Exptl. Biol. Med.* **99**, 707-709.

Masri, M. S., Murray, C. W., and De Eds, F. 1959. Studies on flavonoids and thymus involution. *Proc. Soc. Exptl. Biol. Med.* **101**, 818-819.

Matsuno, T. 1956. Components of citrus species. IV. Constituents of orange oil. *J. Pharm. Soc. Japan* **76**, 1136-1138; *Chem. Abstr.* **51**, 3512b (1957).

Matsuno, T. 1958. Isolation of a new flavonoid glycoside, fortunellin. *Yakugaku Zasshi* **78**, 1311; *Chem. Abstr.* **53**, 6222a (1959).

Matsuno, T. 1959a. Components of citrus species. V. Components of *Citrus medica* var. Sarcodactylus. *Yakugaku Zasshi* **79**, 540-541; *Chem. Abstr.* **53**, 18193h (1959).

Matsuno, T. 1959b. Components of citrus species. VI. Components of Tanikawa Buntan and Itoshima Buntan. *Yakugaku Zasshi* **79**, 547-549; *Chem. Abstr.* **53**, 18193h (1959).

Matsuno, T., and Amano, Y. 1968. Components of the peels of shaddocks. *Yakugaku Zasshi* **88**, 787-789; *Chem. Abstr.* **69**, 65157a (1968).

Matsuura, S. 1957a. Ring closure of chalcones to flavanones in the presence of concentrated phosphoric acid. V. Constitution of ponkanetin. *Yakugaku Zasshi* **77**, 328-329; *Chem. Abstr.* **51**, 11339b (1957).

Matsuura, S. 1957b. Structure of ponkanetin. *Yakugaku Zasshi* **77**, 702; *Chem. Abstr.* **51**, 16448c (1957).

Matsuura, T., Kamikawa, T., and Kubota, T. 1961. Natural furan derivatives. VII. Relationship between the furan ring and six and seven-membered lactone rings in obacunone. *Tetrahedron* **12**, 269-276; *Chem. Abstr.* **55**, 27252d (1961).

Matusis, I. I. 1965. Are bioflavonoids essential factors of nutrition? *Vopr. Pitaniya* **24**(3), 3-13; *Chem. Abstr.* **63**, 4728f (1965).

Matzik, B. 1962. Proportion of β-carotene in the total carotenoids of commercial concentrates of citrus juices. *Fruchtsaft-Ind.* **7**, 317-319.

Mazur, Y., Weizmann, A., and Sondheimer, F. 1958. Steroids and triterpenoids of citrus fruit. III. The structure of citrostadienol, a natural 4-α-methylsterol. *J. Am. Chem. Soc.* **80**, 6293-6296.

Mehlitz, A., and Minas, T. 1964. Bitter orange peel oil. I. Terpenes. *Ind. Obst.- Gemuesever-wert.* **49**, 215-221.

Mehlitz, A., and Minas, T. 1965a. Bitter orange peel oil. II. Terpenoids. *Riechstoffe, Aromen, Koerperpflegemittel* **15**, 204-211.

Mehlitz, A., and Minas, T. 1965b. Bitter orange peel oil. III. Oxygen-containing components by gas chromatography. *Riechstoffe, Aromen, Koerperpflegemittel* **15**, 365-374.

Melera, A., Schaffner, K., Arigoni, D., and Jeger, O. 1957. On the constitution of limonin. I. The course of the alkaline hydrolysis of limonin and limonol. *Helv. Chim. Acta* **40**, 1420-1437.

Menagarishvili, A. D. 1962. The content of boron and manganese in the soils of Georgian SSR and their effect on the soil. *Mikroelementy i Estestv. Radioaktivn. Pochv. Rostovsk. Gos. Univ., Materialy 3-go [Tret'ego] Mezhvuz. Soveshch., 1961* pp. 31-34; *Chem. Abstr.* **60**, 2287g (1964).

Menkin, V. 1959. Anti-inflammatory activity of some water soluble bioflavonoids. *Am. J. Physiol.* **196**, 1205-1210.

Miller, E. V., and Glass, C. S. 1957. Observations on regreening of late varieties of oranges. *Food Res.* **22**, 689-693.

Minutilli, F., and Albanese, F. 1958. Quantitative determination of carbohydrates in fruit juices by ion-exchange chromatography. *Rass. Chim.* **10**, 14-16; *Chem. Abstr.* **54**, 3780h (1960).

Miura, H., Haginuma, S., and Mizuta, T. 1965. Quality of pectin in *Citrus unshiu* (mandarin orange) and *C. natsudaidai* (bitter orange). *Shokuryo Kenkyusho Kenkyu Hokoku* **19**, 290-300; *Chem. Abstr.* **65**, 20511e (1966).

Mizelle, J. W., Dunlap, W. J., Hagen, R. E., Wender, S. H., Lime, B. J., Albach, R. F., and Griffiths, F. P. 1965. Isolation and identification of some flavanone rutinosides of the grapefruit. *Anal. Biochem.* **12**, 316-324.

Mizelle, J. W., Dunlap, W. J., and Wender, S. H. 1967. Isolation and identification of two isomeric naringenin rhamnodiglucosides from grapefruit. *Phytochemistry* **6**, 1305-1307.

Money, R. W. 1964. Analytical data of some common fruits. Potassium and phosphorus contents. *J. Sci. Food Agr.* **15**, 594-596.

Money, R. W. 1966. The potassium, phosphorus and nitrogen contents of whole citrus fruits and juices. *J. Assoc. Public Analysts* **4**, 41-44.

Monselise, S. P., and Goren, R. 1965. Changes in composition and enzymic activity in flavedo of Shamouti oranges during the color break period as influenced by application of gibberellin and 2-chloroethyltrimethylammonium chloride. *Phyton (Buenos Aires)* **22**(1), 61-66; *Chem. Abstr.* **63**, 8969g (1965).

Monselise, S. P., and Halevy, A. H. 1961. Detection of lycopene in pink orange fruit. *Science* **133**, 1478.

Monselise, S. P., and Turrell, F. M. 1959. Tenderness, climate, and citrus fruit. *Science* **129**, 639-640.

Monselise, S. P., Goren, R., and Costo, J. 1967. Hormone-inhibition balance of some citrus tissues. *Israel J. Agr. Res.* **17**(1), 35-45; *Chem. Abstr.* **67**, 79639e (1967).

Moore, E. L., Wenzel, F. W., and Martin, F. G. 1967. Factors affecting the flavor of frozen concentrated orange juice. *Proc. Florida State Hort. Soc.* **80**, 250-253.

Moorthy, N.V.N., Kapur, N. S., and Srivastava, H. C. 1962. Role of precooling on the storage behaviour of sweet oranges. *Food Sci. (Mysore)* **11**, 249-251.

Morgan, R. H. 1963. Combined acids as an index of citrus juice content. *Food Process Packaging* **32**, 163-167.

Morgan, R. H. 1966. Serine as an index of orange content of soft drinks. *J. Assoc. Public Analysts* **4**, 73-80.

Morikawa, E., Tadishi, I., and Takiguchi, Y. 1968. Naringinase having a very weak juice clarifying action. Japanese Patent 68 28,936; *Chem. Abstr.* **70**, 95481g (1969).

Morioka, T., and Ueyanagi, F. 1966. Konatsu mikan *(Citrus tamurana)*. II. Effects of storage on vitamin C, sugar, and organic acid contents of Konatsu mikan. *Kochi Daigaku Gakujutsu Kenkyu Hokoku, Shizen Kagaku* **15**, 143-147; *Chem. Abstr.* **68**, 2085j (1968).

Moshonas, M. G. 1967. Isolation of piperitenone and 6-methyl-5-hepten-2-one from orange oil. *J. Food Sci.* **32**, 206-207.

Moshonas, M. G., and Lund, E. D. 1969a. Isolation and identification of a series of α,β-unsaturated aldehydes from Valencia orange peel oil. *J. Agr. Food Chem.* **17**, 802-804.

Moshonas, M. G., and Lund, E. D. 1969b. Aldehydes, ketones, and esters in Valencia orange peel oil. *J. Food Sci.* **34**, 502-503.

Moss, G. P. 1966. Some aspects of triterpene bitter principle biosynthesis. *Planta Med.* Suppl., 86-96.

Mucci, P., Sternieri, E., and Bertolini, A. 1966. Effectiveness of bioflavonoids in modifying leukopenia induced in rats by antimitotic agents and x-irradiation. *Biochim. Biol. Sper.* **5**(1), 16-25; *Chem. Abstr.* **66**, 102314k (1967).

Mukherjee, B. D., Moslem Ali, A.K.M., Fazl-i-Rubbi, S., Rahman, H., and Khan, N. A. 1958. Utilization of some rare varieties of East Pakistan limes. *Pakistan J. Sci. Ind. Res.* **1**, 228-230.

Mukherjee, S. K., and Bose, A. N. 1962a. Mandarin orange peel oil. I. Factors affecting the essential oil content of the peel. *J. Proc. Inst. Chemists (India)* **34**, 3-6.

Mukherjee, S. K., and Bose, A. N. 1962b. Mandarin orange peel oil. II. Physico-chemical characteristics of Sikkim (mandarin/orange) oil. *J. Proc. Inst. Chemists (India)* **34**, 233-238.

Muller, P. A. 1966. Bergamot and bergamot oil. *Perfumery Essent. Oil Record* **57**(1), 18-25.

Murdock, D. I. 1968. Diacetyl test as a quality control tool in processing frozen concentrated orange juice. *Food Technol.* **22**, 90-94.

Murdock, D. I., Hunter, G.L.K., Bueck, W. A., and Brent, J. A. 1967. Relationship of bacterial contamination in orange oil recovery system to quality of finished product. *Proc. Florida State Hort. Soc.* **80**, 267-280.

Murphy, J. W., Toube, T., and Cross, A. D. 1968. Spectra and stereochemistry. XXIX. The structure of zapoterin. *Tetrahedron Letters 1968* (49), 5153-5156.

Nakabayashi, T. 1960. Extraction of hesperidin from the citrus plant. Japanese Patent 11,200; *Chem. Abstr.* **55**, 9798i (1961).

Nakabayashi, T. 1961a. Citrus flavonoids. V. Chemical structure of flavonoid pigment, fortunellin, in the peel of the fruit of kumquat. *Nippon Nogeikagaku Kaishi* **35**, 45-48; *Chem. Abstr.* **59**, 14232g (1963).

Nakabayashi, T. 1961b. Citrus flavonoids. VI. Chemical structure of the disaccharides of neohesperidin, naringin and poncirin. *Nippon Nogeikagaku Kaishi* **35**, 942-945; *Chem. Abstr.* **60**, 11043h (1964).

Nakabayashi, T. 1961c. Citrus Flavonoids. VII. Flavonoid glycosides in sour orange. *Nippon Nogeikagaku Kaishi* **35**, 945-949; *Chem. Abstr.* **60**, 11044b (1964).

Nakabayashi, T. 1962a. Naringin in bitter orange fruits. *Nippon Shokuhin Kogyo Gakkaishi* **9**, 24-25; *Chem. Abstr.* **59**, 13273e (1963).

Nakabayashi, T. 1962b. Chemistry of flavonoids in oranges. *Nippon Shokuhin Kogyo Gakkaishi* **9**, 28-38; *Chem Abstr.* **59**, 13273g (1963).

Nakabayashi, T. 1962c. Removal of bitter substances in bitter orange with naringinase. *Nippon Shokuhin Kogyo Gakkaishi* **9**, 284-289; *Chem. Abstr.* **59**, 13274a (1963).

Nakabayashi, T., and Kamiya, S. 1959. Citrus flavonoids. II. Substances which turned yellow in an alkaline condition in mandarin orange syrup. *J. Util. Agr. Prod.* **6**, 149-155; *Chem. Abstr.* **56**, 6433h (1962).

Nakabayashi, T., Mizukami, Y., Honda, H., and Furui, H. 1959. Citrus flavonoids. III. Muddiness of canned mandarin syrup. *J. Util. Agr. Prod.* **6**, 261-266; *Chem. Abstr.* **56**, 6434a (1962).

Neurath, G., and Luettich, W. 1968. Natural occurrence of ethyl citrates. *Z. Lebensm.-Untersuch.-Forsch.* **136**, 284-289.

Newhall, W. F., and Ting, S. V. 1965. Isolation and identification of α-tocopherol, a vitamin E factor, from orange flavedo. *J. Agr. Food Chem.* **13**, 281-282.

Newhall, W. F., and Ting, S. V. 1967. Degradation of hesperetin and naringenin to phloroglucinol. *J. Agr. Food Chem.* **15**, 776-777.

Nigam, I. C., Dhingra, D. R., and Gupta, G. N. 1958. Essential oil from the peels of *Citrus Macrocarpa. Indian Perfumer* **2**(2), 36-38; *Chem. Abstr.* **55**, 18020d (1961).

Nikonov, G. K., and Molodozhnikov, M. M. 1964. *Poncirus Trifoliata* and *Citrus meyeri*—new sources for obtaining limonene [limonin]. *Med. Prom. SSSR* **18**(7), 24-26; *Chem. Abstr.* **61**, 14466g (1964).

Nishiura, M. 1964. Citrus breeding and bud selection in Japan. *Proc. Florida State Hort. Soc.* **77**, 79-83.

Nishiura, M., Esaki, S., and Kamiya, S. 1969. Flavonoids in Citrus and Poncirus. I. Distribution of flavonoid glycosides in Citrus and Poncirus. *Agr. Biol. Chem. (Tokyo)* **33**, 1109-1118.

Nobile, L., and Pozzo-Balbi, T. 1968. Composition of the essential oil from Citrus trifoliata fruit rinds. I. *Ann. Chim. (Rome)* **58**, 968-981.

Nomura, D. 1950. Coumarin derivatives found in the fruit of *Citrus aurantium* natsudaidai. *Kagaku No Ryoiki* **4**, 561-564; *Chem. Abstr.* **45**, 7112g (1951).

Nomura, D. 1953. Manufacture of natsudaidai juice. VII. Determination of vitamin P (flavonoid) in the juice. *J. Ferment. Technol.* **31**, 271-275; *Chem. Abstr.* **48**, 4722i (1954).

Nomura, D. 1963. Limonoid contents in the fruits of *Citrus natsudaidai. Nippon Shokuhin Kogyo Gakkaishi* **10**, 381-382; *Chem. Abstr.* **63**, 7349a (1965).

Nomura, D. 1965. Naringinase produced by *Concothyrium diplodeilla.* I. Properties of naringinase and the removal of co-existing pectinase from the enzyme preparation. *Enzymologia* **29**, 272-282; *Chem. Abstr.* **64**, 8560b (1966).

Nomura, D. 1966a. Isolimonin in *Citrus natsudaidai* seeds. *Bull. Fac. Agr., Yamaguti Univ.* **17**, 891-894.

Nomura, D. 1966b. Studies on the decomposition of limonoid by enzymes. II. A search for limonoid decomposition enzyme. *Bull. Fac. Agr., Yamaguti Univ.* **17**, 903-910.

Nomura, D., and Akiyama, K. 1964. Perfect utilization of citrus fruits. IV. Some properties of commercial naringinase preparation and practical uses for debittering of natsudaidai products. *Nippon Shokuhin Kogyo Gakkaishi* **71**, 267-271; *Chem. Abstr.* **64**, 20091e (1966).

Nomura, D., and Santo, T. 1965. New method for quantitative determination of limonoid in *Citrus natsudaidai* juice. *Nippon Shokuhin Kogyo Gakkaishi* **12**, 100-101; *Chem. Abstr.* **64**, 18316c (1966).

Nomura, D., Akiyama, K., Shinmoto, S., and Kushiyama, Y. 1963. Perfect utilization of citrus fruits. I. Canning of *Citrus natsudaidai* fruits which were debittered by naringinase. *Nippon Shokuhin Kogyo Gakkaishi* **10**, 115-118; *Chem. Abstr.* **63**, 1155b (1965).

Nordby, H. E., and Nagy, S. 1969. Fatty acid profiles of citrus juice and seed lipids. *Phytochemistry* **8**, 2027-2038.

Norman, S., and Craft, C. C. 1968. Effect of ethylene on production of volatiles by lemons. *HortScience* **3**, 66-68.

Norman, S., Craft, C. C., and Davis, P. L. 1967. Volatiles from injured and uninjured Valencia oranges at different temperatures *J. Food Sci.* **32**, 656-659.

Nursten, H. E., and Williams, A. A. 1967. Fruit aromas: A survey of components identified. *Chem. & Ind. (London)* pp. 486-497.

Oashi, S. 1964. Paper chromatographic determination of naringin in the naringinase-treated Natsumikan orange. *Nippon Shokuhin Kogyo Gakkaishi* **11**, 376-380; *Chem. Abstr.* **64**, 14785b (1966).

Oberbacher, M. F. 1967. Citrus tissue culture as a means of studying the metabolism of carotenoids and chlorophyll. *Proc. Florida State Hort. Soc.* **80**, 254-257.

Oberbacher, M. F., and Vines, H. M. 1963. Spectrophotometric assay of ascorbic acid oxidase. *Nature* **197**, 1203-1204.

Oberbacher, M. F., Ting, S. V., and Deszyck, E. J. 1960. Internal color and carotenoid pigments of Burgundy grapefruit. *Proc. Am. Soc. Hort. Sci.* **75**, 262-265.

Oberbacher, M. F., Grierson, W., and Deszyck, E. J. 1961. Handling Florida lemons for the fresh fruit market. *Proc. Am. Soc. Hort. Sci.* **77**, 225-230.

Ohta, Y., and Hirose, Y., 1966. Constituents of cold-pressed peel oil of *Citrus natsudaidai* Hayata. *Agr. Biol. Chem. (Tokyo)* **30**, 1196-1201.

Okada, R., and Nakamura, T. 1967. Summer orange. III. Detection of neutral flavor substances by gas-liquid chromatography. *Nippon Shokuhin Kogyo Gakkaishi* **14**, 137-140; *Chem. Abstr.* **68**, 6099q (1968).

Okada, S., Kishi, K., Higashihara, M., and Fukumoto, J. 1963a. Flavonoid-hydrolysing enzymes. I. Crystallization of naringinase I and hesperidinase I and their actions. *Nippon Nogeikagaku Kaishi* **37**, 84-89; *Chem. Abstr.* **62**, 12081d (1965).

Okada, S., Kishi, K., Higashihara, M., and Fukumoto, J. 1963b. Flavonoid-hydrolysing enzymes. II. Substrate specificities of naringinase I and hesperidinase I. *Nippon Nogeikagaku Kaishi* **37**, 142-145; *Chem. Abstr.* **62**, 12081g (1965).

Okada, S., Kishi, K., Itaya, K., and Fukumoto, J. 1963c. Flavonoid-hydrolysing enzymes. III. Purification of prunin and hesperetin-7-glucoside hydrolysing enzyme. *Nippon Nogeikagaku Kaishi* **37**, 146-150; *Chem. Abstr.* **62**, 12081h (1965).

Okada, S., Yano, M., and Fukumoto, J. 1964a. Flavonoid-hydrolysing enzymes. V. Inhibition of naringin-hydrolysing enzymes by glucose. *Nippon Nogeikagaku Kaishi* **38**, 242-245; *Chem. Abstr.* **62**, 15004c (1965).

Okada, S., Yano, M., and Fukumoto, J. 1964b. Application of naringin-hydrolysing enzymes. A new method of naringin and prunin determination and its application. *Hakko Kyokaishi* **22**, 371-374; *Chem. Abstr.* **63**, 16672h (1965).

Okada, S., Yano, M., and Fukumoto, J. 1965. New method of naringin determination and its application. *Kagaku To Kogyo (Osaka)* **39**, 96-101; *Chem. Abstr.* **64**, 13293b (1966).

Olsen, R. W., and Hill, E. C. 1964. Debittering of concentrated grapefruit juice with naringinase. *Proc. Florida State Hort. Soc.* **77**, 321-325.

Omura, H., Ishizaki, K., Osajima, Y., and Yamafuji, K. 1963a. Naringinase activity of some enzyme preparations. *Kyushu Daigaku Nogakubu Gakugei Zasshi* **20**, 169-177; *Chem. Abstr.* **61**, 4639e (1964).

Omura, H., Chikano, T., Hatano, S., and Yamafuji, K. 1963b. Pectic enzyme in naringinase preparations. *Kyushu Daigaki Nogakabu Gakugei Zasshi* **20**, 309-319; *Chem. Abstr.* **61**, 3358a (1964).

Omura, H., Chikano, T., Ishizaki, K., and Yamafuji, K. 1963c. Removal of pectinase activity in naringinase preparations. *Kyushu Daigaku Nogakubu Gakugei Zasshi* **20**, 321-327; *Chem. Abstr.* **61**, 3358b (1964).

Omura, H., Yasukochi, T., and Yamafuji, K. 1966. Influence of sugars on naringinase action. *Kyushu Daigaku Nogakubu Gakugei Zasshi* **22** 181-190; *Chem. Abstr.* **65**, 12486h (1966).

Osborn, R. A. 1964. Chemical composition of fruit and fruit juices. *J. Assoc. Offic. Agr. Chemists* **47**, 1068-1086.

Parekh, C. M., Pruthi, J. S., Lal, G., and Subrahmanyan, V. 1961. Chemistry and technology of citrus essential oils–a review. *Food Sci. (Mysore)* **10**, 339-342.

Pennisi, L., and Di Giacomo, A. 1965. Lemon oils from several cultivation regions in Italy. *Riv. Ital. Essenze-Profumi, Piante Offic., Aromi-Saponi, Cosmet.-Aerosol.* **47**, 370-371; *Chem. Abstr.* **67**, 36342k (1967).

Peyron, L. 1963. Fluorescent substances present in the oleiferous cells of citrus fruits. *Compt. Rend.* **257**, 235-238.

Peyron, L. 1966. Some little known essential oils of potential interest in perfumery. *Soap, Perfumery Cosmetics* **39**, 633-643.

Phillips, R. L., and Meagher, W. R. 1966. Physiological effects and chemical residues resulting from 2,4-D and 2,4,5-T sprays used for control of preharvest fruit drop in 'Pineapple' oranges. *Proc. Florida State Hort. Soc.* **79**, 75-79.

Pinkas, J., Lavie, D., and Chorin, M. 1968. Fungistatic constituents in citrus varieties resistent to the Mal-secco disease. *Phytochemistry* **7**, 169-174.

Pisano, J. J., Oates, J. A., Karmen, A., Sjverdsma, A., and Udenfriend, S. 1961. Identification of synephrine in human urine. *J. Biol. Chem.* **236**, 898-901.

Platt, W. C. 1967. Lemon flavoring oil. U.S. Patent 3,347,687.

Porretta, A., Casoli, U., and Dall'Aglio, G. 1966. Anthocyanins in orange juice. *Ind. Conserve (Parma)* **41**, 175-179.

Pretorius, A. 1966. Influence of trace elements on the effectiveness of calcium arsenate sprays. *S. African Citrus J.* **395**, 5-7.

Primo, E., and Mallent, D. 1966. Detection of adulteration in citrus juices. VII. Method for the characterization of natural and synthetic carotenoids. *Rev. Agroquim. Tecnol. Alimentos* **6**, 215-220.

Primo, E., and Royo, J. 1965a. Detection of adulteration in citrus juices. V. Mineral composition of single strength orange juices manufactured in Spain. *Rev. Agroquim. Tecnol. Alimentos* **5**, 216-224.

Primo, E., and Royo, J. 1965b. Detection of adulteration in juices. VI. Mineral composition of juices from industrial orange varieties. *Rev. Agroquim. Tecnol. Alimentos* **5**, 471-481.

Primo, E., and Royo, J. 1967. Detection of adulteration in citrus juices. IX. Mineral composition of the serum of single strength orange juices manufactured in Spain and the United States. *Rev. Agroquim. Tecnol. Alimentos* **7**, 364-375.

Primo, E., and Royo, J. 1968. Detection of adulterations in citrus juices. XII. Characteristics of the neutralization curve and their variation as a consequence of adulteration. *Rev. Agroquim. Tecnol. Alimentos* **8**, 352-359.

Primo, E., and Royo, J. 1969. Detection of adulteration in citrus juices. XIV. Mineral composition of the serum from lemon juices of Spanish varieties. *Rev. Agroquim. Tecnol. Alimentos* **9**, 100-105.

Primo, E., Royo, J., Sala, J. M., and Gasque, F. 1962. Index of quality of orange varieties. I. Statement of the problem and comparison between the Navelate and Washington Navel varieties. *Rev. Agroquim. Tecnol. Alimentos* **2**, 235-240.

Primo, E., Sala, J., Gasque, F., and Moreno, R. 1963a. Quality of orange varieties. II. Standardisation of methods. Significance of values obtained for the percentage of juice by the usual methods. *Rev. Agroquim. Tecnol. Alimentos* **3**, 258-266.

Primo, E., Royo, J., and Sala, J. M. 1963b. Quality of orange varieties. IV. Standardization of methods. Statistical significance of vitamin C values. *Rev. Agroquim. Tecnol. Alimentos* **3**, 341-348.

Primo, E., Sanchez, J., and Alberola, J. 1963c. Detection of adulteration in citrus juices. I. Methods for the identification of acids in orange juice both by thin-layer and by gas-liquid partition chromatography. *Rev. Agroquim. Tecnol. Alimentos* **3**, 349-356.

Primo, E., Duran, L., and Flores, J. 1964a. Suitability of "Navelate" oranges for commercial use. *Rev. Agroquim. Tecnol. Alimentos* **4**, 255-256.

Primo, E., Sala, J. M., and Asensi, P. 1964b. Quality of orange varieties. V. Standardisation of the method for estimation of juice content. *Rev. Agroquim. Tecnol. Alimentos* **4**, 361-370.

Primo, E., Sanchez, J., and Alberola, J. 1965a. Detection of adulteration in citrus juices. III. Identification of non-volatile acids in orange juices of American origin. *Rev. Agroquim. Tecnol. Alimentos* **5**, 121-124.

Primo, E., Royo, J., and Sala, J. M. 1965b. New Navel type orange variety suitable for industrialization. *Rev. Agroquim. Tecnol. Alimentos* **5**, 495.

Primo, E., Serra, J., and Montesinos, M. 1967. Manometric technique for determination of pectinesterase in citrus juices. *Rev. Agroquim. Tecnol. Alimentos* **7**, 105-110.

Pritchett, D. E. 1957. Extraction of the bitter principle from Navel orange juice. U.S. Patent 2,816,033.

Pritchett, D. E. 1962. Changes in Valencia orange composition during marketing. *Calif. Citrograph* **48**, 29-30.

Pruthi, J. S., Rao, N.S.S., and Lal, G. 1960. Some technological aspects of manufacture of mandarin orange concentrate. *Food Sci. (Mysore)* **9**, 169-174.

Pruthi, J. S., Parekh, C. M., and Lal, G. 1961. Indian mandarin oils. *Food Sci. (Mysore)* **10**, 345-358.

Purcell, A. E., and Schultz, E. F. 1964. Influence of fruit age on lycopene concentration in colored grapefruit. *Proc. Am. Soc. Hort. Sci.* **85**, 183-188.

Purcell, A. E., and Stephens, T. S. 1959. Determining the effect of reciprocal grafts of red and white varieties of grapefruit on the accumulation of carotenoids. *Proc. Am. Soc. Hort. Sci.* **74**, 328-332.

Purcell, A. E., Carra, J. H., and de Gruy, I. V. 1963. Development of chromoplasts and carotenoids in colored grapefruit. *J. Rio Grande Valley Hort. Soc.* **17**, 123-127; *Chem. Abstr.* **65**, 11246f (1966).

Purcell, A. E., Young, R. H., Schultz, E. F., and Meredith, F. I. 1968. Effect of artificial climate on the internal fruit color of Redblush grapefruit. *Proc. Am. Soc. Hort. Sci.* **92**, 170-178.

Ragab, M.H.H. 1962a. An enzyme system reducing L-dehydro-ascorbic acid. V. Occurrence of the enzyme in particular kinds of citrus fruits. *Z. Lebensm.-Untersuch.- Forsch.* **116**, 397-404.

Ragab, M.H.H. 1962b. An enzyme system reducing L-dehydro-ascorbic acid. VI. Various inhibitors of the enzyme and their influence on the stability of ascorbic acid in citrus fruits. *Z. Lebensm.-Undersuch.- Forsch.* **116**, 492-496.

Rahman, A., and Khan, A. 1962. Isolation of citronin from the peel of *Citrus sinensis. Rec. Trav. Chim.* **81**, 102-106; *Chem. Abstr.* **56**, 15472h (1962).

Randhawa, G. S., Singh, J. P., and Seshadri, V. S. 1958. A promising new bud mutant in Foster grapefruit *(Citrus paradisi). Indian J. Hort.* **15**, 59-60.

Rasmussen, G. K. 1964. Seasonal changes in the organic acid content of Valencia orange fruit in Florida. *Proc. Am. Soc. Hort. Sci.* **84**, 181-187.

Rasmussen, G. K., Peynado, A., Hilgeman, R., Furr, J. R., and Cahoon, G. 1966. The organic acid content of Valencia oranges from four locations in the United States. *Proc. Am. Soc. Hort. Sci.* **89**, 206-210.

Rasquinho, L.M.A. 1965. Practical approach to the qualitative characterisation of essential oils by programmed temperature gas chromatography. *J. Gas Chromatog.* **3**, 340-344.

Ravina, A. 1964. Action of magnesium chelates upon the conjunctiva during certain senescence manifestations. *Presse Med.* **72**, 3185-3186; *Chem. Abstr.* **62**, 16804e (1965).

Reitz, H. J., and Hunziker, R. R. 1961. A nitrogen rate and arsenic spray experiment on Marsh grapefruit in the Indian River area. *Proc. Florida State Hort. Soc.* **74**, 62-67.

Reitz, H. J., and Koo, R.C.J. 1960. Effect of nitrogen and potassium fertilization on yield, fruit quality, and leaf analysis of Valencia orange. *Proc. Am. Soc. Hort. Sci.* **75**, 244-252.

Reuther, W. 1962. Citrus in southern Africa. *Calif. Citrograph* **48**, 31-35, 44, and 58-61.

Reuther, W., Rasmussen, G. K., Hilgeman, R. H., Cahoon, G. A., and Cooper, W. C. 1969. Comparison of maturation and composition of Valencia oranges in some major subtropical zones of the United States. *J. Am. Soc. Hort. Sci.* **94**, 144-157.

Riehl, L. A. 1967. Characterization of petroleum oils for the control of pests of citrus. *J. Agr. Food Chem.* **15**, 878-882.

Rispoli, G., and Di Giacomo, A. 1962. Colour preservation in citrus fruit beverages by means of natural constituents. *Essenze Deriv. Agrumari* **32**, 187-192.

Rispoli, G., and Di Giacomo, A. 1965. Sesquiterpene constituents from Sicilian lemon essential oils. *Riv. Ital. Essenze-Profumi -Piante Offic.-Aromi-Saponi-Cosmet.* **47**, 650-655; *Chem. Abstr.* **64**, 17350f (1966).

Rispoli, G., Di Giacomo, A., and Tracuzzi, M. L. 1963. Interpretation of the determination results of "10% distillate" in citrus oils analysis. *Riv. Ital. Essenze-Profumi-Piante Offic.-Aromi-Saponi-Cosmet.* **45**, 445-449; *Chem. Abstr.* **60**, 7867g (1964).

Rispoli, G., Di Giacomo, A., and Tracuzzi, M. L. 1965. Lemon essence aging control through infrared spectrophotometry. *Riv. Ital. Essenze-Profumi-Piante Offic.-Aromi-Saponi-Cosmet.* **47**, 118-122; *Chem. Abstr.* **64**, 1897f (1966).

Robbins, R. C. 1966. Effect of vitamin C and flavonoids on blood cell aggregation and capillary resistance. *Intern. Z. Vitaminforsch.* **36**, 10-15; *Chem. Abstr.* **65**, 2701f (1966).

Robbins, R. C. 1967. Effects of flavonoids on survival time of rats fed thrombogenic or atherogenic regimens. *J. Atherosclerosis Res.* **7**, 3-10; *Chem. Abstr.* **66**, 64142t (1967).

Rockland, L. B., Beavens, E. A., and Underwood, J. C. 1957. Removal of bitter principles from citrus fruits. U.S. Patent 2,816,835.

Rodrigues, J., and Subramanyam, H. 1966. Effect of preharvest spray of plant growth regulators on size, composition and storage behaviour of Coorg mandarins. *J. Sci. Food Agr.* **17**, 425-427.

Rodrigues, J., Dalal, V. B., Subramanyam, H., Aiyappa, K. M., and Srivastava, H. C. 1963. Effect of preharvest sprays of plant growth regulators on Coorg mandarins and their storage with and without wax coating. *Food Sci. (Mysore)* **12**, 336-340.

Rolle, L. A., and Vandercook, C. E. 1963. Lemon juice composition. III. Characterization of California–Arizona lemon juice by use of a multiple regression analysis. *J. Assoc. Offic. Agr. Chemists* **46**, 362-365.

Romanenko, E. V. 1964. Extraction of Vitamin P from mandarin orange juice production residues. *Konservn. i Ovoshchesushil'. Prom.* **19**, 19-22; *Chem. Abstr.* **60**, 16423f (1964).

Romanenko, E. V. 1966. Determination of hesperidin in vitamin P preparations. *Prikl. Biokhim. i Mikrobiol.* **2**, 308-312; *Chem. Abstr.* **65**, 5304a (1966).

Rotondaro, F. 1964. Spectrophotometric characterisation of some authentic oil of bergamot and a few related oils. *J. Assoc. Offic. Agr. Chemists* **47**, 611-617.

Rouse, A. H. 1967a. Evaluation of pectins from Florida's citrus peels and cores. *Citrus Ind.* **48**(6), 9-10.

Rouse, A. H. 1967b. Characteristics of oranges from 5-year-old trees. *Proc. Florida State Hort. Soc.* **80**, 222-227.

Rouse, A. H., and Atkins, C. D. 1955. Pectinesterase and pectin in commercial citrus juices as determined by methods used at the Citrus Experiment Station. *Florida, Univ., Agr. Expt. Sta., (Gainesville), Bull.* **570**.

Rouse, A. H., and Knorr, L. C. 1969. Seasonal changes in pectinesterase activity, pectins, and citric acid of Florida lemons. *Food Technol.* **23**, 829-831.

Rouse, A. H., Atkins, C. D., and Moore, E. L. 1958. Chemical characteristics of citrus juices from freeze-damaged fruit. *Proc. Florida State Hort. Soc.* **71**, 216-220.

Rouse, A. H., Atkins, C. D., and Moore, E. L. 1962a. Seasonal changes occurring in the pectinesterase activity and pectic constituents of the component parts of citrus fruits. I. Valencia oranges. *J. Food Sci.* **27**, 419-425.

Rouse, A. H., Atkins, C. D., and Moore, E. L. 1962b. Occurrence and evaluation of pectin in component parts of Valencia oranges during maturation. *Proc. Florida State Hort. Soc.* **75**, 307-311.

Rouse, A. H., Atkins, C. D., and Moore, E. L. 1963. Effect of sub-freezing temperatures on component parts of citrus fruits with particular reference to the pectic constituents. *Proc. Florida State Hort. Soc.* **76**, 295-301.

Rouse, A. H., Atkins, C. D., and Moore, E. L. 1964a. Seasonal changes occurring in the pectinesterase activity and pectic constituents of the component parts of citrus fruits. II. Pineapple oranges. *J. Food Sci.* **29**, 34-39.

Rouse, A. H., Atkins, C. D., and Moore, E. L. 1964b. Evaluation of pectin in component parts of Pineapple oranges during maturation. *Proc. Florida State Hort. Soc.* **77**, 271-274.

Rouse, A. H., Atkins, C. D., and Moore, E. L. 1964c. Evaluation of pectin in component parts of Silver Cluster grapefruit during maturation. *Proc. Florida State Hort. Soc.* **77**, 274-278.

Rouse, A. H., Moore, E. L., and Atkins, C. D. 1965a. Characteristics of oranges from three-year-old trees. *Proc. Florida State Hort. Soc.* **78**, 283-288.

Rouse, A. H., Atkins, C. D., and Moore, E. L. 1965b. Seasonal changes occurring in the pectinesterase activity and pectic constituents of the component parts of citrus fruits. III. Silver Cluster grapefruit. *Food Technol.* **19**, 673-676.

Row, L. R., and Sastry, P. G. 1962a. Glycosides of citrus fruits. I. Oranges. *Indian J. Appl. Chem.* **25**, 86-89; *Chem. Abstr.* **58**, 9412d (1963).

Row, L. R., and Sastry, G. P. 1962b. Chemical investigation of *Citrus mitis.* IV. Isolation of β-sitosterol D-glucoside. *J. Sci. Ind. Res. (India)* **21B**, 343-344; *Chem. Abstr.* **57**, 16728d (1962).

Row, L. R., and Sastry, G. P. 1963. Chemical investigation of *Citrus mitis.* V. Some reactions of citromitin. *Indian J. Chem.* **1**, 207-209; *Chem. Abstr.* **59**, 9861h (1963).

Rowell, K. M., and Beisel, C. G. 1963. Isolation of gram quantities of a rhamnoglucoside of apigenin from grapefruit. *J. Food Sci.* **28**, 195-197.

Rowell, K. M., and Winter, D. H. 1959. Determination of flavanones in citrus bioflavonoids by potassium borohydride reduction. *J. Am. Pharm. Assoc., Sci. Ed.* **48**, 746-749.

Royo, J. 1962. Proposed method for determining an index for the internal quality of oranges. *Fruits (Paris)* **17**, 457-464.

Royo, J., and Aranda, A. 1967. Detection of adulteration in orange juices and beverages. *Rev. Agroquim. Tecnol. Alimentos* **7**, 376-380.

Rudol'fi, T. A., and Sharapova, R. I. 1965. Determination of the quality of bergamot oil. *Tr. Vses. Nauchn.-Issled. Inst. Sintetich. i Natural'n. Dushistnykh Veshchestv* **7**, 167-170; *Chem. Abstr.* **66**, 13992n (1967).

Rymal, K. S., Wolford, R. W., Ahmed, E. M., and Dennison, R. A. 1968. Changes in volatile flavor constituents of canned single-strength orange juice as influenced by storage temperature. *Food Technol.* **22**, 1592-1595.

Saha, S. K., and Chatterjee, A. 1957. Isolation of alloimperatorin and β-sitosterol from the fruits of *Aegle marmelos*. *J. Indian Chem. Soc. (India)* **34**, 228-230.

Samish, Z., and Ganz, D. 1950. Some observations on bitterness in Shamouti oranges. *Canner* **110**(23), 7-9; (24), 36 and 37; (25), 22 and 24.

Sanchez, J., Alberola, J., and Garcia, I. 1964. Detection of adulterations in citrus juices. II. Identification of acids in orange varieties. *Rev. Agroquim. Tecnol. Alimentos* **4**, 371-374.

Sannie, C., and Sosa, A. 1949. Heteroside of *Citrus trifoliata*. *Bull. Soc. Chim. Biol.* **31**, 36-42; *Chem. Abstr.* **43**, 6582g (1949).

Sarin, P. S., and Seshadri, T. R. 1959. Chemical investigation of *Citrus limonum*. *J. Sci. Ind. Res. (India)* **18B**, 129-130.

Sarin, P. S., and Seshadri, T. R. 1960. New components of *Citrus aurantium*. *Tetrahedron* **8**, 64-66.

Sarin, P. S., and Seshadri, T. R. 1961. Special crystalline components of citrus fruits. *Proc. Natl. Inst. Sci. India* **A26** Suppl. I, 162-171.

Sarkar, S. R. 1958. Vitamin C content of citrus fruits in Darjeeling area. *Indian J. Appl. Chem.* **21**, 92-94; *Chem. Abstr.* **53**, 7456d (1959).

Sastry, G. P., and Row, L. R. 1960. Chemical investigations of *Citrus mitis*. *J. Sci. Ind. Res. (India)* **19B**, 500-501.

Sastry, G. P., and Row, L. R. 1961a. Chemical investigations of *Citrus mitis* Blanco. II. Isolation of two new flavanones. *J. Sci. Ind. Res. (India)* **20B**, 187-188.

Sastry, G. P., and Row, L. R. 1961b. Chemical investigations of *Citrus mitis* Blanco. III. Isolation of two new flavanones. *Tetrahedron* **15**, 111-114.

Sastry, G. P., Rao, P. R., Rao, P.V.S., and Row. L. R. 1964. Chemical investigation of *Citrus aurantium*. *Indian J. Chem.* **2**, 462-463; *Chem. Abstr.* **62**, 11766g (1965).

Sawayama, Z., Shimoda, Y., Oku, M., and Matsumoto, K. 1966. Studies on muddiness prevention in canned mandarin orange with enzyme. I. *Rept. Toyo Junior Coll. Food Technol. Toyo Inst. Food Technol.* **7**, 126-134.

Sawyer, R. 1963. Chemical composition of some natural and processed orange juices. *J. Sci. Food Agr.* **14**, 302-310.

Schneider, G., Unkrich, G., and Pfaender, P. 1968. Methoxylated flavones of *Citrus aurantium* subsp. *amara*. *Arch. Pharm. (Weinheim)* **301**, 785-792; *Chem. Abstr.* **70**, 22862e. (1969).

Schormueller, J., Pfrogner, N., and Belitz, H.D. 1965a. Vegetable phosphatases. I. Isolation and fractionation of phosphatases from sweet orange. *Z. Lebensm.-Untersuch.- Forsch.* **127**, 57-61.

Schormueller, J., Pfrogner, N., and Holz, F. 1965b. Vegetable phosphatases. II. Characterization of phosphatases from orange peel. *Z. Lebensm.-Untersuch.- Forsch.* **127**, 325-341.

Schultz, T. H., Teranishi, R., McFadden, W. H., Kilpatrick, P. W., and Corse, J. 1964. Volatiles from oranges. II. Constituents of the juice identified by mass spectra. *J. Food Sci.* **29**, 790-795.

Schultz, T. H., Black, D. R., Bomben, J. L., Mon, T. R., and Teranishi, R. 1967. Volatiles from oranges. VI. Constituents of the essence identified by mass spectra. *J. Food Sci.* **32**, 698-701.

Schwartz, H. M., Biedron, S. I., von Holdt, M. S., and Rehm, S. 1964. Plant esterases. *Phytochemistry* **3**, 189-200.

Schwarz, R. E. 1968. Thin layer chromatographical studies on phenolic markers of the greening virus in various citrus species. *S. African J. Agric. Science* **11**, 797-802.

Scora, R. W., and Newman, J. E. 1967. A phenological study of the essential oils of the peel of Valencia oranges. *Agr. Meteorol.* **4**, 11-26.

Scora, R. W., England, A. B., and Bitters, W. P. 1966. Essential oils of *Poncirus trifoliata* and its selections in relation to classification. *Phytochemistry* **5**, 1139-1146.

Scora, R. W., Cameron, J. W., and England, A. B. 1968. Rind oil components of intergeneric Citrus-Poncirus hybrids and their parents. *Proc. Am. Soc. Hort. Sci.* **92**, 221-226.

Scott, W. C., 1968. Collaborative study of recoverable oil in citrus juices by bromate titration. *J. Assoc. Offic. Anal. Chemists* **51**, 928-931.

Scott, W. C., and Hearn, C. J. 1966. Processing qualities of new citrus fruit hybrids. *Proc. Florida State Hort. Soc.* **79**, 304-307.

Scott, W. C., and Veldhuis, M. K. 1966. Rapid estimation of recoverable oil in citrus juices by bromate titration. *J. Assoc. Offic. Anal. Chemists* **49**, 628-633.

Scott, W. C., Kew, T. J., and Veldhuis, M.K. 1965. Composition of orange juice cloud. *J. Food Sci.* **30**, 833-837.

Sehgal, J. M., Seshadri, T. R., and Vadehra, K. L. 1955. Synthetic experiments in the benzopyrone series. LVII. Synthesis of 4′5,6,7,8-pentamethoxyflavanone and flavone: The constitution of ponkanetin. *Proc. Indian Acad. Sci.* **A42**, 252-254; *Chem. Abstr.* **50**, 7788i (1956).

Sherratt, J. G., and Sinar, R. 1963. Analytical data for grapefruit juice. *J. Assoc. Public Analysts* **1**, 18-20.

Shigeyama, T., and Murakami, H. 1962. Utilization of Fortunella fruits. *Miyazaki Daigaku Nogakubu, Kenkyu Jiho* **7**, 131-134; *Chem Abstr.* **59**, 13273e (1963).

Shimba, R., and Nakayama, A. 1963. Natsudaidai juice. IV. Differences in the qualities and juice compositions of various size fruits. *Nippon Shokuhin Kogyo Gakkaishi* **10**, 456-460; *Chem. Abstr.* **63**, 4869g (1965).

Shimoda, Y., Oku, M., Sawayama, Z., and Matsumoto, K. 1966a. Studies of debittering of natsudaidai (*Citrus natsudaidai* Hoyata) with naringinase enzymes. I. Problems in the naringin analysis. *Rept. Toyo Junior Coll. Food Technol. Toyo Inst. Food Technol.* **7**, 108-116.

Shimoda, Y., Oku, M., Sawayama, Z., and Matsumoto, K. 1966b. Studies of debittering of natsudaidai (*Citrus natsudaidai* Hoyata) with naringinase enzymes. II. Debittering of canned natsudaidai with a one step process. *Rept. Toyo Junior Coll. Food Technol. Toyo Inst. Food Technol.* **7**, 117-125.

Shimoda, Y., Oku., M., Mori, D., and Sawayama, Z. 1968a. Studies on debittering of natsudaidai with naringinase enzymes. III. Effect of naringinase on products manufactured under different conditions. *Rept. Toyo Junior Coll. Food Technol. Toyo Inst. Food Technol.* **8**, 140-147.

Shimoda Y., Oku, M., Mori, D., Sawayama, Z., and Otsuka, S. 1968b. Studies on debittering of natsudaidai with naringinase enzymes. IV. Effect of artificially sweetened products. *Rept. Toyo Junior Coll. Food Technol. Toyo Inst. Food Technol.* **8**, 148-158.

Shimoda, Y., Oku, M., Mori, D., and Sawayama, Z. 1968c. Studies on muddiness prevention in canned mandarin orange with enzymes. II. *Rept. Toyo Junior Coll. Food Technol. Toyo Inst. Food Technol.* **8**, 131-139.

Shinoda, J., and Sato, S. 1931. Synthesis of polyhydroxychalcones, polyhydroxyhydrochalcones and polyhydroxyflavanones. IV. Synthesis of citronetin and a few flavanones. *J. Pharm. Soc. Japan* **51**, 576-582; *Chem Abstr.* **26**, 1916 (1932).

Shioiri, H., and Katayama, O. 1955. Fruit juices. I. Identification of free amino acids by the microbiological method. *Shokuryo Kenkyusho Kenkyu Hokoku* **10**, 161-165; *Chem. Abstr.* **53**, 11703f (1959).

Sholokhova, V. A. 1962. Chemical composition of fruits in lemon productive clones. *Tr., Gos. Nikitsk. Botan. Sad* **36**, 259-263; *Chem. Abstr.* **61**, 3623a (1964).

Sholokhova, V. A., and Domanskaya, E. N. 1962. Influence of gibberellic acid on citrus crops and quality. *Tr., Gos. Nikitsk. Botan. Sad* **36**, 115-124; *Chem. Abstr.* **61**, 6293a (1964).

Siddappa, G. S., and Bhatia, B. S. 1954. Comparative study of the composition and bitterness of loose-jacket oranges grown in India. *J. Sci. Ind. Res. (India)* **13B**, 148-150.

Siddappa, G. S., and Bhatia, B. S. 1959. Effect of method of extraction of juice on the development of bitterness in preserved orange juice. *Food Technol.* **13**, 349-351.

Siddiqi, A. I., and Freedman, S. O. 1963. Identification of chlorogenic acid in castor bean and oranges. *Can. J. Biochem. Physiol.* **41**, 947-952.

Silber, R. L., Becker, M., Cooper, M., Evans, P., Fehder, P., Gray, R., Gresham, P., Rechsteiner, J., and Searles, M. A. 1960. Paper chromatographic identification and estimation of the free amino acids in thirty-two fruits. *Food Res.* **25**, 675-680.

Sinclair, W. B., ed. 1961. The Orange. Univ. of California, Berkeley, California.

Sinclair, W. B., and Jolliffe, V. A. 1958a. Changes in pectic constituents of Valencia oranges during growth and development. *Botan. Gaz.* **119**, 217-223.

Sinclair, W. B., and Jolliffe, V. A. 1958b. Free galacturonic acid in citrus fruits. *Botan. Gaz.* **120**, 117-121.

Sinclair, W. B., and Jolliffe, V. A. 1960. Methods of analysis of soluble carbohydrates and pectic substances of citrus fruits. *Food Res.* **25**, 148-156.

Sinclair, W. B., and Jolliffe, V. A. 1961a. Pectic substances of Valencia oranges at different stages of maturity. *J. Food Sci.* **26**, 125-130.

Sinclair, W. B., and Jolliffe, V. A. 1961b. Chemical changes in the juice vesicles of granulated Valencia oranges. *J. Food Sci.* **26**, 276-282.

Singh, D., and Schroeder, C. A. 1962. Taxonomic and physiological relationships of the so-called mandarin-lime group of citrus. *Proc. Am. Soc. Hort. Sci.* **80**, 291-295.

Singh, I. S. 1964. Varietal differences in fruit quality of tangelos *(Citrus paradisi x C. reticulata). Allahabad Farmer* **38**, 154-157; *Chem. Abstr.* **67**, 10515u (1967).

Sites, J. W., Wander, I. W., and Deszyck, E. J. 1961. The rate and timing of nitrogen for grapefruit on Lakeland fine sand. *Proc. Florida State Hort. Soc.* **74**, 53-57.

Slater, C. A. 1961a. Citrus essential oils. I. Evaluation of natural and terpeneless lemon oils. *J. Sci. Food Agr.* **12**, 257-264.

Slater, C. A. 1961b. Composition of natural lime oil. *Chem. & Ind. (London)* pp. 833-835.

Slater, C. A. 1961c. Citrus essential oils. II. Composition of distilled oil of limes. *J. Sci. Food Agr.* **12**, 732-734.

Slater, C. A. 1963. Citrus essential oils. III. Evaluation of Sicilian natural lemon oils. *J. Sci. Food Agr.* **14**, 58-64.

Slater, C. A., and Watkins, W. T. 1964. Citrus essential oils. IV. Chemical transformations of lime oil. *J. Sci. Food Agr.* **15**, 657-664.

Smith, P. F. 1963a. Quality measurements on selected sizes of Marsh grapefruit from trees differentially fertilized with nitrogen and potash. *Proc. Am. Soc. Hort. Sci.* **83**, 316-321.

Smith, P. F. 1963b. Nitrogen in citrus fertilization. *Citrus Ind.* **44**(1), 9-10.

Smith, P. F. 1967. A comparison of three nitrogen sources on mature Valencia orange trees. *Proc. Florida State Hort. Soc.* **80**, 1-7.

Smith, P. F., and Rasmussen, G. K. 1959. Relation of potassium nutrition to size and quality of Valencia oranges. *Proc. Am. Soc. Hort. Sci.* **74**, 261-265.

Smith, P. F., and Rasmussen, G. K. 1960. Relationship of fruit size, yield and quality of Marsh grapefruit to potash fertilization. *Proc. Florida State Hort. Soc.* **73**, 42-49.

Smith, P. F., and Rasmussen, G. K. 1961a. Effect of nitrogen source, rate, and pH on the production and quality of Marsh grapefruit. *Proc. Florida State Hort. Soc.* **74**, 32-38.

Smith, P. F., and Rasmussen, G. K. 1961b. Effect of potash rate on growth and production of Marsh grapefruit in Florida. *Proc. Am. Soc. Hort. Sci.* **77**, 180-187.

Smith, P. F., Scudder, G. K., and Hrnciar, G. 1963. Twenty years of differential phosphate application on Pineapple oranges. *Proc. Florida State Hort. Soc.* **76**, 7-12.

Smythe, C. V., and Thomas, D. W. 1960. Conversion of flavonoid glycosides. U.S. Patent 2,950,974.

Soine, T. 1964. Naturally occurring coumarins and related physiological activities. *J. Pharm. Sci.* **53**, 231-264.

Soost, R. K., and Cameron, J. W. 1961. Contrasting effecs of acid and nonacid pummelos on the acidity of hybrid citrus progenies. *Hilgardia* **30**, 351-364.

Sperti, G. S. 1968. Orange juice product. U.S. Patent 3,385,711.

Srere, P. A., and Senkin, J. 1966. Citrate condensing enzyme in citrus fruit. *Nature* **212**, 506-507.

Srivas, S. R., Pruthi, J. S., and Siddappa, G. S. 1963. Effect of stage of maturity of fruit and storage temperature on the volatile oil and pectin content of fresh limes *(Citrus aurantifolia, Swingle). Food Sci. (Mysore)* **12**, 340-343.

Srivastava, M. P., and Tandon, R. N. 1966. Free amino acid spectrum of healthy and infected Moshambi fruit. *Naturwissenschaften* **53**, 508-509.

Stanley, W. L. 1962. Citrus oils: Analytical methods and compositional characteristics. *Intern. Fruchtsaft-Union, Ber. Wiss.-Tech. Komm.* **4**, 91-103.

Stanley W. L. 1964. Recent developments in coumarin chemistry. *Aspects of Plant Phenolic Chem., Proc. 3rd Symp., Univ. Toronto, 1963* pp. 79-102.

Stanley W. L. 1965. Recent progress in analysis (characterization) of citrus juices. *Intern. Fruchtsaft-Union, Ber. Wiss.-Tech. Komm* **6**, 207-220.

Stanley, W. L., and Vannier, S. H. 1959. Furocoumarins. U.S. Patent 2,889,337.

Stanley, W. L., and Vannier, S. H. 1967. Psoralens and substituted coumarins from expressed oil of lime. *Phytochemistry* **6**, 585-596.

Stanley, W. L., Lindwall, R. C., and Vannier, S. H. 1958. Specific quantitative colorimetric method of analysis for citral in lemon oil. *J. Agr. Food Chem.* **6**, 858-860.

Stanley, W. L., Ikeda, R. M., Vannier, S. H., and Rolle, L. A. 1961a. Determination of the relative concentrations of the major aldehydes in lemon, orange, and grapefruit oils by gas chromatography. *J. Food Sci.* **26**, 43-48

Stanley, W. L., Ikeda, R. M., and Cook, S. 1961b. Hydrocarbon composition of lemon oils and its relationship to optical rotation. *Food Technol.* **15**, 381-385.

Stanley, W. L., Waiss, A. C., Lundin, R. E., and Vannier, S. H. 1965. Auraptenol, a coumarin compound in bitter (Seville) orange oil. *Tetrahedron* **21**, 89-92.

Stanley, W. L., Vannier, S. H., Petriceks, L., and Lundin, R. E. 1967. Cited by Maier and Metzler (1967b).

Stevens, K. L., Lundin, R. E., and Teranishi, R. 1965. Volatiles from oranges. III. Structure of sinensal. *J. Org. Chem.* **30**, 1690-1692.

Stewart I. 1963. An ephedra alkaloid in citrus juices. *Proc. Florida State Hort. Soc.* **76**, 242-245.

Stewart, I., and Wheaton, T. A. 1964a. *l*-Octopamine in citrus. *Science* **145**, 60-61.

Stewart, I., and Wheaton, T. A. 1964b. Phenolic amines in citrus juice. *Proc. Florida State Hort. Soc.* **77**, 318-320.

Stewart, I., and Wheaton, T. A. 1965. A nitrogen source and rate study on Valencia oranges. *Proc. Florida State Hort. Soc.* **78**, 21-25.

Stewart, I., Leonard, C. D., and Wander, I. W. 1961. Comparison of nitrogen rates and sources for Pineapple oranges. *Proc. Florida State Hort. Soc.* **74**, 75-79.

Stewart, I., Newhall, W. F., and Edwards, G. J. 1963. The isolation and identification of *l*-synephrine in the leaves and fruit of Citrus. *J. Biol. Chem.* **239**, 930-932.

Subbarayan, C., and Cama, H. R. 1965. Carotenoids in Nagpur orange (Citrus species) pulp and peel. *Indian J. Chem.* **3**, 463-465.

Subramanyam, H., Narasimhan, P., and Srivastava, H. C. 1965. Physical and biochemical changes in limes during growth and development. *J. Indian Botan. Soc.* **44**, 105-109.

Sudarsky, J. M., and Fisher, R. A. 1961. Bioflavonoids. U.S. Patent 2,984,601.

Sumitani, M., Fujita, S., Komura, Y., and Hara, T. 1964. Purification of naringinase. Japanese Patent 2983; *Chem. Abstr.* **61**, 1231c (1964).

Sundt, E., Willhalm, B., and Stoll, M. 1964. Analysis of acidic components in saponified bergamot oil. *Helv. Chim. Acta* **47**, 408-413.

Suyama, Y., and Tsusaka, N. 1958. Pectic substances. I. Colorimetric determination of pectin with anthrone. *Meiji Daigaku Nogakubu Kenkyu Hokoku* **8**, 6-8; *Chem. Abstr.* **55**, 16841d (1961).

Swain, T. 1962. Economic importance of flavonoid compounds: Foodstuffs. *In* "The Chemistry of Flavonoid Compounds" (T. A. Geissman, ed.), pp. 513-552. Pergamon Press, Oxford.

Swain T., 1965. Biosynthetic variation and its effect and chemotaxonomy. *Mem. Soc. Botan. France* pp. 176-186; *Chem. Abstr.* **68**, 27486d (1968).

Swift, L. J. 1960. Nobiletin from the peel of the Valencia orange (*Citrus sinensis*, L.). *J. Org. Chem.* **25**, 2067-2068.

Swift, L. J. 1961. Determination of linaloöl and α-terpineol in Florida orange products. *J. Agr. Food Chem.* **9**, 298-301.

Swift, L. J. 1964. Isolation of 5,6,7,3′,4′-pentamethoxyflavone from orange-peel juice. *J. Food Sci.* **29**, 766-767.

Swift, L. J. 1965a. Tetra-*O*-methylscutellarein in orange peel. *J. Org. Chem.* **30**, 2079-2080.

Swift, L. J. 1965b. Proximate analyses of Florida orange peel juice extract for the 1962-63 and 1963-64 seasons. *J. Agr. Food Chem.* **13**, 282-284.

Swift, L. J. 1965c. Flavones of the neutral fraction of the benzene extractables of an orange peel juice. *J. Agr. Food Chem.* **13**, 431-433.

Swift, L. J. 1967. TLC-spectrophotometric analysis for neutral fraction flavones in orange peel juice. *J. Agr. Food Chem.* **15**, 99-101.

Swingle, W. T., and Reece, P. C. 1967. The botany of Citrus and its wild relatives. *In* "The Citrus Industry" (W. Reuther, J. H. Webber, and L. D. Batchelor, eds.), rev. ed., Vol. I, pp. 190-430. Div. Agr. Sci., Univ. of California, Berkeley, California.

Swisher, H. E. 1958. Control of Navel bitter in dehydrated juice products. U.S. Patent 2,834,687.

Takiguchi, Y. 1962. Naringinase produced by *Coniella diplodiella. Takamine Kenkyusho Nempo* **14**, 101-114; *Chem. Abstr.* **58**, 5935h (1963).

Takiguchi, Y. 1965. Mold naringinase. II. Comparison of naringinases produced by various molds. *Nippon Nogeikagaku Kaishi* **39**, 194-198; *Chem. Abstr.* **63**, 18996b (1965).

Talalaj, S. 1966. Essential oil of lime fruits from Ghana. *W. African Pharmacist* 8(3), 46-48; *Chem. Abstr.* **66**, 31928a (1967).

Tarazona, V., Royo, J., and Primo, E. 1959. Hesperidin from orange by-products. *Fruchtsaft.-Ind.* **4**, 147-151.

Tarutani, T., and Manabe, M. 1963. Manufacture of low methoxyl pectin. I. Nature of pectic substances in mandarin orange. *Nippon Shokuhin Kogyo Gakkaishi* **10**, 316-320; *Chem. Abstr.* **63**, 1155g (1965).

Tasaka, T. 1965. Flavonoid content of *Citrus natsudaidai* fruits. *Nippon Shokuhin Kogyo Gakkaishi* **12**, 14-16; *Chem. Abstr.* **64**, 18316d (1966).

Tatum, J. H., Parks, G. L., and Berry, R. E. 1965. Comparison of orange concentrate and foam mat powders using chromatographic procedures. *Proc. Florida State Hort. Soc.* **77**, 307-311.

Taylor, W. H. 1957. Formol titration: An evaluation of its various modifications. *Analyst* **82**, 488-498.

Teranishi, R., Schultz, T. H., McFadden, W. H., Lundin, R. E., and Black, D. R. 1963. Volatiles from oranges. I. Hydrocarbons identified by infrared, nuclear magnetic resonance, and mass spectra. *J. Food Sci.* **28**, 541-545.

Teranishi, R., Lundin, R. E., McFadden, W. H., Mon, T. R., Schultz, T. H., Stevens, K. L., and Wasserman, J. 1966. Volatiles from oranges. Oxygenated compounds identified by infrared, proton magnetic resonance, and mass spectra. *J. Agr. Food Chem.* **14**, 447-449.

Thomas, A. F. 1967. Synthesis and structure of α-sinensal. *Chem. Commun.* pp. 947-949.

Thomas, D. W., Smythe, C. V., and Labbee, M. D. 1958. Enzymatic hydrolysis of naringin, the bitter principle of grapefruit. *Food Res.* **23**, 591-598.

Thommen, H. 1962. On the occurrence of β-apo-8′-carotenol in the juice and skins of fresh oranges. *Naturwissenschaften* **49**, 517-518.

Ting, S. V. 1958. Enzymic hydrolysis of naringin in grapefruit juice. *J. Agr. Food Chem.* **6**, 546-549.

Ting, S. V. 1961. The total carotenoid and carotene content of Florida orange concentrate. *Proc. Florida State Hort. Soc.* **74**, 261-267.

Ting, S. V. 1967. Nitrogen content of Florida orange juice and Florida orange concentrate. *Proc. Florida State Hort. Soc.* **80**, 257-261.

Ting, S. V. 1969. Distribution of soluble components and quality factors in the edible portion of citrus fruits. *J. Am. Soc. Hort. Sci.* **94**, 515-519.

Ting, S. V., and Deszyck, E. J. 1958. The internal color and carotenoid pigments of Florida red and pink grapefruit. *Proc. Am. Soc. Hort. Sci.* **71**, 271-277.

Ting. S. V., and Deszyck, E. J. 1959. Isolation of *L*-quinic acid in citrus fruit. *Nature* **183**, 1404-1405.

Ting, S. V., and Deszyck, E. J. 1961. The carbohydrates in the peel of oranges and grapefruit. *J. Food Sci.* **26**, 146-152.

Ting, S. V., and Newhall, W. J. 1965. Occurrence of a natural antioxidant in citrus fruit. *J. Food Sci.* **30**, 57-62.

Ting, S. V., and Vines, H. M. 1966. Organic acids in the juice vesicles of Florida Hamlin orange and Marsh seedless grapefruit. *Proc. Am. Soc. Hort. Sci.* **88**, 291-297.

Todd, G. W., Bean, R. C., and Propst, B. 1961. Photosynthesis and respiration in developing fruits. II. Comparative rates at various stages of development. *Plant Physiol.* **36**, 69-73.

Tokoroyama, T. 1958a. Natural furan derivatives. III. Occurrence of the furan ring in limonin. *Nippon Kagaku Zasshi* **79**, 314-319; *Chem. Abstr.* **55**, 18698b (1961).

Tokoroyama, T. 1958b. Natural furan derivatives. IV. Occurrence of the furan ring in obacunone. *Nippon Kagaku Zasshi* **79**, 319-322; *Chem. Abstr.* **55**, 18698e (1961).

Tokoroyama, T., and Matsuura, T. 1962. Formation of methyl tetrahydroanhydroepimeroobacunolate from limonin. *Nippon Kagaku Zasshi* **83**, 630-635; *Chem. Abstr.* **59**, 534h (1963).

Tokoroyama, T., Kamikawa, T., and Kubota, T. 1961. Natural furan derivatives. V. The number of double bonds in obacunone. *Bull. Chem. Soc. Japan* **34**, 131-133; *Chem. Abstr.* **55**, 15451g (1961).

Tomas, F., Carpena, O., and Abrisqueta, G. C. 1966. Spectrophotometric determination of hesperidin and citrus flavonoids. *Anales Bromatol. (Madrid)* **18**, 393-400; *Chem. Abstr.* **67**, 63076h (1967).

Trammel, K., and Simanton, W. A. 1966. Properties of spray oils in relation to effect on citrus trees in Florida. *Proc. Florida State Hort. Soc.* **79**, 19-26.

Tseng, K. F. 1938. Nobiletin. I. *J. Chem. Soc.* pp. 1003-1004.

Tsusaka, T. 1965. Application of naringinase in canned summer orange. *Nippon Shokuhin Kogyo Gakkaishi* **12**, 167-173; *Chem. Abstr.* **64**, 18316g (1966).

Tucker, D.P.H., and Reuther, W. 1967. Seasonal trends in composition of processed Valencia and Navel oranges from major climatic zones of California and Arizona. *Proc. Am. Soc. Hort. Sci.* **90**, 529-540.

Ueyanagi, F., and Morioka, T. 1965. Seasonal changes of vitamin C, sugar, and organic acid in summer orange, *Citrus Tamurana. Nippon Shokuhin Kogyo Gakkaishi* **12**, 332-334; *Chem. Abstr.* **64**, 18317a (1966).

U.S. Department of Commerce. 1963. "Vitamin P, Its Properties and Uses" (translated from the Russian). Office Tech. Serv., U.S. Dept. Comm., Washington, D.C.

Vandercook, C. E., and Guerrero, H. C. 1968. Effects of chemical preservatives and storage on constituents used to characterize lemon juice. *J. Assoc. Offic. Anal. Chemists* **51**, 6-10.

Vandercook, C. E., and Guerrero, H. C. 1969. Citrus juice characterization. Analysis of the phosphate fractions. *J. Agr. Food Chem.* **17**, 626-628.

Vandercook, C. E., and Rolle, L. A. 1963. Lemon juice composition. II. Characterization of California-Arizona lemon juice by its polyphenolic content. *J. Assoc. Offic. Agr. Chemists* **46**, 359-362.

Vandercook, C. E., and Stephenson, R. G. 1966. Lemon juice composition. Identification of the major phenolic constituents and estimation by paper chromatography. *J. Agr. Food Chem.* **14**, 450-454.

Vandercook, C. E., and Yokoyama, H. 1965. Lemon juice composition. IV. Carotenoid and sterol content. *J. Food Sci.* **30**, 865-868.

Vandercook, C. E., Rolle, L. A., and Ikeda, R. M. 1963. Lemon juice composition. I. Characterization of California-Arizona lemon juice by its total amino acid and *l*-malic acid content. *J. Assoc. Offic. Agr. Chemists* **46**, 353-358.

Vandercook, C. E., Rolle, L. A., Postlmayr, H. L., and Utterberg, K. A. 1966. Lemon juice composition. V. Effects of some fruit storage and processing variables on the characterization of lemon juice. *J. Food Sci.* **31**, 58-62.

Vas, K., Nedbalek, M., Scheffer, H., and Kovacs Proszt, G. 1967. Methodological investigations on the determination of some pectic enzymes. *Fruchtsaft.-Ind.* **12**, 164-184.

Veldhuis, M. K., and Hunter, G.L.K. 1967. Nomenclature of ylangene, copaene, and cubebene. *J. Food Sci.* **32**, 697.

Venturella, P., Bellino, A., and Cusmano, S. 1961. Flavonoid components of citrus. I. Isolation of 5,6,7,8,3′,4′-hexamethoxyflavone (nobiletin) from *Citrus deliciosa* Tenore. *Ann. Chim. (Rome)* **51**, 101-115; *Chem. Abstr.* **55**, 19912h (1961).

Venturella, P., Bellino, A., and Cusmano, S. 1964. Substances in citrus fruits. V. Isolation of the pigments from *Citrus aurantium* (varieties Alphonsii and Foetifera). *Gazz. Chim. Ital.* **96**, 475-482; *Chem. Abstr.* **65**, 9341h (1966).

Venturella, P., Bellino, A., and Cusmano, S. 1965. Citrus components. Isolation of pigments from *Citrus aurantium* var. Foetifera. *Atti Accad. Sci., Lettere, Arti Palermo, Pt. I [4]* **24**, 101-104; *Chem. Abstr.* **64**, 13085h (1966).

Vernin, G. 1967. Applications of infrared spectroscopy to the study of essential oils and their terpenic and aromatic components. *France Parfums* **10**(2), 35-54.

Vincent, Y. 1962. Chemurgy of citrus peel. *Fruits (Paris)* **17**, 451-455.

Vines, H. M. 1968. Citrus enzymes. II. Mitochondrial and cytoplasmic malic dehydrogenase from grapefruit juice vesicles. *Proc. Am. Soc. Hort. Sci.* **92**, 179-184.

Vines, H. M., and Metcalf, J. F. 1967. Seasonal changes in oxidation and phosphorylation in mitochondrial preparations from grapefruit. *Proc. Am. Soc. Hort. Sci.* **90**, 86-92.

Vines, H. M., and Oberbacher, M. F. 1962. Ascorbic acid oxidase in citrus. *Proc. Florida State Hort. Soc.* **75**, 283-286.

Vines, H. M., and Oberbacher, M. F. 1963. Citrus fruit enzymes. I. Ascorbic acid oxidase in oranges. *Plant Physiol.* **38**, 333-337.

Vines, H. M., and Oberbacher, M. F. 1965. Response of oxidation and phosphorylation in citrus mitochondria to arsenate. *Nature* **206**, 319-320.

Vogel, G., and Stroecker, H. 1966. Effect of pharmacologically active principles, especially flavonoids on lymph flow. *Arzneimittel-Forsch.* **16**, 1630-1634; *Chem. Abstr.* **66**, 45231e (1967).

Vogin, E. E. 1960. Review of bioflavonoids. *Am. J. Pharm.* **132**, 363-382.

Vogin, E. E., and Rossi, G. V. 1961. Bioflavonoids in experimental ulceration. *J. Pharm. Sci.* **50**, 14-17.

Wagner, D., and Monselise, J. J. 1963. Determination of some components of citrus and tomato juices produced in Israel. *Israel J. Technol.* **1**, 33-37; *Chem. Abstr.* **60**, 6140a (1964).

Wagner, H., Aurnhammer, G., Hörhammer, L., and Farkas, L. 1969. Synthesis and structural determination of narirutin, a 5,7,4′-trihydroxyflavanone-7-rutinoside from *Citrus sinensis* (L.) Osb. *Chem. Ber.* **102**, 2089-2092.

Walther, K., Rimpler, H., and Leuckert, C. 1966. Occurrence of flavonoids in the secretory cavities of citrus fruits. *Planta Med.* **14**, 453-459; *Chem. Abstr.* **66**, 52946v (1967).

Wan, L. K. 1942. Components of the fruit, *Pseudaegle trifoliata.* II. Alkaline decomposition of neohesperidin. *J. Pharm. Soc. Japan* **62**, 466-468; *Chem. Abstr.* **45**, 4239e (1951).

Weerakoon, A. H. 1960. Composition of the seed fat of Ceylon sweet orange. *J. Sci. Food Agr.* **11**, 273-276.

Weichselbaum, T. E., and Margraf, H. W. 1961. Effect of citrus bioflavonoids on metabolism of hydrocortisone in man. *Proc. Soc. Exptl. Biol. Med.* **107**, 128-131.

Weizmann, A., Meisels, A., and Mazur, Y. 1955. Steroids and terpenoids of grapefruit. I. *J. Org. Chem.* **20**, 1173-1177.

Westbrook, G. F., and Stenstrom, E. C. 1959. A study of the degrees Brix and Brix-acid ratios of tangerines utilized by Florida citrus processors for the seasons 1953-54 through 1956-57 and 1958-59. *Proc. Florida State Hort. Soc.* **72**, 290-297.

Westbrook, G. F., and Stenstrom, E. C. 1963. The degrees Brix and Brix-acid ratios of grapefruit utilized by Florida citrus processors for seasons 1958-59 through 1961-62. *Proc. Florida State Hort. Soc.* **76**, 258-264.

Westbrook, G. F., and Stenstrom, E. C. 1964. Degrees Brix and Brix-acid ratios of oranges utilized by Florida citrus processors for the seasons 1958-59 through 1963-64. *Proc. Florida State Hort. Soc.* **77**, 278-283.

Wheaton, T. A., and Stewart, I. 1965a. Quantitative analysis of phenolic amines using ion-exchange chromatography. *Anal. Biochem.* **12**, 585-592.

Wheaton, T. A., and Stewart, I. 1965b. Feruoylputrescine: Isolation and identification from citrus leaves and fruit. *Nature* **206**, 620-621.

Wheaton, T. A., and Stewart, I. 1969. Biosynthesis of synephrine in citrus. *Phytochemistry* **8**, 85-92.

Williams, B. L., Goad, L. J., and Goodwin, T. W. 1967. The sterols of grapefruit peel. *Phytochemistry* **6**, 1137-1145.

Wilson, K. W., and Crutchfield, C. A. 1968. Spectrophotometric determination of limonin in orange juice. *J. Agr. Food Chem.* **16**, 118-124.

Winterstein, A., Studer, A., and Ruegg, R. 1960. New knowledge in carotenoid research. *Chem. Ber.* **93**, 2951-2965.

Wolford, R. W., Attaway, J. A. 1967. Analysis of recovered natural orange flavor enhancement materials using gas chromatography. *J. Agr. Food Chem.* **15**, 369-377.

Wolford, R. W., Alberding, G. E., and Attaway, J. A. 1962. Analysis of recovered natural orange essence by gas chromatography. *J. Agr. Food Chem.* **10**, 297-301.

Wolford, R. W., Attaway, J. A., Alberding, G. E., and Atkins, C. D. 1963. Analysis of the flavor and aroma constituents of Florida orange juices by gas chromatography. *J. Food Sci.* **28**, 320-328.

Woodruff, R. E., and Olson, E. O. 1960. Effects of rootstocks on physical characteristics and chemical composition of fruit of six citrus varieties in Texas. *J. Rio Grande Valley Hort. Soc.* **14**, 77-84; *Chem. Abstr.* **60**, 11049a (1964).

Woods, R. M. 1966. Cited by Charley (1966).

Wucherpfennig, K., and Franke, I. 1966. Distribution of amino acids in the juice and peel of oranges. *Fruchtsaft-Ind.* **11**, 60-65.

Yamamoto, R., and Oshima, Y. 1931. A new glycoside, citronin, from the peel of lemon ponderosa. *J. Agr. Chem. Soc. Japan* **7**, 312-319; *Chem. Abstr.* **26**, 1295 (1932).

Yamanishi, T., Kobayashi, A., Mikumo, Y., Nakasone, Y., Kita, Y., and Hattori. S. 1968. Composition of peel oil of *Citrus unshu. Agr. Biol. Chem. (Tokyo)* **32**, 593-598.

Yokoyama, F., Levi, L., Laughton, P. M., and Stanley, W. L. 1961. Determination of citral in citrus extracts and citrus oils by conventional and modern chemical methods of analysis. *J. Assoc. Offic. Agr. Chemists* **44**, 535-541.

Yokoyama, H. 1965. Collaborative studies on the characterization of lemon juice. *J. Assoc. Offic. Agr. Chemists* **48**, 530-533.

Yokoyama, H. 1966. Collaborative study of the determination of *l*-malic acid in lemon juice. *J. Assoc. Offic. Anal. Chemists* **49**, 621-623.

Yokoyama, H., and Vandercook, C. E. 1967. Citrus carotenoids. I. Comparison of carotenoids of mature-green and yellow lemons. *J. Food Sci.* **32**, 42-48.

Yokoyama, H., and White, M. J. 1965a. Citrus carotenoids. II. Structure of citranaxanthin, a new carotenoid ketone. *J. Org. Chem.* **30**, 2481-2482.

Yokoyama, H., and White, M. J. 1965b. Citrus carotenoids. IV. The isolation and structure of sintaxanthin. *J. Org. Chem.* **30**, 3994-3996.

Yokoyama, H., and White, M. J. 1966a. Citrus carotenoids. V. The isolation of 8′-hydroxy-8′,9′-dihydrocitranaxanthin. *J. Org. Chem.* **31**, 3452-3454.

Yokoyama, H., and White, M. J. 1966b. Citrus carotenoids. VI. Carotenoid pigments in the flavedo of Sinton citrangequat. *Phytochemistry* **5**, 1159-1173.

Yokoyama, H., and White, M. J. 1967. Carotenoids in the flavedo of Marsh seedless grapefruit. *J. Agr. Food Chem.* **15**, 693-696.

Yokoyama, H., and White, M. J. 1968a. Occurrence of ergosterol in citrus. *Phytochemistry* **7**, 493-494.

Yokoyama, H., and White, M. J. 1968b. Citrus carotenoids. VIII. Isolation of semi-β-carotenone and β-carotenone from Citrus relatives. *Phytochemistry* **7**, 1031-1034.

Yokoyama, H., and White, M. J. 1968c. Causes of lemon bronzing. *Calif. Citrograph* **53**, 283-284.

Yokoyama, H., White, M. J., and Vandercook, C. E. 1965. Citrus carotenoids. III. The structure of reticulataxanthin. *J. Org. Chem.* **30**, 2482-2483.

Young, L. B., and Erickson, L. C. 1961. Influence of temperature on color change in Valencia oranges. *Proc. Am. Soc. Hort. Sci.* **78**, 197-200.

Young, R. E., and Biale, J. B. 1968. Carbon dioxide effects on fruit. III. The fixation of ^{14}C-labelled carbon dioxide in lemons in an atmosphere enriched with carbon dioxide. *Planta* **81**, 253-263.

Young, T. W., and Koo, R.C.J. 1967. Effects of nitrogen and potassium fertilization on Persian limes on Lakeland fine sand. *Proc. Florida State Hort. Soc.* **80**, 337-342.

Zaganiaris, S. 1959. Chemical study of bitter orange juices. *Chim. Ind. (Paris)* **81**, 544-548.

Zemplen, G., Tettamanti, A. K., and Farago, S. 1938. Biose of hesperidin and neohesperidin. *Chem. Ber.* **71B**, 2511-2520.

Zitko, V., and Bishop, C. T. 1965. Fractionation of pectins from sunflowers, sugar beets, apples, and citrus fruits. *Can. J. Chem.* **43**, 3206-3214.

Zuber, H. 1964. Purification and properties of a new carboxypeptidase from citrus fruit. *Nature* **201**, 613.

Zuber, H. 1968. Purification and properties of citrus fruit carboxypeptidase. *Hoppe-Seyler's Z. Physiol. Chem.* **349**, 1337-1352.

AUTHOR INDEX

Numbers in italics refer to the pages on which the complete references are listed.

A

Abraham, E. P., 61, *169*
Abrisqueta, G. C., 120, *208*
Ahmad, M. N., 102, *169*
Ahmed, E. M., 95, *203*
Aichinger, F., 147, *169*
Aiyappa, K. M., 12, *201*
Akiyama, K., 143, *197, 198*
Albach, R. F., 121, 125, 142, 166, *169, 183, 195*
Albanese, F., 24, *195*
Alberding, G. E., 88, 94, *170, 210*
Alberola, J., 24, 32, 33, 160, *169, 200, 203*
Alvarez, B. M., 52, 54, *169*
Amano, Y., 131, *194*
Ammerman, C. B., 57, *169*
Andersen, L., 134, 144, *176*
Anderson, C. A., 15, *169*
Arakawa, H., 117, *169*
Arakawa, N., 144, *187*
Aranda, A., 45, *202*
Aratake, S., 145, *169*
Arigoni, D., 150, 152, 154, 156, *169, 195*
Arita, H., 155, *185*
Armstrong, M. D., 55, *182*
Arnaud, R., 146, *169*
Arnott, S., 150, *169, 170*
Arrington, L. F., 57, *169*
Asensi, P., 8, *200*
Ashoor, S. H. M., 93, 95, *170*
Aspinall, G. O., 29, *170*
Aspridis, J. A., 6, 158, *180*
Atkins, C. D., 14, 22, 23, 25, 26, 60, 88, 94, 158, 159, *202, 210*
Attaway, J. A., 88, 91, 93, 94, 95, 99, 100, 101, 103, *170, 210*
Aurnhammer, G., 126, *210*
Avalle, N., 146, *170*
Averna, V., 52, *170*
Aversa, M. C., 78, *179*
Axelrod, B., 61, *170*

B

Babin, R., 120, 146, *170, 192*
Bacharach, A. L., 147, *170*
Bailey, G. F., 69, 72, 74, 128, *177*
Baker, R. A., 28, 57, *170*
Balls, A. K., 157, 162, *171*
Banerjee, S., 127, *182*
Barabas, L. J., 93, 99, 100, 101, *170*
Bar Akiva, A., 63, *171*
Barr, B. K., 67, *171*
Barron, R. W., 136, *194*
Bartholomew, E. T., 1, *171*
Barton, D. H. R., 150, 152, 154, 155, 156, 158, 161, *169, 171*
Barua, A. D., 27, 83, 102, *174, 175*
Basker, H. B., 102, 140, *171, 192*
Basson, W. J., 7, 9, 13, *194*
Bauernfeind, J. C., 75, 78, *171*
Baumann, G., 54, *182*
Bavetta, L. A., 147, *171*
Bean, R. C., 24, 35, 63, 67, *171, 208*
Beavens, E. A., 162, *201*
Beavieux, J., 146, *192*
Becker, M., 52, 54, *205*
Beisel, C. G., 6, 126, 137, *171, 202*
Belitz, H. D., 61, *203*
Bellenot, P., 6, 8, 139, *173*
Bellino, A., 109, 130, 131, *171, 209*
Bellomo, A., 53, *171*
Ben-Aziz, A., 134, *171*
Benk, E., 6, 7, 30, 45, 46, 54, 75, 78, 79, 102, 129, *171, 172*

Bernal, A. A., 159, *180*
Bernhard, R. A., 85, 88, 89, 93, 96, 110, *170, 172, 176*
Bergmann, R., 6, 7, 75, *172*
Berry, R. E., 97, 128, 149, *172, 207*
Bertolini, A., 148, *196*
Bessho, Y., 24, 27, 135, 137, 139, 140, 143, *172, 190, 191, 194*
Beverly, G. O., 161, 163, *193*
Bezanger-Beauquesne, L., 146, *172*
Bhambota, J. R., 12, *190*
Bhatia, B. S., 158, 159, 162, *205*
Bhatty, M. K., 102, *169*
Biale, J. B., 66, *211*
Biedron, S. I., 60, *203*
Biggs, R. H., 67, *187*
Birdsall, J. J., 6, 39, 40, 45, 47, *172*
Bishop, C. T., 29, 30, *211*
Bissett, D. W., 22, *172*
Bitters, W. P., 88, 93, 99, 100, *204*
Black, D. R., 85, 88, 94, *193, 203, 208*
Blazquez, C. H., 7, *172*
Blomquist, V. H., 37, *180*
Boehm, K., 146, *172*
Böhme, H., 107, 128, 149, *172*
Bogin, E., 33, 35, 65, *172*
Boll, P. M., 118, *176*
Bomben, J. L., 88, 94, *203*
Bonnell, J. M., 145, *172*
Bonner, J., 59, *188*
Born, R., 128, *173*
Bose, A. N., 12, 102, *192, 196*
Boswell, S. B., 18, 20, *176, 177*
Boudene, C., 47, *183*
Bouillant, M. L., 125, *175*
Bouma, D., 15, 16, *173*
Bowden, R. P., 9, 10, 14, 68, 78, 158, 159, *173*
Bram, B., 141, *173*
Braunshweker, H., 117, *184*
Braverman, J. B. S., 120, 140, *173, 191*
Brent, J. A., 94, 97, 102, *189, 196*
Brewer, R. F., 20, 48, *173*
Brogden, W. B., 88, 91, 93, 96, 97, 98, 99, *186, 187*
Bruemmer, J. H., 28, 57, 63, *170, 173*
Brunk, H. D., 31, *192*
Bueck, W. A., 94, *196*
Buffa, A., 6, 8, 139, *173*
Buigues, N. M., 88, 97, 101, 110, *193*
Bunnell, R. H., 75, 78, *171*
Burckhart, O., 46, *183*
Burke, B. A., 156, *173*
Burnett, R. H., 6, *174*
Burns, R. M., 19, *176, 185*
Burr, H. K., 166, *183*
Burstain, I. G., 156, *181*
Buslig, B. S., 93, 100, *170*
Buttery, G., 166, *183*

C

Caglioti, L., 150, 152, 156, *169*
Cahoon, G. A., 20, 21, 34, *177, 201*
Calabro, G., 88, 93, 102, *177*
Calapaj, R., 103, 109, 123, 128, 130, 131, *173, 177*
Call, T. G., 146, *173*
Calvarano, I., 7, 41, 45, 52, 53, 54, 68, 78, 102, *173, 174*
Calvarano, M., 41, 45, 78, 102, 107, 129, *173, 174*
Calvert, D. V., 14, *174*
Cama, H. R., 75, *207*
Cameron, J. W., 6, 35, 73, 88, 93, 99, 100, *174, 204, 206*
Cananzi, V., 7, *174*
Caporale, G., 109, *174*
Cardinali, L. R., 6, *174*
Carpena, O., 102, 120, *174, 208*
Carra, J. H., 73, *200*
Cary, P. R., 15, 16, *174*
Casas, A., 24, *169*
Casoli, U., 51, 52, 128, 174, *199*
Castro, C. E., 46, *174*
Catsouras, G. C., 78, *174*
Cavoli, A. M., 102, *174*
Chaliha, B. P., 6, 24, 27, 52, 83, 102, 123, 130, 131, 132, 135, 136, 137, 149, 159, *174, 175*
Chan, W. R., 156, *173*
Chandler, B. V., 9, 127, 154, 156, 157, 158, 159, 160, 162, 163, 164, *175, 188*
Chang, H-C., 6, *176*
Charley, V. L. S., 53, 146, 147, *175*
Chatt, E. M., 43, *175*
Chatterjee, A., 111, *203*
Cheraskin, E., 146, *180*
Chikano, T., 141, *199*
Chopin, J., 106, 122, 124, 125, 149, *175*

Chorin, M., 131, 134, *171, 199*
Christiansen, K., 118, *176*
Chu, L-C., 6, *176*
Cieri, U. R., 107, 109, 110, *176*
Cingolani, E., 109, *174*
Clark, J. R., 85, *176*
Clark, R. B., 33, 35, 62, 65, *176*
Clayton, E. M., Jr., 146, *176*
Clements, R. L., 33, 51, 52, 54, 56, *176*
Clementson, C. A. B., 134, 144, *176*
Coates, M. E., 147, *170*
Coffin, D. E., 54, 55, 56, *176*
Coggins, C. W., 18, 19, 67, 68, 165, *176, 185, 189, 192*
Cohen, A., 146, *176*
Cohen, L., 146, *176*
Cohen, M., 9, *176*
Colburn, B., 9, *177*
Coleman, R. L., 90, *177*
Connolly, J. D., 154, 156, *177*
Cook, S., 101, *206*
Cooper, M., 52, 54, *205*
Cooper, W. C., 19, 20, 21, 68, *177, 201*
Corey, E. J., 150, 152, 156, *169*
Corigliano, F., 95, 102, *177*
Corse, J., 88, 94, *203*
Costo, J., 165, *195*
Couch, J. F., 133, *190*
Couch, J. R., 147, *178*
Coussin, B. R., 51, *177*
Coustun, F., 146, *192*
Cover, A. R., 99, *193*
Cox, J. E., 6, *177*
Craig, J. W. T., 29, *170*
Craft, C. C., 90, 97, *198*
Cree, C. B., 21, *188*
Cross, A. D., 156, *196*
Crupi, F., 93, *178*
Crutchfield, C. A., 164, *210*
Curl, A. L., 69, 72, 74, 75, 128, *177*
Cusmano, S., 109, 130, 131, *171, 209*

D

Dalal, V. B., 11, 12, *177, 201*
Dall'Aglio, G., 128, *199*
D'Amore, G., 88, 93, 95, 102, 109, 123, 128, 130, 131, *177*
Danziger, M. T., 5, *177*
D'Arcy, A., 125, *175*
Dastoli, F. R., 139, *177*
Datta, S., 157, *178*
Davidek, J., 113, 144, *178*
Davie, A. W., 150, *169, 170*
Davis, P. L., 6, 9, 11, 20, 90, 121, *178, 184, 198*
Davis, T. T., 31, *189*
Davis, W. B., 119, *178*
Dawes, S. N., 7, 45, *178*
Day, J. C., 85, *193*
De, S., 12, *192*
Deacon, L. E., 147, *178*
Dean, H. A., 17, *178*
De Eds, F., 122, 146, *178, 194*
De Fossard, R. A., 15, *178*
de Gruy, I. V., 73, *200*
Dellamonica, G., 106, 124, *175*
Dennison, R. A., 95, *203*
Denny, R., 156, *181*
Derse, P. H., 6, 39, 40, 45, 47, *172*
Deszyck, E. J., 6, 11, 12, 15, 17, 20, 24, 29, 32, 73, 136, *178, 198, 205, 208*
Dev, S., 150, 152, 156, *169*
Deyoe, C. W., 147, *178*
Dhingra, D. R., 111, *197*
Diemair, W., 33, *178*
Di Giacomo, A., 7, 52, 53, 54, 68, 78, 93, 102, 107, 113, 130, 143, 144, 145, *178, 179, 191, 199, 201*
Dodge, F. D., 107, *179*
Doig, A. R., 139, *177*
Domanskaya, E. N., 19, *205*
Dougherty, M. H., 103, *170, 179*
Drawert, F., 121, *179*
Dreyer, D. L., 109, 111, 123, 138, 149, 150, 152, 154, 155, 156, 158, 159, 161, 163, 166, *179, 193*
Drummond, F. E., 148, *179*
D'Souza, S., 11, *177*
Dugger, W. M., 18, *192*
Dunlap, J. A., 9, 11, *185*
Dunlap, W. J., 119, 121, 125, 126, 127, 132, 135, 137, 138, 141, 142, 146, *179, 180, 183, 195*
Dupuis, C., 93, 98, 102, *186*
Duran, L., 159, *200*
Durix, A., 122, 124, 125, *175*

E

Eaks, I.L., 12, 18, 39, *176, 180*

Easley, J. F., 57, *169*
Edwards, G. J., 55, 88, 102, 103, 120, 139, *170, 184, 189, 206*
El Ashiry, G. M., 146, *180*
El Tobshy, Z., 165, *180*
El Zhorkani, A. S., 6, *180*
Embleton, T. W., 15, 16, 17, 21, 47, 166, *180, 188, 191*
Emerson, O. H., 150, 160, 161, 162, *180*
England, A. B., 6, 88, 93, 99, 100, *174, 204*
Erickson, L. C., 33, 35, 65, 68, 73, 172, *180, 211*
Ershoff, B. H., 148, *180*
Esaki, S., 129, 130, 131, 132, 137, 142, *188, 197*
Evans, P., 52, 54, *205*
Exarchos, C. D., 6, 158, *180*

F

Fabianek, J., 146, *180*
Fagerson, I. S., 88, 89, *180*
Farago, S., 116, *211*
Farid, S., 109, 130, *180*
Farkas, L., 123, 126, *180, 210*
Fazl-i-Rubbi, S., 7, *180, 196*
Fehder, P., 52, 54, *205*
Feldhaus, W. D., 146, *176*
Fel'dman, A. L., 162, *194*
Feldman, A. W., 105, *180*
Feldman, J. R., 118, 148, *180*
Feliu, A. R., 159, *180*
Fernandez, B., 88, 89, *180*
Fernandez-Flores, E., 37, *180*
Ferrini, P. G., 150, 152, 156, *169*
Fischer, I. R., 97, *181*
Fisher, J. F., 107, 109, 121, 137, *181*
Fisher, R. A., 146, *207*
Fishman, G. M., 7, 120, 142, *181*
Flath, R. A., 94, *181*
Flavian, S., 154, *181*
Fletcher, W. A., 9, 22, *181*
Flores, J., 159, *200*
Floyd, K. M., 6, 45, *181*
Fontanelli, L., 121, *181*
Foote, C. S., 156, *181*
Franguelli, L., 83, *181*
Frank, N. A., 148, *181*
Franke, I., 54, *211*
Freedman, L., 146, 147, *181*
Freedman, S. O., 32, 106, *181, 205*
Friz, M., 148, *182*
Fujita, A., 155, *185*
Fujita, S., 141, *207*
Fukuda, M., 7, *187*
Fukumoto, J., 119, 142, 143, *182, 198*
Furui, H., 140, *197*
Furr, J. R., 20, 34, *177, 201*

G

Gamkrelidze, I. D., 16, *182*
Ganz, D., 160, 162, *203*
Garber, M. J., 19, 20, 48, 67, *173, 176, 192*
Garcia, I., 32, 33, *203*
Gardner, F. E., 6, 8, 9, *177, 182*
Gasque, F., 7, 8, 159, *200*
Gee, M., 29, *193*
Gentili, B., 105, 106, 109, 116, 118, 119, 120, 122, 123, 124, 125, 126, 127, 132, 133, 137, 138, 148, *182, 186*
Gerngross, O., 120, 124, 127, *182*
Gerritz, H. W., 101, *182*
Gershtein, L. A., 12, *182*
Ghosh, B. P., 127, *182*
Gierschner, K., 54, *182*
Giss, G., 147, *169*
Gjessing, L., 55, *182*
Glass, C. S., 11, *195*
Glazier, E. R., 150, 152, 156, *169*
Gloppe, K., 129, *190*
Goad, L. J., 149, 152, *182, 210*
Goldschmidt, E. E., 62, *182*
Goodwin, T. W., 149, 152, *182, 210*
Goren, R., 19, 41, 49, 62, 120, 134, 135, 165, *182, 183, 195*
Goretti, G., 89, 102, *183*
Gori, C., 110, *183*
Gorin, P. A. J., 116, *183*
Gounelle, H., 47, *183*
Grant, E. R., 163, *193*
Graves, H. B., 20, 48, *192*
Gray, R., 52, 54, *205*
Gresham, P., 52, 54, *205*
Grierson, W., 6, 12, *183, 198*
Griffiths, F. P., 119, 121, 125, 137, 141, 142, *183, 195*
Griffiths, J. T., 97, 102, *189*
Guadagni, D. G., 166, *183*
Guenther, F., 46, *183*

Guenther, H., 98, *183*
Guerrero, H. C., 37, 47, 57, *209*
Guilbert, N., 146, *172*
Guillemet, F. B., 20, 48, *173*
Guislain, R., 146, *172*
Gumanitskaya, M. N., 120, 142, *181*
Gupta, G. N., 111, *197*
Gutfinger, T., 144, *183*

H

Haase-Sajak, E., 6, 74, 78, 79, 119, 127, 145, *190*
Hänsel, R., 148, *183*
Haginuma, S., 27, *195*
Hama, M., 142, *188*
Hagen, R. E., 119, 121, 125, 135, 137, 138, 141, 142, *179, 183*
Hales, T. A., 9, *185*
Halevy, A., 73, *195*
Hamell, M., 118, 148, *180*
Handa, K. L., 154, 156, *177*
Hanson, J., 163, *183*
Hara, T., 141, *183, 207*
Harborne, J. B., 146, *183*
Hardegger, E., 117, *184*
Harding, P. L., 6, 9, 10, 11, 13, 20, 22, *184, 192, 193*
Harrell, J. E., 6, *181*
Hart, B. F., 148, *184*
Hart, S. H., 7, 45, *184*
Hasegawa, S., 161, *194*
Haskell, G., 134, *184*
Hatanaka, C., 30, *184*
Hatano, S., 141, *199*
Hattori, S., 13, 93, 95, 144, *184, 211*
Hearn, C. J., 6, *204*
Heimann, W., 121, *179*
Heintze, K., 146, *184*
Hendershott, C. H., 133, *184*
Hendrick, D. V., 88, 94, *170*
Hendrickson, R., 8, 81, 83, 84, 91, 93, 94, 96, 97, 100, 102, 119, 120, 121, 131, 139, 145, *184, 189*
Henry, W. H., 19, *177*
Hera, E., 161, *194*
Herrman, K., 113, *184*
Herzog, P., 136, *184*
Hess, D., 23, *190*
Hield, H. Z., 11, 18, 19, 67, *176, 185, 193*
Higashihara, M., 142, *198*
Higby, R. H., 160, 162, 164, *185*
Higby, W. K., 78, *185*
Hilgeman, R. H., 9, 11, 20, 21, 34, *177, 185, 201*
Hill, E. C., 52, 94, 119, 136, 142, *185, 192, 194, 198*
Hirose, Y., 88, 99, 155, *185, 198*
Hirota, I., 30, 145, *185*
Hobson, G. F., 60, *185*
Hodgson, R. W., 1, 2, 3, *185*
Hoelscher, C. E., 17, *178*
Hörhammer, L., 113, 121, 126, *185, 210*
Holeman, E. H., 43, *185*
Hollies, M., 9, 22, *181*
Holz, F., 61, *203*
Hom, D., 147, *171*
Honda, H., 140, *197*
Hopkins, G. A., 45, 46, *185*
Horanic, G. E., 8, 9, *177, 182*
Horie, T., 132, *185*
Horowitz, R. M., 105, 106, 109, 113, 116, 118, 119, 120, 122, 123, 124, 125, 126, 127, 129, 132, 133, 134, 135, 137, 138, 146, 148, *179, 182, 185, 186*
Hrnciar, G., 15, *206*
Huet, R., 7, 8, 90, 93, 98, 101, 102, 113, 135, 136, 139, 146, 159, *186*
Huffaker, R. C., 35, 64, *186*
Huffman, C. E., 97, 102, *189*
Huffmann, M. N., 146, *188*
Huggart, R. L., 94, 136, *185, 194*
Hulme, B., 43, 45, 49, *186*
Hunter, G. L. K., 88, 91, 93, 94, 96, 97, 98, 99, *186, 187, 196, 209*
Hunziker, R. R., 14, 17, *174, 201*
Hutchins, P. C., 8, *182*
Hutchinson, D. J., 8, *182*

I

Ichikawa, N., 131, *187*
Igarashi, O., 144, *187*
Iizuka, H., 141, *187*
Ikeda, R. M., 36, 53, 54, 88, 93, 96, 98, 101, *187, 206, 209*
Inagaki, C., 144, *187*
Inoue, H., 7, *187*
Ishizaki, K., 141, *198, 199*
Ismail, M. A., 52, 57, 67, *187*

Itaya, K., 142, *198*
Ito, S., 18, *188*
Ito, T., 141, *187*
Izumi, Y., 18, *188*

J

Jahn, O. L., 67, *188*
Jang, R., 59, *188*
Janku, I., 126, *188*
Jansen, E. F., 59, *188*
Jeffery, J. D'A., 61, *169*
Jeger, O., 150, 152, 154, 156, *169, 195*
Jennings, W. G., 166, *188*
Johnson, A. R., 37, *180*
Jolliffe, V. A., 23, 24, 25, 33, 60, *205*
Jones, C., 147, *171*
Jones, R. W., 146, *188*
Jones, W. W., 15, 16, 17, 21, 47, 49, 166, *180, 188, 191*
Joseph, G. H., 6, 39, 40, *188*
Jutras, P. J., 13, *190*

K

Kadota, R., 7, 93, *188*
Kamikawa, T., 152, *188, 191, 194, 208*
Kamiya, S., 120, 129, 130, 131, 132, 137, 142, *188, 197*
Kapur, N. S., 11, *196*
Karimullah, 102, *169*
Kariyone, T., 107, 132, *188*
Karmen, A., 55, *199*
Karrer, W., 129, *188*
Katayama, O., 52, *204*
Kefford, J. F., 1, 2, 8, 9, 43, 74, 106, 133, 135, 139, 140, 144, 149, 154, 156, 157, 158, 159, 160, 163, 164, *175, 188*
Kemper, W. C., 11, *193*
Kennedy, B. M., 13, *188*
Kertesz, Z. I., 60, *188*
Kesterson, J. W., 8, 81, 83, 84, 91, 93, 94, 96, 97, 100, 102, 119, 120, 121, 131, 139, 145, 148, *181, 184, 189*
Kew, T. J., 27, 28, 57, 81, 90, 121, *181, 204*
Khadr, A., 166, *189*
Khalifah, R. A., 11, 165, *189, 192*
Khan, A., 127, *200*
Khan, N. A., 7, 27, 131, *180, 189, 196*
Khuda, M. Q., 27, 131, *189*
Kilburn, R. W., 31, *189*
Kilpatrick, P. W., 88, 94, *203*
Kirchner, J. G., 97, *189*
Kishi, K., 140, 141, 142, *189, 190, 198*
Kita, Y., 93, 95, *190, 211*
Kitchel, R. L., 6, 44, *171*
Knorr, L. C., 6, 25, 60, *202*
Koaze, Y., 141, *183*
Kobayashi, A., 93, 95, *190, 211*
Koch, J., 6, 23, 74, 78, 79, 119, 127, 145, *190*
Kodama, M., 24, 27, 93, 139, 140, 143, *190, 191, 193, 194*
Koeppen, B. H., 118, 140, *190*
Kohli, R. R., 12, *190*
Kolle, F., 129, *190*
Komatsu, S., 107, *190*
Komura, Y., 141, *207*
Kondo, K., 155, *185*
Konishi, T., 144, *184*
Koo, R. C. J., 14, 15, 16, 20, 47, *190, 201, 211*
Koorama, M., 135, 137, 139, *172*
Kordan, H. A., 138, *190*
Kovacs Proszt, G., 59, *209*
Kovats, E., 88, 89, 93, 95, 99, *190, 191*
Kretchman, D. W., 13, *190*
Krewson, C. F., 132, *190*
Krummel, G., 7, *190*
Krupey, J., 106, *181*
Kubo, R., 107, *190*
Kubo, S., 24, 27, 135, 137, 139, 140, 143, *172, 190, 191, 194*
Kubota, T., 152, *188, 191, 194, 208*
Kugler, E., 88, 89, 93, 95, *191*
Kunjukutty, N., 57, *191*
Kunkar, A., 7, 54, 102, 106, 128, *191*
Kushii, M., 93, *193*
Kushiyama, Y., 143, *198*
Kuykendall, J. R., 11, *189*
Kwietny, A., 120, 121, *191*

L

Labanauskas, C. K., 17, 18, 47, *191, 192*
Labbee, M. D., 119, 141, 142, *208*
Labruto, G., 107, *191*
Laencina, J., 102, *174, 183*
La Face, F., 7, 102, 109, *191*

Lal, G., 60, 102, *199, 200*
Lang, K., 33, *191*
Langendorf, H., 33, *191*
Latz, H. W., 109, 110, *191*
Laughton, P. M., 93, 98, 103, *191, 192, 211*
Lavie, D., 131, 134, *199*
Lavollay, J., 146, *191*
Leach, E. H., 103, *191*
Lebreton, P., 106, 124, *175*
Lecomte, J., 148, *191*
Lee, R. E., 148, *191*
Leger, H., 146, *192*
Leland, H. V., 51, 52, 54, *176*
Lenz, F., 15, 158, *175, 178*
Leonard, C. D., 14, 20, 48, *192, 206*
Leuckert, C., 123, 133, *210*
Levi, L., 93, 98, 103, *191, 192, 211*
Levy, A., 154, *181*
Lewis, L. N., 18, 67, 68, 165, *176, 189, 192*
Lewis, W. M., 54, 55, *192*
Lezhava, V. V., 16, *192*
Liberti, A., 102, *183*
Lifshitz, A., 102, *192*
Lime, B. J., 7, 119, 121, 125, 137, 141, 142, *183, 192, 195*
Lin, Y-C., 6, *176*
Lindwall, R. C., 103, *206*
Lisle, D. B., 41, *192*
Lloyd, J. P. F., 103, *191*
Lockett, M. F., 146, *192*
Lodh, S. B., 12, *192*
Lombard, P. B., 31, *192*
Long, S. K., 52, *192*
Long, W. G., 6, 9, 11, 13, 22, *192, 193*
Loori, J. J., 99, *193*
Lopiekes, D. V., 139, *177*
Lo Presti, V., 130, 145, *178*
Ludin, A., 7, *193*
Luettich, W., 88, *197*
Lund, E. D., 88, 90, 94, *177, 196*
Lundin, R. E., 88, 94, 107, 109, 126, *181, 206, 208*

M

Mabry, T., 125, *175*
McCarty, C. D., 11, *193*
McConnell, B., 146, *193*
McCornack, A. A., 14, 16, 20, *190*
McCready, R. M., 29, 59, *193*
McCrindle, R., 154, 156, *177*
McDuff, O. R., 159, *193*
McFadden, W. H., 85, 88, 94, 101, *193, 203, 208*
McGraw, J. Y., 148, *193*
McIntyre, G. A., 15, *173*
MacLeod, W. D., 88, 97, 101, 110, *193*
MacRill, J. R., 6, 39, 40, *188*
Madsen, B. C., 109, 110, *191*
Maekawa, K., 93, *193*
Magnus, K. E., 156, *173*
Mahanta, D., 27, 83, 102, *174, 175*
Maier, V. P., 109, 125, 138, 149, 158, 159, 161, 163, *193, 194*
Maier, V. M., 105, 106, 109, 111, 124, 125, 132, 138, 166, *193*
Maleki, M., 7, 136, *194*
Mallent, D., 79, *199*
Manabe, M., 27, *207*
Manabe, T., 24, 27, 135, 137, 139, 140, 143, *172, 190, 191, 194*
Mannheim, C. H., 5, 60, *177, 194*
Manwaring, D. G., 127, *194*
Maraulja, M. D., 136, *194*
Margileth, D. A., 161, 163, *193*
Margraf, H. W., 146, *210*
Mariani, E., 83, *181*
Markh, A. T., 162, *194*
Marloth, R. H., 7, 9, 13, *194*
Marnay, C., 47, *183*
Martin, F. G., 167, *196*
Masias, E., 12, *180*
Masri, M. S., 122, *194*
Masumura, M., 132, *185*
Matsuda, Z., 107, *190*
Matsumoto, K., 119, 143, *203, 204*
Matsumoto, T., 152, *191*
Matsuno, T., 107, 109, 116, 131, 132, 133, *188, 194*
Matsuura, S., 131, *194*
Matsuura, T., 152, *191, 194, 208*
Matusis, I. I., 148, *195*
Mazur, Y., 149, 150, *195, 210*
Meagher, W. R., 19, *199*
Mehlitz, A., 102, *195*
Meisels, A., 150, *210*
Melera, A., 150, 152, 154, 156, *169, 195*
Menagarishvili, A. D., 17, *195*
Menarchery, M., 57, *191*
Menkin, V., 146, *195*

Meredith, F. I., 73, *200*
Merritt, A. J., 146, 147, *181*
Metcalf, J. F., 36, 65, *209*
Metzler, D. M., 105, 106, 109, 111, 124, 125, 133, 138, 166, *193*
Mikumo, Y., 93, 95, *211*
Miller, E. V., 11, *195*
Miller, J. M., 97, *189*
Minas, T., 102, *195*
Minutilli, F., 24, *195*
Mitamura, M., 93, *193*
Miura, H., 27, *195*
Mizelle, J. W., 121, 125, 127, 132, *183, 195*
Mizukami, Y., 140, *197*
Mizuta, T., 27, *195*
Molodozhnikov, M. M., 111, 150, 158, *197*
Mon, T. R., 88, 94, *203, 208*
Money, R. W., 44, 46, 49, *195*
Monselise, J. J., 7, 119, 145, *210*
Monselise, S. P., 19, 20, 41, 49, 62, 73, 120, 134, 136, 137, 165, *171, 182, 183, 184, 195*
Montesinos, M., 59, *200*
Moore, E. L., 14, 22, 25, 26, 60, 136, 158, 159, 167, *196, 202*
Moorthy, N. V. N., 11, *196*
Moreno, R., 8, *200*
Morgan, R. H., 43, 53, *196*
Morgenstern, L., 138, *190*
Mori, D., 143, *204*
Morikawa, E., 141, *196*
Morioka, T., 12, *196, 209*
Morries, P., 43, 45, 49, *186*
Moshonas, M. G., 88, 90, 94, 96, 97, 98, 99, *172, 177, 187, 196*
Moslem, Ali, A. K. M., 7, *180, 196*
Moss, G. P., 152, 156, *196*
Mucci, P., 148, *196*
Mukherjee, A. K., 127, *182*
Mukherjee, B. D., 7, *180, 196*
Mukherjee, S. K., 12, 102, *192, 196*
Muller, P. A., 102, *196*
Murakami, H., 7, 84, *204*
Murdock, D. I., 94, 95, *196*
Murphy, J. W., 156, *196*
Murray, C. W., 122, *194*

N

Nagy, S., 81, 84, *198*
Nakabayashi, T., 113, 116, 120, 129, 133, 137, 139, 140, 141, 143, 145, *196, 197*
Nakagawa, Y., 146, *180*
Nakamura, T., 7, 93, *188, 198*
Nakasone, Y., 93, 95, *211*
Nakatani, Y., 93, 95, *190*
Nakayama, A., 13, 135, *204*
Nakazaki, M., 117, *169*
Narasimhan, P., 12, *207*
Nawar, W. W., 88, 89, *180*
Nedbalek, M., 59, *209*
Neurath, G., 88, *197*
Neumann, J., 146, *191*
Newhall, W. F., 55, 144, 148, *197, 206, 208*
Newman, J. E., 21, 94, *204*
Newton, G. G. F., 61, *169*
Nicholas, H. J., 157, *178*
Nigam, I. C., 111, *197*
Nikonov, G. K., 111, 150, 158, *197*
Nimni, M. E., 147, *171*
Nishiura, M., 7, 129, 130, 131, 132, 137, *197*
Nobile, L., 102, *197*
Nogradi, M., 123, *180*
Nomura, D., 107, 129, 141, 143, 152, 157, 158, 162, 163, *197, 198*
Nordby, H. E., 81, 84, 107, 109, 121, *181, 198*
Norman, S., 90, 97, *198*
Nota, G., 89, *183*
Nursten, H. E., 88, *198*

O

Oashi, S., 121, *198*
Oates, J. A., 55, *199*
Oberbacher, M. F., 12, 18, 36, 61, 62, 67, 73, 91, 166, *170, 187, 198, 209, 210*
Ohta, T., 144, *187*
Ohta, Y., 88, 99, *198*
Okada, R., 93, *198*
Okada, S., 119, 142, 143, *182, 198*
Okamoto, Y., 13, *184*
Okano, S., 166, *183*
Oku, M., 119, 143, *203, 204*
Okumara, F. S., 132, *185*
Olsen, R. W., 119, 142, *198*
Olson, F. O., 9, 73, *174, 211*
Omura, H., 141, 142, *198, 199*
Ono, Y., 107, *190*

Oostinga, I., 46, *183*
Osadca, M., 78, *171*
Osajima, Y., 141, *198*
Osborn, R. A., 45, *199*
Oshima, Y., 127, 132, *211*
Otsuka, S., 143, *204*
Overton, K. H., 154, *177*
Ozawa, J., 30, *184*
Ozawa, S., 107, *190*

P

Page, A. L., 16, *180*
Parekh, C. M., 102, *199, 200*
Parks, G. L., 88, 128, 149, *187, 207*
Patterson, J. A., 146, *173*
Pennisi, L., 53, 93, 102, *178, 179, 199*
Perlin, A. S., 116, *183*
Petriceks, L., 126, *206*
Peynado, A., 20, 34, *177, 201*
Peyron, L., 102, 106, 109, 111, 124, 133, *199*
Pfaender, P., 130, *203*
Pfeifer, K., 33, *178*
Pfrogner, N., 61, *203*
Phillips, R. L., 19, *199*
Phythyon, J. M., 146, *176*
Pieringer, A. P., 93, 99, 100, 101, *170*
Pietsch, G., 107, *172*
Pinkas, J., 131, 134, *199*
Pisano, J. J., 55, *199*
Platt, R. G., 166, *188*
Platt, W. C., 98, *199*
Porretta, A., 128, *199*
Porter, G. G., 63, 67, *171*
Postlmayr, H. L., 8, 12, 29, 36, 37, 53, 137, 139, 149, *209*
Pounden, W. D., 148, *181*
Pozzo-Balbi, T., 102, *197*
Pradhan, S. K., 150, 152, 154, 155, 156, 158, 161, *169, 171*
Pretorius, A., 18, *199*
Primo, E., 7, 8, 24, 32, 37, 41, 45, 46, 59, 79, 136, 145, 159, 160, *169, 199, 200, 207*
Pritchett, D. E., 11, 163, *200*
Propst, B., 67, *208*
Pruthi, J. S., 27, 60, 99, 102, *199, 200, 206*
Purcell, A. E., 73, *200*

R

Raciti, G., 102, *179*
Ragab, M. H. H., 6, 63, *200*
Rahman, A., 127, *200*
Rahman, H., 7, 27, 131, *180, 189, 196*
Randhawa, G. S., 6, *201*
Rao, N. S. S., 60, *200*
Rao, P. R., 6, 24, 27, 52, 123, 128, 130, 131, 132, 135, 136, 137, 149, 159, *175, 203*
Rao, P. V. S., 128, *203*
Rasmussen, G. K., 15, 16, 21, 33, 34, 47, 68, *177, 201, 205, 206*
Rasquinho, L. M. A., 101, *201*
Ravina, A., 146, *201*
Rechsteiner, J., 52, 54, *205*
Redman, G. H., 166, *169*
Reece, P. C., 1, 2, 3, 6, *182, 207*
Rehm, S., 60, *203*
Reich, H., 146, *188*
Reichert, I., 134, *171*
Reitz, H. J., 9, 14, 16, 17, *174, 176, 201*
Renda, N., 120, 124, 127, *182*
Reuther, W., 21, 159, *201, 209*
Rickards, R. W., 127, *194*
Riehl, L. A., 17, *201*
Rimpler, H., 123, 133, *210*
Ringsdorf, W. M., 146, *180*
Rispoli, G., 7, 52, 53, 54, 68, 78, 93, 102, 113, 143, 144, *178, 179, 201*
Robbins, R. C., 146, 147, *201*
Robertson, J. M., 150, *169, 170*
Rockland, L. B., 162, *201*
Rodney, D. R., 9, *185*
Rodrigues, J., 12, 19, *201*
Rogers, G. R., 6, 45, *181*
Rolle, L. A., 8, 12, 29, 36, 37, 53, 54, 88, 93, 96, 98, 109, 110, 119, 120, 137, 139, 144, 145, 149, *187, 201, 206, 209*
Romaneko, E. V., 120, 146, *201*
Rossi, G. V., 148, *210*
Rotondaro, F., 109, *202*
Rouse, A. H., 6, 14, 22, 23, 25, 26, 60, 136, 158, 159, *202*
Roux, B., 122, 124, 125, 149, *175*
Row, L. R., 116, 128, 130, 131, 149, 158, *202, 203*
Rowell, K. M., 120, 126, 137, *202*

Royo, J., 5, 7, 37, 41, 45, 46, 136, 145, 159, *199, 200, 202, 207*
Rudol'fi, T. A., 110, *203*
Ruegg, R., 73, 74, *210*
Rymal, K. S., 95, *203*

S

Sadin, S., 146, *181*
Sagiv, J., 63, *171*
Saha, S. K., 111, *203*
Sala, J. M., 7, 8, 41, 45, 159, *200*
Samish, Z., 7, 51, 160, 162, *177, 193, 203*
Sanchez, J., 32, 33, *200, 203*
Sankunny, T. R., 57, *191*
Sannie, C., 132, *203*
Sarin, P. S., 106, 113, 130, 132, 136, *203*
Santo, T., 163, *197*
Sarkar, S. R., 6, *203*
Sarkissian, S., 7, 136, *194*
Sastry, G. P., 6, 24, 27, 52, 116, 123, 128, 130, 131, 132, 135, 136, 137, 149, 158, 159, *175, 202, 203*
Sato, S., 127, 132, *204*
Sawayama, Z., 119, 143, *203, 204*
Sawyer, R., 30, 39, 41, 45, *203*
Schaffner, K., 150, 152, 154, 156, *169, 195*
Scheffer, H., 59, *209*
Schelstraete, M., 13, *188*
Schenck, G. O., 156, *181*
Schmitt, R. A., 46, *174*
Schneider, E., 107, *172*
Schneider, G., 130, *203*
Schormueller, J., 61, *203*
Schroeder, C. A., 6, *205*
Schulman, R., 106, *181*
Schulte-Elte, K-H., 156, *181*
Schultz, E. F., 73, *200*
Schultz, T. H., 88, 94, *203, 208*
Schwartz, H. M., 60, *203*
Schwarz, R. E., 105, *204*
Scora, R. W., 21, 88, 93, 94, 99, 100, *204*
Scott, W. C., 6, 27, 28, 57, 81, 90, 101, *204*
Scudder, G. K., 15, *206*
Searles, M. A., 52, 54, *205*
Seegmiller, C. G., 59, *193*
Sehgal, J. M., 131, *204*
Sehon, A. H., 106, *181*
Seibold, H., 78, *172*
Seiler, F. E. E., 6, *174*
Seiler, R. R., 97, 102, *189*
Senkin, J., 66, *206*
Sergi, G., 103, *173*
Serra, J., 59, *200*
Seshadri, T. R., 106, 113, 130, 131, 132, 136, *203, 204*
Seshadri, V. S., 6, *201*
Sharapova, R. I., 110, *203*
Sharp, F. O., 11, *185*
Sharples, G. C., 11, *185*
Shen, C-P., 6, *176*
Sherratt, J. G., 45, *204*
Shigeo, F., 132, *185*
Shigeyama, T., 7, 84, *204*
Shiizaki, T., 13, *184*
Shimba, R., 13, 135, *204*
Shimoda, Y., 119, 143, *203, 204*
Shimoo, H., 132, *185*
Shinmoto, S., 143, *198*
Shinoda, J. 127, 132, *204*
Shioiri, H., 52, *204*
Shirley, R. L., 57, *169*
Sholokhova, V. A., 7, 19, *205*
Siddappa, G. S., 27, 83, 99, 102, 158, 159, 162, *174, 175, 205, 206*
Siddigi, A. I., 32, 106, *181, 205*
Silber, R. L., 52, 54, *205*
Silvestri, S., 121, *181*
Sim, G. A., 150, *169, 170*
Simanton, W. A., 17, *208*
Sinar, R., 45, *204*
Sinclair, J. B., 165, *180*
Sinclair, W. B., 1, 23, 24, 25, 33, 60, *171, 205*
Singh, D., 6, *205*
Singh, I. S., 6, *205*
Singh, J. P., 6, *201*
Sites, J. W., 15, *205*
Siv, S., 60, *194*
Sjverdsma, A., 55, *199*
Skakum, W., 103, *191*
Slater, C. A., 85, 89, 93, 99, 101, 102, 109, *205*
Smit, C. J. B., 140, *190*
Smoot, J. J., 68, *177*
Smith, P. F., 13, 14, 15, 16, 47, 49, *205, 206*
Smythe, C. V., 119, 141, 142, *206, 208*
Soine, T., 106, *206*
Sokoloff, B., 146, *193*

Solomons, G. L., 141, *173*
Sondheimer, F., 149, *195*
Soost, R. K., 6, 35, 73, *174, 206*
Sosa, A., 132, *203*
Soule, M. J., 6, 13, *192*
Sperti, G. S., 163, *206*
Spitler, E. M., 93, 98, *187*
Srere, P. A., 66, *206*
Srivas, S. R., 27, 99, *206*
Srivastava, H. C., 11, 12, *177, 196, 201, 207*
Srivastava, M. P., 52, *206*
Stainsby, W. J., 43, 45, 49, *186*
Stanley, W. L., 88, 89, 91, 93, 94, 96, 98, 101, 103, 106, 109, 110, 126, 181, *187, 206, 211*
Steers, C. W., 148, *180*
Stein, H., 54, *172*
Steinberg, B. M., 63, *171*
Stenstrom, E. C., 6, 20, *210*
Stepak, Y., 102, *192*
Stephens, T. S., 73, *200*
Stephenson, R. G., 105, 109, 121, 123, 136, *209*
Sternhell, S., 150, 152, 154, 155, 156, 158, 161, *169, 171*
Sternieri, E., 148, *196*
Stevens, J. W., 6, 39, 40, *188*
Stevens, K. L., 88, 94, *206, 208*
Stewart, I., 14, 55, 56, *192, 206, 210*
Still, F., 97, *181*
Stoll, M., 88, 107, *207*
Stout, M. G., 146, *188*
Stroecker, H., 148, *210*
Studer, A., 73, 74, *210*
Subbarayan, C., 75, *207*
Subrahmanyan, V., 102, *199*
Subramanyam, H., 11, 12, 19, *177, 201, 207*
Sudarsky, J. M., 146, *207*
Sumitani, M., 141, *207*
Sunday, M. B., 6, 9, 10, 11, 13, 20, 22, 67, *184, 188, 192, 193*
Sundt, E., 88, 107, *207*
Sutherland, F. H., 20, 48, *173*
Suyama, Y., 30, *207*
Swain, T., 113, 133, *207*
Swift, L. J., 5, 94, 128, *207*
Swingle, W. T., 1, 2, 3, *207*
Swisher, H. E., 163, *207*

T

Tabata, S., 123, *179*
Tadishi, I., 141, *196*
Takiguchi, H., 141, *187*
Takiguchi, Y., 141, *196, 207*
Talalaj, S., 102, *207*
Tanaka, S., 107, *190*
Tandon, R. N., 52, *206*
Tarazona, V., 136, 145, *207*
Tarutani, T., 27, *207*
Tasaka, T., 135, *207*
Tatum, J. H., 128, 149, *207*
Taylor, D. R., 156, *173*
Taylor, W. H., 53, *208*
Templeton, J. F., 150, 152, 154, 155, 156, 158, 161, *169, 171*
Teply, L. J., 6, 39, 40, 45, 47, *172*
Teranishi, R., 85, 88, 94, *181, 193, 203, 206, 208*
Tettamanti, A. K., 116, *211*
Thomas, A. F., 94, *208*
Thomas, D. W., 119, 141, 142, *206, 208*
Thommen, H., 74, *208*
Ting, S. V., 6, 10, 11, 17, 20, 24, 29, 32, 34, 35, 49, 73, 78, 136, 141, 144, 148, *178, 183, 197, 198, 208*
Tita, S., 7, 52, 68, *179*
Tobinaga, S., 150, 152, 156, *169*
Todd, G. W., 35, 67, *171, 208*
Tokorayama, T., 152, *191, 208*
Tomas, F., 120, *208*
Toube, T., 156, *196*
Tracuzzi, M. L., 102, *178, 201*
Trammel, K., 17, *208*
Treiber, H., 79, *172*
Trupin, N., 146, *172*
Tseng, K. F., 131, *208*
Tsusaka, N., 30, *207*
Tsusaka, T., 143, *208*
Tucker, D. M., 7, *192*
Tucker, D. P. H., 21, *209*
Turrell, F. M., 20, *195*

U

Ueyanagi, F., 12, *196, 209*
Udenfriend, S., 55, *199*
Underwood, J. C., 162, *201*

Unkrich, G., 130, *203*
Utterberg, K. A., 8, 12, 29, 36, 37, 53, 137, 139, 149, *209*

V

Vadehra, K. L., 131, *204*
Vandercook, C. E., 8, 12, 29, 36, 37, 47, 53, 54, 57, 72, 74, 77, 78, 105, 109, 110, 119, 120, 121, 123, 136, 137, 139, 144, 145, 149, *201, 209, 211*
Vannier, S. H., 88, 93, 96, 98, 103, 106, 109, 110, *187, 206*
Van Walleghem, P. A., 57, *169*
Vas, K., 59, *209*
Veldhuis, M. K. 27, 28, 57, 81, 88, 90, 101, *204, 209*
Venturella, P., 109, 130, 131, *171, 209*
Vernin, G., 101, *209*
Vincent, Y., 145, *209*
Vines, H. M., 18, 32, 34, 35, 36, 61, 62, 65, *198, 208, 209, 210*
Vogel, G., 147, 148, *169, 210*
Vogin, E. E., 146, 148, *210*
Volcker, P. E., 128, 149, *172*
von Holdt, M. S., 60, *203*

W

Wagner, C. J., 97, *172*
Wagner, D., 7, 119, 145, *210*
Wagner, H., 113, 121, 126, *185, 210*
Waiss, A. C., 107, 109, *181, 206*
Walker, D. R., 133, *184*
Walkley, V. T., 45, 46, *185*
Wallace, A., 33, 35, 62, 64, 65, 166, *172, 176, 186, 189*
Wallach, J., 125, *175*
Walther, K., 123, 133, *210*
Wan, L. K., 132, *210*
Wander, I. W., 14, 15, *192, 205, 206*
Ward, W. W., 118, 148, *180*
Wasserman, J., 88, *208*
Watkins, W. T., 99, *205*
Watson, D. G., 150, *169, 170*
Weerakoon, A. H., 83, *210*
Weichselbaum, T. E., 146, *210*
Weizmann, A., 149, 150, *195, 210*
Wender, S. H., 119, 121, 125, 126, 127, 132, 135, 137, 138, 141, 142, 146, *179, 180, 183, 195*
Wenzel, F. W., 94, 167, *185, 196*
Westbrook, G. G., 6, 20, *210*
Wexler, S., 156, *181*
Wheaton, T. A., 14, 52, 55, 56, *192, 206, 210*
White, M. J., 69, 72, 77, 149, *211*
Whyte, J. L., *29, 170*
Wildfeuer, I., 7, 102, *172*
Wilkes, P. S., 6, *181*
Willhalm, B., 88, 107, *207*
Williams, A. A., 88, *198*
Williams, B. L., 149, 152, *182, 210*
Wilson, K. W., 164, *210*
Winter, D. H., 120, *202*
Winterstein, A., 73, 74, *210*
Wolff, I., 79, *172*
Wolford, R. W., 52, 88, 91, 94, 95, 103, *170, 187, 203, 210*
Woodruff, R. E., 9, *211*
Woods, R. M., 146, *211*
Wucherpfennig, K., 54, *211*
Wuesthoff, M. T., 156, *181*

Y

Yamafugi, K., 141, 142, *198, 199*
Yamamoto, H., 7, *187*
Yamamoto, R., 127, 132, *211*
Yamanishi, T., 93, 95, *190, 211*
Yamashita, T., 131, *187*
Yasukochi, T., 141, 142, *199*
Yano, M., 119, 142, *198*
Yokoyama, F., 93, 98, 103, *211*
Yokoyama, H., 37, 53, 69, 72, 74, 77, 78, 121, 145, 149, *209, 211*
Yoshioka, H., 125, *175*
Young, L. B., 68, 73, *211*
Young, R. E., 66, *211*
Young, R. H., 73, *200*
Young, T. W., 15, 16, *211*
Yunus, M., 27, 131, *189*

Z

Zaganiaris, S., 6, *211*
Zemplen, G., 116, *211*
Ziegler, A., 121, *179*
Ziemelis, G., 163, *175*
Zimmermann, G., 121, 144, *183, 191*
Zitko, V., 29, 30, *211*
Zocolillo, L., 89, *183*
Zuber, H., 64, *211*
Zubovics, Z., 123, *180*

SUBJECT INDEX

A

Acacetin, 116
  chemical structure of, 114
  in citrus fruits, as glycoside, 130
Acetal, in citrus volatile flavors, 87
Acetaldehyde, in citrus volatile flavorings, 87, 90
Acetone, in citrus volatile flavorings, 87
Acetylmethylcarbinol, 95
Acidity of citrus fruits, 31, 36
Acids of citrus fruits, 31-37
  analytical aspects of, 36-37
  biosynthesis sites and mechanisms of, 35-36
  maturation effects on, 33-35
  in orange juice, 32
  in volatile flavors, 88
Aconitic acid, in citrus fruits, 32, 33
Adipic acid, in citrus fruits, 32, 33
Aegele marmelos, coumarins in, 111
Aeglopsis chevalieri, coumarins in, 111
Agricultural chemicals, citrus flavonoids as, 148
Alanine, in citrus fruits, 50, 52
Albedo
  definition of, 4
  of oranges
    composition of, 28
  pectin in, 28
Aldehydes, in citrus volatiles, 87, 91, 92, 96
  as analytical method, 103
Adulteration of citrus juices, *see under* Citrus juices and individual types
Algeria, information sources on citrus fruits of, 6
Alkaline earth metals, in citrus juices, 46
Alloimperatorin, in trifoliate orange seeds, 111
Alloisoleucine, in citrus fruits, 50
Amino acids in citrus fruits, 50-52
  in adulteration detection, 36-37, 53-54
  in citrus juices, 50-51
α-Aminoadipic acid, in citrus fruits, 52
γ-Aminobutyric acid, in citrus fruits, 50, 52, 54
Ammonia, in citrus fruits, 54
Antheraxanthin, in citrus fruits, 71
Anthocyanins, in citrus fruits, 113, 127-128
Antihistaminic compounds, in citrus fruits, 55
Antiinflammatory agents, citrus flavonoids as, 147
Antioxidants, citrus flavonoids as, 143-144
Apigenin
  chemical structure of, 114
  in citrus fruits, 122, 126, 138
    as glycosides, 124-125
Apo-β-carotenals
  in citrus fruits, 72, 74, 75
  separation of, 79
  structure of, 76
Apo-violaxanthals
  in citrus fruits, 75
  structure of, 76
Arachidic acid, in citrus seed oil, 83
Arginine, in citrus fruits, 50, 52
Aromadendrene, in citrus volatile flavors, 86, 99
Arsenate sprays
  arsenic in orange juice and, 18
  effect on
    general composition of fruit, 17-18, 36
    flavonoids of fruits, 136-137
Ascorbic acid, in citrus fruits, 39-41
Ascorbic acid oxidase, in citrus fruits, 61-62
Ash, of citrus fruits, 43, 44, 46
Asparagine, in citrus fruits, 50, 52

Aspartic acid, in citrus fruits, 50, 52
Aspergillus spp., naringinase from, 141, 142
Auranetin
  chemical structure of, 114
  in citrus fruits, 130, 136
Aurapten, in citrus fruits, 107, 108
Auraptene, *see* Meranzin
Auraptenol, in citrus fruits, 107, 108
Auraptin, *see* Isoimperatorin
Aureusidin, 6-rhamnoglucoside of, artifact in lemons, 124
Aurones
  in citrus fruits, 113
  structural formula of, 115
Auroxanthin, in citrus fruits, 71
Australia, information sources on citrus fruits of, 6
Auxin, from citrus fruit, unidentified, 165

B

Barbituric acid reagent, for citrus oil analysis, 103
Benzaldehyde, in citrus volatile flavors, 87
Benzoic acid, in citrus fruits, 32
Benzyl acetate, in citrus volatile flavors, 88
Benzyl alcohol, in citrus volatile flavors, 87
Bergamot
  botanical classification and varieties of, 3
  composition of information sources for, 7
  flavonoids in, 129
  hesperidin reported in, 129
Bergamot oil
  coumarins in, 107-110
  flavonoid in, 130
  mutagenic property of, 110
  nootkatone in, 97
  properties, information sources for, 210
α-Bergamotene, in citrus volatile flavors, 86, 98
Bergamottin, in citrus fruits, 107, 108, 110
Bergapten
  in citrus fruits and seeds, 107, 108, 111
  as photosensitizer, 110
Bergaptol, in citrus fruits, 107, 108, 111, 138
Betaine, in citrus fruits and products, 54-55
Bioflavonoids, *see* Flavonoids
Biondo di Sicilia oranges, amino acids in, 50-51
Biotin, in citrus fruits, 39, 40
β-Bisabolene, in citrus volatile flavors, 86, 98
Bisdehydrolycopene, in citrus fruits, 73
Bitter orange
  botanical classification of, 2
  coumarins in, 107, 108
  enzymes in, 62, 63
  flavonoids in, 122, 129-130, 134-137
    bitterness and, 139
  inorganic constituents of, 44
  oil
    analysis of, 101
    properties, information sources for, 102
  peel
    amino acids in, 52
    inorganic constituents in, 44
  taxonomy of, 130
Bitterness problems in citrus products
  flavonoids and, 138-140
  limonoids and, 159-163
Bittersweet orange, botanical classification of, 2
Blood orange
  amino acids in, 52, 54
  anthocyanins of, 127-128
  pigments in, 73
Borneo lemon, taxonomy of, 3, 110-111
Borneol, in citrus volatile flavors, 87
Bornyl acetate, in citrus volatile flavors, 88
Borohydride, in colorimetric method for flavanones, 120
Boron
  effect on fruit yield, 17
  in fruit, from borax treatment, 47
Brazil, information sources on citrus fruits of, 6
Bromate method, for oil content of citrus juice, 101, 103
Bromination, use in limonin assay, 164
Bronzing of lemons, carotenoid ketones in, 77
Butanols, in citrus volatile flavors, 86
2-Butanone, in citrus volatile flavors, 87
2-Butyl acetate, in citrus volatile flavors, 88
Butyraldehyde, in citrus volatile flavors, 87
Byakangelican, in citrus fruits, 108

C

Cadinenes, in citrus volatile flavors, 86

Caffeic acid, in citrus fruits, 106
Calacorene, in citrus volatile flavors, 86, 99
Calamenene, in citrus volatile flavors, 86, 99
Calamondin fruit
  acidity of, 31
  botanical classification and varieties of, 3
  flavonoids of, 130, 131
  nitrogen bases in, 55
  nucleic acids in, 57
  oil, citropten in, 111
  seed oil, lipids of, 83
  sitosterols in, 149
Calcium, in citrus fruits, 44, 47
California, information sources on citrus fruits of, 6
Campestrol
  in grapefruit, 150
  structure of, 151
Camphene, in citrus volatile flavors, 86, 96
Camphor, in citrus volatile flavors, 87
Canthaxanthin, separation from other pigments, 79
Capillary fragility, citrus flavonoids and, 146, 147
Carbazole, use in pectin determination, 23
Carbohydrates in citrus fruits, 23-30
  in juices, 24
  polysaccharides, 29-30
    separation techniques for, 29-30
  soluble type, methods for, 24
  structural types, 24-30
    changes during maturation, 24-27
Carbonyl compounds, in citrus volatile flavors, 87
Carboxylic ester hydrolase, in citrus fruits, 60
Carboxypeptidase C, in citrus fruits, 64
3-Carene, in citrus volatile flavors, 86
Carotenes, in citrus juices, as index of quality, 78
α-Carotene, in citrus fruits, 69, 70
β-Carotene, in citrus fruits, 40, 68, 69, 70, 73
γ-Carotene, in citrus fruits, 70
δ-Carotene, in citrus fruits, 70
ζ-Carotene, in citrus fruits, 69, 70, 74
η-Carotene, in citrus fruits, 70, 74
β-Carotene-5,6-monoepoxide, in citrus fruits, 74
Carotenoids
  as adulterants, separation techniques for, 79
  in citrus fruits, 40, 68-72
Carotenoid esters, as index of quality of citrus juices, 78
Carotenoid ketones, in citrus fruits and "bronzing" pigments, 77
β-Carotenone, in citrus fruits, 77
Carveol, in citrus volatile flavors, 87
Carvone, in citrus volatile flavors, 87, 99
β-Caryophyllene, in citrus volatile flavors, 86, 98, 101
Caryophyllene oxide, in citrus volatile flavors, 87
Catalase, in citrus fruits, 62
Catechins, 113
CCC, effect on fruit composition, 19
Chalcone
  in citrus fruits, 113
  structural formula of, 115
Chemical oxygen demand (COD), of orange juice, 103
China, information sources on citrus fruits of, 6
Chlorine, in citrus fruits, 44
Chloramine value, of amino acids in citrus products, 54
2-Chloroethyltrimethylammonium chloride, *see* CCC
Chlorogenic acid
  as allergen, 106
  in citrus fruits, 32, 33, 106
Chlorophenoxyacetic acid, effect on fruit composition, 19
Chlorophyll, in citrus fruits, maturation effects on, 67-68
Chlorophyll a, in citrus fruits, 67
Chlorophyll b, in citrus fruits, 67
2-Chlorotrimethylammonium chloride, effect on flavonoids of fruit, 137
Choline, in citrus fruits, 40, 54
Cholinesterases, in citrus fruits, 60-61
Chromatography, in analysis of flavonoids, 121
Chrysoeriol
  chemical structure of, 114
  in citrus fruits, 122
Cineoles, in citrus volatile flavors, 87, 93, 99
Cirantin, *see* Hesperidin
Citral, in citrus volatile flavors, 91, 92, 97,

98, 99, 103
as reference compound in oil analysis, 103
toxicology of, 103
Citramalic acid, in citrus fruits, 32, 33
Citranaxanthin
in citrus fruits, 71, 77
structure of, 75
Citrange
botanical classification and varieties of, 3
nitrogen bases in, 55
Cystine, in citrus fruits, 50
Citrangequat
botanical classification and varieties of, 3
pigments in, 70-72, 75-77
Citrangor, botanical classification and varieties of, 3
Citrate condensing enzyme, in citrus fruits, 65-66
β-Citraurin, in citrus fruits, 74
Citric acid
in citrus fruits, 32, 33
changes with maturity, 34
lemon varieties, 36
juice adulteration by, detection, 37
Citrofolioside, *see* Poncirin
Citromitin
chemical structure of, 114
in citrus fruits, 131, 136
Citron
botanical classification and varieties of, 2
oil, analysis of, 101
Citronellal, in citrus volatile flavor, 86, 100
Citronellol, in citrus volatile flavors, 87, 95
Citronellyl esters, in citrus volatile flavors, 88, 93
"Citronin," reported isolation of, 127, 132
Citropten, in citrus fruits, 107, 110, 111
Citrostadienol, in grapefruit peel, 149, 150
Citrumelo, botanical classification and varieties of, 3
Citrus acetylesterase, in preparation of deacetyl-cephalosporin C derivatives, 61
Citrus fruits
acidity related to taste of, 31
acids in, 31-37
carbohydrates in, 23-30
classification of, 2
coumarins in, 106-111
enzymes in, 58-66
flavonoids of, 113-148, 166
freezing effects on, 22
general composition of, 5-22
climatic factors and, 20-22
fruit size and, 13-14
horticultural spray effect on, 17-20
maturity and storage effects on, 10-13
nutrient status of tree and, 14-17
minor elements, 16-17
nitrogen, 14-15
phosphorus, 15-16
potassium, 16
plant growth regulator effect on, 18-20
rootstock effects on, 8-9
sources of information on, 6-7
hybrids of, *see* Hybrids of citrus fruits
information sources for fruit from various countries, 6-7
inorganic constituents of, 43-48
limonoids in, 150
nitrogen compounds in, 49-57
nucleic acids, in, 57
phenolic acids in, 105-106
phloroglucinol and derivatives in, 105
polyphenolic compounds in, 105-114
pigments in, 67-79
proteins in, 56-57
research needs for, 165-166
rhamnoglucosides in, 116-118
seeds of, *see* Seeds of citrus fruits
seed oils, lipids of, 81-84
steroids in, 149-150
storage
methods of extension, 10
triterpenoids in, 149-150
vitamins in, 39-42
volatile flavoring constituents of, 85-103
world production of, 1
Citrus hybrids, *see* Hybrids of citrus fruits
Citrus juice(s)
adulteration detection
by amino acid content, 53-54
by flavonoids, 119, 144-145
by mineral analyses, 43, 46
by "pentose equivalent," 30
by phenolic bases, 56
by pigments, 77-79
amino acids in, 50-51
betaine content as quality index, 54-55
composition, processing effects on, 5, 8
inorganic constituents in, 43-48

oils
extraction of, 89-90
rapid analysis of, 101, 103
Citrus oils, *see also* Volatile flavoring constituents; individual fruit oils
analytical methods for, 101-103
coumarins in detection of adulteration, 110
extraction methods for, 89-90
nonvolatile compounds in, 90
volatile flavoring constituents of, 85-103
Citrus sudachi
botanical classification of, 3
flavonoids of, 123, 132
Citrus tamurana, composition as affected by maturity, 12
Citrus waste, hesperidin recovery from 145-146
Clementine mandarin
carotenoids in, 74, 77
composition of, information sources for, 6
hesperidin reported in, 129
seed oil, lipids of, 82
Cloud, *see under* Orange juice
Colombia, information sources on citrus fruits of, 6
Coniella diplodiella, naringinase from, 141
Copaenes, in citrus volatile flavors, 86
Core of citrus fruits
definition of, 4
pectin in, 26
p-Coumaric acid
in citrus fruits, 106, 137
in "metabolic grid," 138
Coumarin, structure of, 109
Coumarins
in citrus fruits, 90, 106-111
as aid in taxonomic classification, 10-11
Cryptoflavin, in citrus fruits, 75
Cryptoflavinlike pigment, in citrus fruits, 71
Cryptoxanthin, in citrus fruits, 69, 70, 74, 75, 79
Cubebenes, in citrus volatile flavors, 86
Cuminaldehyde, in citrus volatile flavors, 87
Cyanidin reaction, for flavanones, 120
Cyanidin, chemical structure of, 114
Cyanidin-3-glucoside, in blood oranges, 127
Cycloartenol
in grapefruit, 149
structure of, 151
Cycloeucalenol
in grapefruit, 149
structure of, 151
"Curing" of lemons and limes, 10
p-Cymene, in citrus volatile flavors, 86, 92, 99
Cysteic acid, in citrus fruits, 50

D

Davis method for flavonoid analysis, 119-120
Deacetylnomilin, 154
as isolimonin, 152
tastelessness of, 157
Decanal
in citrus volatile flavors, 87, 91, 92, 96, 98
as reference for aldehyde content of oils, 103
Decanoic acid, in citrus volatile flavors, 88
Decanol, in citrus volatile flavors, 87, 92
γ-Decanolactone, in citrus volatile flavors, 88
2-Decanone, in citrus volatile flavors, 87
Decyl acetate, in citrus volatile flavors, 88, 99
L-Dehydroascorbic acid reductase, in citrus fruits, 62-63
Delphinidin, chemical structure of, 114
Delphinidin-3-glucoside, in blood oranges, 127
3-Demethoxysudachitin, in citrus fruits, 123, 132
5-*O*-Demethylcitromitin, in mandarin orange, 131, 136
5-Demethylnobiletin, in sour orange, 136
5-*O*-Demethyltangeretin, in rough lemon, 132, 136
Deoxylimonin
occurrence of, 154
structure of, 153
tastelessness of, 157
Deoxyribonucleic acid (DNA), in Calamondin fruit, 57
Diacetyl, in citrus volatile flavors, 87, 94-95
1, 2-Dialkylacroleins, in citrus volatiie flavors, 87, 94
O-Dianisidine, as reagent for analysis of citrus oils, 103
2,4-Dichlorophenoxyacetic acid, effect on

fruit composition, 19
Didymin, chemical structure of, 114
Diepoxycryptoxanthins, in citrus fruits, 71
Diethyl citrate, in citrus volatile flavors, 88
Dihydrochalcones, 113
sweetness of, 118
Dihydrokaempferol
chemical structure of, 114
in citrus fruits, 125, 126, 138
4′, 5-Dihydroxy-3′, 6, 7, 8-tetramethoxyflavone, as fungistat found in mandarins, 134
5, 8-Dihydroxy-3, 3′, 4′, 7-tetramethoxyflavone, in orange peel oil, 128
4′, 5-Dihydroxy-6, 7, 8-trimethoxyflavone, as fungistat found in mandarins, 134
p-Dimethylaminobenzaldehyde, as chromogenic spray for limonoids, 163, 164
2, 2-Dimethyl-5-(1-methyl-1-propenyl)-tetrahydrofuran, in citrus volatile flavors, 87
Dimethyl-*p*-phenylenediamine, in colorimetric method for hesperidin, 120
p,α-Dimethystyrene, in citrus volatile flavors, 86, 99
N-Dimethyltyramine, *see* Hordenine
Dinitrophenylhydrazone, use in limonin assay, 164
Diosmetin
chemical structure of, 114
in citrus fruits
as glucoside, 127
as *C*-glycosides, 124-125
Diosmin
chemical structure of, 114
in citrus fruits, 122, 123, 132
Dipentene, in citrus volatile flavors, 86
Djenkolic acid, in citrus fruits, 52
Dodecanal, in citrus volatile flavors, 87, 91, 92
Dodecanoic acid, in citrus volatile flavors, 88
Dodecanol, in citrus volatile flavors, 87
2-Dodecenal, in citrus volatile flavors, 87
Dolce Romana low-acid lime, carotenoids in, 75
Duncan grapefruit
general composition of,
fruit size and, 13
sprays and, 17-18
seed oil lipids of, 82

E

East Indian Rangpur, botanical classification of, 3
Edema reduction, flavonoids in, 147
Egypt, information sources on citrus fruits of, 6
Electrophoresis, of citrus proteins, 56
Elemenes, in citrus volatile flavors, 86, 93, 95, 96
Elemol, in citrus volatile flavors, 87
Endocarp, definition of, 4
Enzymes in citrus fruits, 58-66
carboxypeptidase, 64
esterases, 60-61
nitrate reductase, 63
oxidoreductases, 61-63
pectolyzing type, 58-59
phosphatases, 61
of tricarboxylic acid cycle, 64-66
Epoxides, in citrus volatile flavors, 87
5, 6-Epoxy-β-carotene, in citrus fruits, 70
5, 6-Epoxycryptoxanthin, in citrus fruits, 71
5, 6-Epoxylutein, in citrus fruits, 69, 71
Ergosterol, in the Rangpur, 149
Eriocitrin
chemical structure of, 114
in citrus fruits, 122, 123
Eriodictyol
chemical structure of, 114
in citrus fruits, 105, 122, 123, 125, 126, 136, 138
determination of, 121
as fine chemical, 148
pharmacological activity of, 122
Esculetin, in citrus fruits, 109, 111
Essence, definition of, 90
Essential oils, *see* Citrus oils and individual oils
Esters, in citrus volatile flavors, 88, 92
Esterases, in citrus fruits, 60-61
Ethanol, in citrus volatile flavors, 86, 90
Ethanolamine, in citrus fruits, 54
Ethers, in citrus volatile flavors, 87
Ethyl acetate, in citrus volatile flavors, 88, 90, 91
2-Ethylbutyraldehyde, in citrus volatile flavors, 87

Ethyl butyrate
in citrus volatile flavors, 88, 91, 94
as reference compound in oil analysis, 103
Ethylene
effect on bitterness of orange juice, 162
effect on lemon volatiles, 97
Ethyl formate, in citrus volatile flavors, 88, 91
Ethyl heptanoate, in citrus volatile flavors, 88, 91
Ethyl hexanoate, in citrus volatile flavors, 88, 91
Ethyl 3-hydroxyhexanoate, in citrus volatile flavors, 88
Ethyl isovalerate, in citrus volatile flavors, 88, 91
Ethyl octanoate, in citrus volatile flavors, 88, 91
Ethyl propionate, in citrus volatile flavors, 88, 91
Ethyl valerate, in citrus volatile flavors, 88, 91
Eureka lemon
acid biosynthesis, 35-36
amino acids in, 50-51
carotenoids in, 70-71, 74
enzymes in, 65
mineral composition of, 47
pigments in, 70-72, 74
vitamins in, 39-40
Evodol
occurrence and biogenesis of, 155-156
structure of, 155
taste of, 157

F

Farnesene, in citrus volatile flavors, 86
Fatty acids of citrus fruits, 81
Fenchol, in citrus volatile flavors, 87, 93
Ferric hydroxamates of limonoids, use for assay, 164
Ferulic acid, in citrus fruits, 106
Feruloylputrescine
in citrus fruits, 55-56, 106
structural formula of, 56
Flavanones, 113
analytical methods for, 118-122
structural formula of, 115
Flavedo
carboxypeptidase in, 64
definition of, 3
pigments in, 67, 69
Flavones
in citrus fruits, 113
structural formula of, 115
Flavonoids of citrus fruits, 113-148
as agricultural chemicals, 148
analytical methods for, 118-122
chromatographic, 121
colorimetric, 119-120
Davis procedure, 119-120
spectrophotometric, 120-121
anthocyanins, 113, 127-128
as antioxidants, 143-144
as authenticity standards, 144-145
biochemical aspects of, 133-138
biogenesis of, 137-138
in bitterness problems, 138-140
as by-products, 145-148
crystallization problems from, 140
enzymes for hydrolysis of, 140-142
extraction procedures for, 121
as fine chemicals, 148
flavanones, 113, 126-127
flavones, 113, 126-127
flavon-3-ols, 113
glycosides of, 118
grapefruit, 125-126
horticultural variable effects on, 135-137
importance of, 113
lemons, 122-125
mandarins and related fruit, 130-131
occurrence of, 113
oranges, 126-129
permethoxylated compounds, 128-129
as pharmaceutical chemicals, 146-148
negative findings on activity of, 147-148
physiological role in fruits, 133-135
pummelos and related fruits, 131-132
reviews on, 113, 115
rhamnoglucosides, 116-118
sour orange, 129-130
structure and nomenclature of, 114
taxonomic significance of, 129, 134-135
technological aspects of, 138-145
types of, 113
Flavon-3-ols
in citrus fruits, 113
structural formula of, 115

Flavoprotein aerodehydrogenase, in citrus fruits, 63
Flavor of citrus products, need for research on, 166-167
Flavoring, volatile constituents in, *see* Volatile flavoring constituents
Flavoxanthin, 69
 in citrus fruits, 71
Flavylium ion. (*See also* Anthocyanins)
 structural formula of, 115
Florida, information sources on citrus fruits of, 6
Fluoride, effect on fruit composition, 19-20, 47-48
Folic acid, in citrus fruits, 40
Formol titration, of amino acids in citrus products, 53
Fortunellin
 chemical structure of, 114
 in kumquats, 133
 sugar moiety in, 116
Freezing, effect on fruit composition, 22
Friedelin
 in grapefruit, 150
 structure of, 151
Fruit, size effects on general composition, 13-14
Fruit drinks
 citrus flavonoids as preservatives for, 144
 orange juice in, identification by flavonoids, 119, 145
Furfuraldehyde, in citrus volatile flavors, 87

G

Galactose, in orange juice, 24
Galacturonic acid, in citrus fruits, 32, 33, 60
 in method for pectin, 23
Gas chromatography, of citrus oils, 85, 89, 101
Gedunin, photogedunin from, 156
Gentisic acid
 in citrus fruits, 105
 as diagnostic test for greening virus, 105-106
Geranial, in citrus volatile flavors, 87, 98
Geraniol, in citrus volatile flavors, 86, 92
5-Geranoxy-7-methoxycoumarin, in citrus fruits, 107, 110, 111
Geranyl esters, in citrus volatile flavors, 88, 93, 95, 98
German market, information sources on citrus fruits of, 6
Gibberellic acid, effect on general composition of fruit, 18-20
Gibberellin
 effect on flavonoids, 137
 effect on pigments, 67-68
Gibberellinlike substances, from citrus fruit, unidentified, 165
Glucosidase, flavonoid hydrolysis by, 142
Glucosides, in lemons, 123
Glucosuria, from phloroglucinol, 105
Glutamic acid, in citrus fruits, 50, 52
Glycine, in citrus fruits, 50, 52, 54
Grandifolione
 occurrence of, 154
 structure of, 153
Grapefruit
 acidity of, 31, 36
 acids of, 35
  maturation effects on, 36
 amino acids in, 50-51
 bitterness of
  change upon maturation, 137-138
  by naringin determination, 119
  naringinase in debittering processes, 141, 142
 botanical classification and varieties of, 2
 citrostadienol in, 149
 cores, pectin in, 26
 coumarins in, 107-109, 111, 138
 enzymes in, 60, 62, 63, 64, 65
 feruloylputrescine in, 55
 flavonoids in, 116-118, 125-126, 133, 136-138
  determination of, 120
  maturation effects on, 135, 137
  "metabolic grid" of 125, 138
 flavor, nootkatone and aldehydes in, 97
 friedelin in, 150
 general composition of,
  climatic factors and, 20
  fruit size and, 13
  information sources for, 6
  plant growth regulators and, 19
  sprays and, 17-18
  tree nutrient status and, 15, 16
 inorganic constituents of, 44, 47
 limonoids in, 154, 158
  role in bitterness, 159
 pectin in, 26-27

peel
structural polysaccharides of, 29
sugars in, 24
phenolic acids in, 106, 166
phloroglucinol and derivatives in, 105, 138
pigments in, 69-73
polyphenols of, 125, 138
ripening process of, 10
seed oil lipids of, 82, 84
β-sitosterol in, 149
sterols in, 149-150
Grapefruit juice
acidity of, 31
acids in, 33
amino acids in, 50-51, 54
debittering of, 142
flavonoids in, 120
bitterness from, 139
crystal formation from, 140
determination of, 119, 121
inorganic constituents of, 44, 47
lipids of, 81
β-sitosterol in, 149
volatiles of, 96
Grapefruit oil
analysis of, 103
compounds in, 92-93, 96-97, 100
coumarins in, 107
"curing" of, 96
flavones in, 126
properties, information sources of, 102
nootkatone in, 97
Greece, information sources on citrus fruits of, 6
Greening virus
effect on juice flavor, 159
gentisic acid as diagnostic compound for, 105-106

H

Hamlin orange
acids in, 34
general composition of
sprays and, 17
tree nutrient status and, 16
juice, acids of, 32
seed oil, lipids of, 82
volatile flavor compounds in, 91
Hemicellulase, use to obtain flavonoids, 122
Hemicellulose, method for, 23
Heptadecanal, in citrus volatile flavors, 87
Heptamethoxyflavone
in grapefruit peel oil, 126
in oranges, 128, 129
Heptanal, in citrus volatile flavors, 87, 92
Heptanol, in citrus volatile flavors, 87
3-Hepten-1-ol, in citrus volatile flavors, 87
Herniarin, in citrus fruits, 109
Hesperidin, 144
analytical methods for, 119-121
as by-product from citrus fruits, 145-146
in citrus fruits, 41, 123, 126, 129-130, 132, 135-136
crystallization in citrus products, 140
effect on fungi, 134
extraction of, 121
as fine chemical, 148
neohesperidin and, 116, 130
pharmacological properties of, 126, 147
physiological effects on fruits, 134
structural studies on, 116-118
structure of, 114, 117
taxonomic significance of, 130
Hesperidinase
flavonoid hydrolysis by, 142
use in turbidity prevention, 143
Hesperetin
chemical structure of, 114
in citrus fruits, 126, 136
as fine chemical, 148
Hexadecanal, in citrus volatile flavors, 87
Hexanal, in citrus volatile flavors, 87
Hexanol, in citrus volatile flavors, 86
2-Hexenal, in citrus volatile flavors, 87
Hexenols, in citrus volatile flavors, 87
Histidine, in citrus fruits, 50
Hordenine, in mandarin juice, 55
Horticultural sprays
effect on,
general composition of fruit, 17-20
flavonoids, 136-137
inorganic constituents, 47
pigments, 68
Humulenes, in citrus volatile flavors, 86
Hybrids of citrus fruits
botanical classification and varieties of, 2
general composition of, 10-12
pigments in, 75, 77

5-Hydroxyauranetin, in sour orange, 136
2-Hydroxy-2-butanone, in citrus volatile flavors, 87
3-Hydroxy-α-carotene, in citrus fruits, 70
Hydroxy-ζ-carotene, in citrus fruits, 70, 75
8-Hydroxy-7',8',-dihydrocitranaxanthin, in citrangequat, 77
Hydroxy epoxy-α-carotenes, in citrus fruits, 71
Hydroxylamine titration of citrus oils, 103
3-Hydroxysintaxanthin, in citrus fruits, 72, 77
5-Hydroxytetramethoxyflavone, in bergamot oil, 130
Hyuganatsu tangelo
  botanical classification of, 3
  composition of, information sources for, 7

I

Ichang lemon
  botanical classification of, 3
  flavonoids in, 129
Ichangin
  bitter taste of, 157
  occurrence of, 154
  structure of, 153
Imperatorin, in citrus fruits and seeds, 108, 111
India, information sources on citrus fruits of, 6
Indoleacetic acid oxidase, in citrus fruits, 62
Infrared spectroscopy, of citrus oils, 101
Inorganic constituents of citrus fruits, 43-48
  analytical interest in, 43
Inositol, in citrus fruits, 40, 105
  as juice quality index, 41-42
Ion exchange chromatography, of amino acids, 52
Iron, in citrus fruits, 44
Isosakuranetin
  chemical structure of, 114
  in citrus fruits, 126
    as glycosides, 125, 126
Isocitric acid, in citrus fruits, 32, 33
Isoflavones, 113
Isohexanoic acid, in citrus volatile flavors, 88
Isoimperatorin, in citrus fruits, 107, 108
Isoleucine, in citrus fruits, 50, 54
Isolimocitrol
  chemical structure of, 114
  in citrus fruits,
    as glucoside, 123
Isolimonin, *see* Deacetylnomilin
Isonicotinic acid hydrazide reagent, in analysis of citrus oils, 103
Isopimpinellin, in citrus fruits, 107, 108, 111
Isopulegol, in citrus volatile flavors, 86
Isorhamnetin
  chemical structure of, 114
  in citrus fruits, 122, 125, 126, 138
    as glycoside, 130
Isoterpinolene, in citrus volatile flavors, 86
Israel, information sources on citrus fruits of, 7
Italy, information sources on citrus fruits of, 7

J

Jamaica, information sources on citrus fruits of, 7
Jambhiri, *see* Rough lemon
Japan, information sources on citrus fruits of, 7
Jasmone, in citrus volatile flavors, 87
Joppa orange
  composition of
    fruit size and, 14
    maturity and storage effects on, 10
  pigments in, 68
Juice, definition of, 4
Juices, *see* Citrus juices
"Juice oils," extraction methods for, 89-90
Juice sacs of citrus fruits, pectin in, 26

K

Kaempferol
  chemical structure of, 114
  in citrus fruits, 125, 126, 138
Kagzi nimbu, composition as affected by maturity, 12-13
Key lime, seed oil, properties of, 82
Kumquat
  amino acids in, 52
  botanical classification of, 2
  composition of, information source for, 7

fortunellin in, 116, 133
seeds, composition of, 84
Kusaie Rangpur,
botanical classification of, 3
flavonoids of, 130

L

Lactic acid, in citrus fruits, 32
Lauric acid, in citrus seed oil, 83
Lebanon, information sources on citrus fruits of, 7
Lemon
acidity of, 31
acids of, 33,
analysis, 36-37
biosynthesis, 35-36
differences in varieties, 36
amino acids in, 35-36, 50-52
botanical classification and varieties of, 2
"bronzing" in, 74, 77
coumarins in, 108, 110, 111
"curing" of, 10
enzymes in, 62, 64, 65-66
flavonoids in, 122-125, 129, 133
biogenesis of, 137, 138
O-glucosides and aglycones, 122-124
C-glycosides, 124-125, 133
horticultural variables and, 136
recovery from, 145
general composition of,
climatic factors and, 20
information sources for, 6
maturity and storage effects on, 12-13
plant growth regulators and, 18, 19
tree nutrient status and, 16
hybrid of, botanical classification and varieties of, 3
inorganic constituents of, 44
limonoids in, 152, 154
nitrogen bases in, 55
pectin in, 27
phenolic acids in, 106
phenolic amine absence in, 55
pigments, 70-72, 74
seed oil lipids of, 82, 84
sitosterols in, 149
tocopherol in, 144
vitamins in, 39-40
Lemon juice
acids in, 33
as index of quality, 36
amino acids in 50-51, 54
as index of quality, 53, 54
coumarins in, 110, 123
inorganic constituents of, 44, 46, 47
flavonoids in, 105, 123, 137
bitterness and, 139
determination of, 121
as index of quality, 119, 144-145
lipids of, 81
phloroglucinol in, 105
pigments in, as index of quality, 78
polyphenols in, storage effects on, 137
β-sitosterol in, 149
sugars in, 24
vitamins in, 40
Lemon oil
analysis of, 103
compounds in, 85, 97-98, 101
coumarins in, 110
from peel, compounds in, 92-93, 97-98
properties, information sources for, 102
Lemon peel
pectin in, 26
vitamins in, 40
Leucine, in citrus fruits, 51, 54
Leucoanthocyanins, 113
Lignoceric acid, in citrus seed oil, 83
Lime
acidity of, 31
acids in, 33
botanical classification and varieties of, 2
coumarins in, 108
"curing" of, 10
enzymes in, 62
general composition of
information sources for, 6
maturity and storage effects on, 12-13
tree nutrient status and, 15
limonin in, 158
pectin in, 26, 27
peel, pectin in, 26
seed oil, 81, 82
Lime juice
amino acids in, 54
lipids of, 81, 84
proline absence from, 52
Lime oil
analysis of, 103

components of, 89, 97, 101
peel oil, 92-93, 99
coumarins in, 107-110
properties, information sources for, 102
West Indian distilled type, 99
Limetta, enzymes in, 65
Limettin *see* Citropten
Limocitrin
chemical structure of, 114
in citrus fruits, 122, 123
structure elucidation of, 123
Limocitrol
chemical structure of, 114
in citrus fruits, 122, 123
structure elucidation of, 123
d-Limonene
in analysis of lemon oil, 101
in citrus volatile flavors, 86, 91, 92, 95-97, 100
biogenetic aspects, 100
in rapid analysis of citrus oils, 101, 103
Limonene oxides, in citrus volatile flavors, 87
Limonexic acid
from limonin, 156
structure of, 155
Limonin
A-ring lactone of, 161
biogenesis and biodegradation of, 150, 152, 154-157, 161
radioactive tracers in study of, 157
in citrus fruits, 150
D-ring monolactone of, 161
extraction of, 162-163
physiological activity of, 150
relationship to simarubolides, 156
structure of, 153
elucidation of, 150
Limonin D-ring lactone hydrolase, isolation of, 161
Limonoids in citrus fruits 81, 150, 152-164
analytical methods for, 163-164
biogenesis and biodegradation of, 150, 152, 154-157
horticultural aspects, 157-159
role in bitterness of citrus products, 81, 139, 159-163, 166
technological aspects, 159-163
Linalool
in citrus volatile flavors, 86, 93, 94, 95, 96, 98, 100
of leaf oils, 100
Linalool oxides
biogenetic aspects of, 100
in citrus volatile flavors, 87, 96
Linalyl esters, in citrus volatile flavors, 88, 93, 98
Linalyl pyrophosphate, biogenesis from isopropenylpyrophosphate, 100
Linarin, 116
Linoleic acid, in citrus seed oils, 82-83, 84
Linolenic acid, in citrus seed oils, 82-83, 84
Lipids in citrus fruits, 81-84
Lisbon lemon, amino acids in, 50-51
Longifolene, in citrus volatile flavors, 86
Lonicerin, in sour oranges, 129
Low-acid lemon, composition of, information source for, 6
Lutein, in citrus fruits, 71, 74
Lutein-5, 6-epoxide, 69, 71
Luteolin
chemical structure of, 114
in citrus fruits, 122
Luteoxanthins, in citrus fruits, 69, 71
Lycopene, in grapefruit, 73
Lysine, in citrus fruits, 51

M

Magnesium, in citrus fruits, 44
Magnesol column chromatography, in separation of flavonoids, 125, 126
Malsecco, flavones in resistance to, 133-134, 148
Malic acid
in citrus fruits, 32, 33, 34, 36
as index of juice quality, 37
determination (official method), 37
Malic dehydrogenase, in citrus fruits, 65, 66
Malonic acid, in citrus fruits, 32, 33
Mandarin oil
from peel, compounds in, 89, 92-93, 95-96, 101
properties, information sources for, 102
Mandarin oranges
amino acids in, 52
botanical classification of, 2
composition of
information sources for, 6
maturity and storage effects on, 10-12

plant growth regulators and, 19
tree nutrient status and, 16
coumarins in, 108
flavonoids in, 128, 129, 134, 136
crystallization problems from, 140
as fungistats, 134
juice
nitrogen bases in, 55
pigments in, 78-79
limonin in, 158
role in bitterness, 159-160
nonbitter juices from, 162
pectin, 27, 30
pigments in, 74-75
products of, turbidity prevention in, 143
seed oil lipids of, 82, 84
sitosterols in, 149
tocopherol in, 144
Manganese
effect on fruit yield, 16-17
in juice, from foliar sprays, 47
Marsh grapefruit
acids in, 35
amino acids in, 50-51
enzymes in, 62
peel oil
maturity effects on, 100
volatile compounds of, 96
pigments in, 69
carotenoids, 70-71
seed oil lipids of, 82
Maturation, effects on
acids, 33-35
carbohydrates, 24-27
flavonoids, 134-137
general composition of fruit, 10-13
limonoids, 156-157
pigments, 67-68
seed oils, 84
volatiles, 91, 94, 97, 100
Mediterranean mandarin, botanical classification of, 2
Membrane of citrus fruits, pectin in, 26
p-Menthadienols, in citrus volatile flavors, 87, 95
p-Mentha-1,8-dien-7-yl acetates, in citrus volatile flavors, 88
p-Mentha-8-ene-1,2-diol, in citrus volatile flavors, 87
p-Mentha-1-en-9-ol, in citrus volatile flavors, 87
p-Mentha-1-en-9-ol acetate, in citrus volatile flavors, 88
Menthone, in citrus volatile flavors, 87
Meranzin, in citrus fruits, 107, 108
Methanol, in citrus volatile flavors, 86
Methionine, in citrus fruits, 51
Methionine sulfoxide, in citrus fruits, 51
Methyl-8′-apo-β-carotenate, separation from other pigments, 79
3-Methylbutanol, in citrus volatile flavors, 86
3-Methyl-3-buten-2-ol, in citrus volatile flavors, 86
Methyl butyrate, in citrus volatile flavors, 88
24-Methylene cycloartenol, in grapefruit, 149
24-Methylene lophenol
in grapefruit, 150
structure of, 151
Methyl 2-ethylhexanoate, in citrus volatile flavors, 88
Methylheptanol, in citrus volatile flavors, 87
Methylheptanone, in citrus volatile flavors, 87
Methylheptenones, in citrus volatile flavors, 87, 94
Methyl hexanoate, in citrus volatile flavors, 88
Methyl 3-hydroxyhexanoate, in citrus volatile flavors, 88
Methyl isovalerate, in citrus volatile flavors, 88
Methyl *N*-methyl anthranilate, in citrus volatile flavors, 88, 93, 95
4-Methylpentanol, in citrus volatile flavors, 86
4-Methyl-2-pentanone, in citrus volatile flavors, 87
N-Methyltyramine, in citrus fruits, 55
Meyer lemon
botanical classification of, 3
carotenoids in, 70-71, 75
composition as affected by maturity, 13
flavonoids in, 129
nitrogen bases in, 55
pigments in, 70-72
Mexico, information sources on citrus fruits of, 7
Minerals, in citrus fruits, 43-48
Minneola tangelo

botanical classification of, 3
carotenoids in, 77
Moisture status, effect on fruit composition, 20
Morocco, information sources on citrus fruits of, 7
Murcott tangor
botanical classification of, 3
composition of
information source for, 6
maturity and storage effects on, 11
flavonoids of, 131
nitrogen bases in, 55
peel oil, compounds in, 96
pigments of, 68
seed oil lipids of, 83
Murraya exotica, carotenoid ketones in, 77
Mutatochrome, in citrus fruits, 70
Mutatoxanthin, in citrus fruits, 71
Myrcene, in citrus volatile flavors, 86, 90, 92, 100
Myristic acid, in citrus seed oil, 83

N

NAD and NADP, in citrus fruits, 63
α-Naphthaleneacetic acid, effect on fruit composition, 19
β-Naphthoxyacetic acid, effect on fruit composition, 19
Naringenin
in anti-inflammatory studies, 147
chemical structure of, 114
in citrus fruits, 126
as rhamnoglucoside, 125, 127
determination of, 121
as fine chemical, 148
physiological role in fruit, 133-134
Naringin, 144
analytical methods for, 119-121
biogenesis of, 137
bitterness from, 139-140, 159
as by-product from citrus fruits, 145-146
chemical structure of, 114
in citrus fruits, 125, 126, 129, 131, 135, 136
crystallization of in citrus products, 140
Davis method for, 119
enzymatic hydrolysis of, 141-142
extraction of, 121
as fine chemical, 148
as fungistat, 133-134
sugar moiety of, 116
Naringinase
composition of, 121, 141
in debittering of grapefruit products, 140-142
of Japanese citrus products, 143
in modification of Davis method, 119
properties of, 141-142
Narirutin
chemical structure of, 114
in citrus fruits, 132
Natsudaidai
amino acids in, 52
botanical classification of, 3
coumarins in, 107, 108
debittering of, 143
flavonoids in, 121, 129, 135, 137
bitterness and, 139-140
crystallization and, 140
general composition of
fruit size and, 13
information source for, 7
juice, debittering of, 143, 162
limonin in seeds of, 157, 158
pectin in, 27, 30
peel oil
compounds in, 99
darkening of, 144
sugars in, 24
Navel orange
amino acids in, 50-51, 54
flavonoids of, 126, 135
general composition
climatic factors and, 21
fruit size and, 13-14
information source for, 6-7
plant growth regulators and, 18-20
tree nutrient status and, 15, 16
inorganic constituents of, 43, 47
juice of,
debittering of, 162-163
phosphatase in, 61
quality related to limonin content, 159-161
limonoids in, 150
pigments in, 67, 69-72
proteins in, 56

vitamins in, 39-40
Neo-β-carotene-U, in citrus fruits, 70, 74
Neohesperidin
chemical formula of, 114
in citrus fruits, 125, 129-132, 135, 136, 137
hesperidin and, 116, 130
structural studies on, 116-118
structure of, 117
Neohesperidose
as sugar in neohesperidin, 116
structure of, 116
Neoisopulegol, in citrus volatile flavors, 87
Neoponcirin, chemical structure of, 114
trans-Neoxanthin
in citrus fruits, 69, 71, 74, 75
structure of, 76
Neral, in citrus volatile flavors, 87, 91, 98
Nerol, in citrus volatile flavors, 86, 93, 98
Nerolidol, in citrus volatile flavors, 87
Neryl esters, in citrus volatile flavors, 88, 93, 95, 98
Neurosporene, in citrus fruits, 70
New Zealand, information sources on citrus fruits of, 7
Nicotinic acid
in citrus fruits, 39, 40
as juice quality index, 41-42
Nimbin, autoxidation of, 156
Nitrogen, in citrus fruits, 44
Nitrogen bases, in citrus fruits, 54-56
Nitrogen compounds in citrus fruits, 49-57
amino acids, 50-54
analytical applications, 53-54
nitrogen bases, 54-56
nucleic acids, 57
proteins, 56-57
Nitrogen status of tree, effect on general composition of fruit, 14-15
Nobiletin
chemical structure of, 114
in citrus fruits, 128, 129, 130
from drug (Chinese) "Chen pi," 131
in edema reduction, 147
as fungistat, 134
Nomilin
bitter taste of, 157
in citrus fruits, 152, 154
structure of, 153
Nonanal, in citrus volatile flavors, 87, 92, 96, 98
Nonane, in citrus volatile flavors, 86
Nonanoic acid, in citrus volatile flavors, 88
Nonanols, in citrus volatile flavors, 87, 100
Nonyl acetate, in citrus volatile flavors, 88, 93
Nootkatone
in citrus volatile flavors, 87, 96
properties of, 97
grapefruit flavor and, 97
Nucleic acids, in citrus fruits, 57

O

Obacunone
in citrus fruits, 152, 154
limonin from, 161
structure of, 153
tastelessness of, 157
Obacunol, structure and occurrence of, 152
Obtusifoliol
in grapefruit, 149
structure of, 151
Ocimene, in citrus volatile flavors, 86
of leaf oils, 100
Octanal
in citrus volatile flavors, 87, 91, 92, 96, 98, 103
as reference compound in oil analysis, 103
Octanols, in citrus volatile flavors, 86, 90, 92
2-Octanone, in citrus volatile flavors, 87
2-Octenal, in citrus volatile flavors, 87
5-Octen-2-one, in citrus volatile flavors, 87
Octopamine
in citrus fruits, 55
structural formula of, 56
Octyl acetate, in citrus volatile flvors, 88, 93
Octyl butyrate, in citrus volatile flavors, 8
Octyl isovalerate, in citrus volatile flavors, 88
Octyl octanoate, in citrus volatile flavors, 88
Oils of seeds, *see* Seed oils of citrus fruits
Oil sprays, effect on fruit general composition, 17
Oleic acid, in citrus seed oils, 82-83, 84
Optical rotation of citrus oils, 101
Orange (sweet orange)
acidity of, 31
acids in, 31, 33
maturation effects on, 33

sites of formation in, 35
albedo, general composition and pectin in, 28
amino acids in, 50-51, 52
aroma, compounds contributing to, 91
ascorbic acid in, 39-41
bitterness in
rootstock effects on, 9
test for (in juice making), 164
botanical classification and varieties of, 2
cores, pectin in, 26
coumarins in, 107-111
enzymes in, 58-66
flavonoids in, 116, 124, 126-129
anthocyanins, 127-128
determination of, 119-121
hesperidin and neohesperidin, 116, 122
horticultural variables in, 135-137
permethoxylated flavonoids, 128-129
simple flavanones and flavones, 126-127
recovery of, 145-146
general composition of,
climatic factors and, 20-22
fruit size and, 13, 14
information sources for, 6
maturity and storage effects on, 10-12
plant growth regulators, 18-20
sprays and, 17-20
tree nutrient status and, 14-17
gentisic acid in, 105
index of quality for, 5
inorganic constituents of, 43-48
limonoids in, 158
enzymes for degrading in, 154
nitrogen in, 49
nitrogen bases in, 55, 56
nitrogen compounds in, 49-57
pectin in, 24-30
phenolic acids in, 105-106
phloroglucinol and derivatives in, 105
pigments in, 67-69
product purity, serine as index of, 53
proteins in, 56-57
pulp, general composition and pectin in, 28
rag, general composition and pectin in, 28
ripening process in, 10
seed oil lipids of, 82-84
tocopherols in, 144
vitamins in, 40
volatile flavoring compounds of, 90-95
orange juice, 94-95
peel, *see* Orange oil
whole orange, 90-91
Orange blossom oil, volatile components of, 100-101
Orange juice
acids in, 32, 33, 37
adulteration
detection of, 37, 41, 46, 53
carotenes as index of, 78
vitamins as index of, 41-42
amino acids in, 52, 53-54
arsenic in, from arsenate sprays, 18
betaine in, 54
as quality index, 55
bitterness
nomilin-limonin system in, 157, 159
precursor theory of, 160-161
rootstock effects on, 157-158
technological aspects, 159-163
variety and maturity effects on, 158
cloud
composition of, 27-29
lipids in, 81
COD values of, 103
debittering of, 162
enzymes in, 61
flavonoids in, 120, 135, 136
determination of, 119
flavor assessment method for, 167
in fruit drinks, identification by flavonoids, 119, 145
horticultural aspects of, 165-166
inorganic constituents of, 44, 46, 47
limonin removal from, 163
nitrogen content of, 49
pigments in, 69
polyphenols in, as index of quality, 120-121
proteins, 57
β-sitosterol in, 149
sugars in, 24
vitamins in, 40
as index of quality, 41-42
volatiles of, 94-95
xylose in, 30
Orange oil
components of, 89, 91-94, 97, 100, 101
analysis of, 101, 103

coumarins in, 111
permethoxylated flavonoids in, 128-129
properties, information sources for, 102
steroids in, 149
Orange peel
acid in, 33
ascorbic acid in, 41
hesperidin in, 41
structural polysaccharides of, 29
sugars in, 24
vitamins in, 40
Orangelo, botanical classification and varieties of, 3
Ortanique tangor
botanical classification of, 3
composition of, information source for, 7
Orthohydroxycinnamic acids, in citrus fruits, 106
Osthol, in citrus fruits, 108
Otaheite Rangpur
botanical classification of, 3
flavonoids of, 130
Oxalacetic-aspartic transaminase, 66
Oxalic acid, in citrus fruits, 32, 33, 34
Oxidoreductases, in citrus fruits, 61-63
Oxypeucedanin hydrate, in citrus fruits, 108

P

Pakistan, information sources on citrus fruits of, 7
Palmitic acid, in citrus seed oils, 82-83
Palmitoleic acid, in citrus seed oil, 83
Pantothenic acid, in citrus fruits, 39, 40
Papeda, hybrid of, botanical classification and varieties of, 3
Parson Brown orange
acids in juice of, 32
fatty acids in seed oil of, 82
Peach buds and flowers, inhibition by naringenin, 148
Pectic substances
in citrus fruits, 24-27
characterization of, 29-30
methods for, 23, 26-27
table of, 28
in orange components, 28
in orange juice cloud, 28
Pectinesterase, in citrus fruits, 60
Pectinmethylesterase, in citrus fruits, 58-59
Pectolyzing enzymes in citrus fruits, 58-59
Peel of citrus fruits
definition of, 4
flavonoid recovery from, 146
pectin in, 24-30
Peel juice, definition of, 4
Peel oil, extraction methods for, 90
Pentadecanal, in citrus volatile flavors, 87
Pentadecane, in citrus volatile flavors, 86, 91, 98
Pentamethoxyflavone, in citrus fruits, 130
Pentanols, in citrus volatile flavors, 86
1-Penten-3-ol, in citrus volatile flavors, 86
Perillaldehyde, in citrus volatile flavors, 87, 99
Permethoxylated flavonoids, in citrus fruits, 126, 128-129, 130
Peroxidase, in citrus fruits, 62
Persian lime, citrus seed oil, properties of, 82
Pharmaceutical chemicals, citrus flavonoids as, 146-148
Phellandrenes, in citrus volatile flavors, 86
Phellopterin, in citrus fruits, 108
Phenolic acids, in citrus fruits, 105-106
Phenolic amines
in citrus fruits, 55-56
as indexes of juice quality, 56
Phenylalanine, in citrus fruits, 51
p-Phenylenediamine, as reagent for analysis of citrus oils, 103
m-Phenylenediamine condensation method for citrus oil analysis, 103
Phenylhydrazine titration of citrus oils, 103
Phlorin, in citrus fruits, 105
Phloroglucinol and derivatives in citrus fruits, 105, 123
Phomopsis citri, substance inhibiting growth of, from orange peel, 165
Phosphatases, in citrus fruits, 61
Phosphoenolpyruvate carboxykinase, in citrus fruits, 64-65
Phosphoenolpyruvate carboxylase, in citrus fruits, 64-65
Phosphoric acid, in citrus fruits, 32, 33
Phosphorus, in citrus fruits, 43, 44, 47
Phosphorus status of tree, effect on general composition of fruit, 15-16
Photogedunin, occurrence of, 156

Phytoene, in citrus fruits, 69, 70, 75
Phytoenol, in citrus fruits, 70, 75
Phytofluene, in citrus fruits, 69, 70, 74, 75
Phytofluenol, in citrus fruits, 70, 75
Phytosterols, from squalene cyclization products, 149
Pigments in citrus fruits, 67-79
  analytical aspects, 77-79
  grapefruit, 69-73
  hybrids, 75, 77
  lemons, 74
  maturation effects on, 67-68
  oranges, 68-69
Pineapple orange
  acids in juice of, 32
  bitterness following 1962 freeze, 158-159
  enzymes in, 60, 62
  fatty acids in seed oil of, 82
  general composition of
    sprays and, 17
    tree nutrient status and, 15-16
  pectin distribution in, 26
  volatiles of, 91
Pinenes
  in analysis of lemon oil, 101
  in citrus volatile flavors, 86, 90, 92, 96, 98
Pinol, in citrus volatile flavor, 87
Piperitenone, in citrus volatile flavors, 87, 94
Plant growth regulators
  in citrus fruits, need for research on, 165
  effect on composition of fruit, 18-20
Polygalacturonase, absence, in citrus fruits, 60
Polyphenolic compounds in citrus fruits, 105-114
  coumarins, 106-111
  as index of juice quality, 37, 119-120, 144-145
  phenolic acids, 105-106
  phloroglucinol and derivatives, 105
Polysaccharides, of citrus fruits, 29-30
Poncirin, 121
  chemical structure of, 114
  in citrus fruits, 125, 126, 131, 137
  sugar moiety of, 116-117
Ponderosa fruit
  botanical classification of, 3
  flavonoids of, 132
  seed lipids of, 83, 84
Ponkan mandarin, flavonoids in, 131
Ponkanetin, *see* Tangeretin
Potassium, in citrus fruits, 43, 44, 46, 47
Potassium status of tree, effect on general composition of fruit, 16
Primo Fiore lemon, amino acids in, 50-51
Proline, in citrus fruits, 51, 52
Prolycopene, in citrus fruits, 71, 75
Propanol, in citrus volatile flavors, 86
Proteins in citrus fruits, 56-57
Protopectin, in orange peel, freezing effect on, 22
Prunin
  in citrus fruits, 137
  determination of, 121
  as interference in Davis test, 119
Psoralen
  derivatives of, in grapefruit, 109
  structure of, 109
Pummelo
  acids of, genetic effects on, 35
  botanical classification and varieties of, 2
  composition of information sources for, 6-7
  flavonoids of, 131-132, 135, 137
    bitterness and, 139
  hybrid (*Citrus tamurana*)
    composition as affected by maturity, 12
  pectin in, 27
  seed oil, lipids of, 82
Putrescine, in citrus fruits, 54
Pyridoxine, in citrus fruits, 40

Q

Quercetin, 144
  chemical structure of, 114
  in citrus fruits, 122, 123, 125, 126, 133, 138
Quinic acid, in citrus fruits, 32, 33, 34

R

Rag and pulp, definition of, 4
Rangpur
  botanical classification and varieties of, 3
  sitosterol and ergosterol from, 149
"Red" grapefruit, pigments in, 69, 73
Research needs for citrus fruits, 165-167
Reticulataxanthin

in citrus fruits, 69, 72, 74, 77
structure of, 76
Retinal, in citrus fruits, 71
L-Rhamnose, in citrus pectin, 29-30
Rhoifolin
chemical structure of, 114
in citrus fruits, 125, 126, 129, 131, 132, 137
pharmacological activity of, 126
Riboflavin, in citrus fruits, 40
Ribonucleic acid, in Calamondin fruit, 57
Rootstock, effects on general composition of citrus fruits, 8-9, 157-158
Rough lemon
amino acids in, 52
botanical classification of, 3
composition of, information source for, 6
flavonoids of, 132, 134, 136
Rubixanthin, in citrus fruits, 70
Rusk citrange
botanical classification of, 3
carotenoids in, 75-76
Russia, information sources on citrus fruits of, 7
Rutaceae, possible limonin degradation products in, 154-155, 166
Rutaevin
occurrence and biogenesis of, 155-156
structure of, 155
taste of, 157
Rutin
chemical structure of, 114
in satsumelo, 132-133
Rutinose
in hesperidin, 116
structure of, 116

S

Sabinene, in citrus volatile flavors, 86
of orange leaves, 100
Sabinene hydrate, in citrus volatile flavors, 87, 95
Sacaton citrumelo, botanical classification of, 3
Sathgudi oranges, composition of, maturity and storage effects on, 11
Satsuma mandarin
botanical classification of, 2
composition of, information source of, 7
pectin hydrolysis products of, 30
Satsumelo
botanical classification of, 3
rutin in, 132-133
Scopoletin, in citrus fruits, 108, 111, 137
Seeds of citrus fruits
coumarins in, 111
limonin in, 158
protein content of, 57
Seed oils of citrus fruits, lipids of, 81-84
Segments, definition of, 4
Selinenes, in citrus volatile flavors, 86, 99
Semi-β-carotene, in citrus fruits, 77
Serine
in citrus fruits, 51, 52
as index of purity of orange products, 53
Sesquicitronellene, in citrus volatile flavors, 86
β-Sesquiphellandrene, in citrus volatile flavors, 86, 93, 95
Sesquiterpenes, in citrus fruits, 85, 86, 91, 92, 95, 96, 98, 99
Seville orange, *see* Bitter orange
Shaddock, *see* Pummelo
Shamouti orange
amino acids in, 50-51
ascorbic acid in, 41
inorganic constituents in, 46
enzymes in, 62
extracts of, coleoptile test on, 165
general composition of, 6
hesperidin in, 135, 137
role of, 134
nitrogen in, 49
nonbitter juices from, 162
pigments in, 73
Simarubolides
bitter taste of, 157
limonin relationship to, 156
Sinapic acid, in citrus fruits, 106, 137
Sinensals, in citrus volatile flavors, 87, 91, 94
Sinensetin
chemical structure of, 114
in citrus fruits, 128, 130
Sinensiachrome, in citrus fruits, 71
Sinensiaxanthin, in citrus fruits, 71
Sintaxanthin
in citrus fruits, 72, 77
structure of, 76

Sinton citrangequat
  botanical classification of, 3
  carotenoids in, 70-71, 75
Sitosterols, in citrus fruits, 149
β-Sitosterol, in citrus fruits, 149, 150
β-Sitosterol glycoside, in orange peel oil, 128
Sodium, in citrus fruits, 44, 46
Sour orange, *see* Bitter orange
Sourness of citrus fruits, 31
South Africa, information sources on citrus fruits of, 7
Spain
  information sources on citrus fruits of, 7
  orange varieties of, acids in juices of, 32, 33
Sprays, horticultural, *see* Horticultural sprays; individual types
Stachydrine, in citrus fruits, 54
Stearic acid, in citrus seed oils, 82-83
Steroids in citrus fruits, 149-150
  biosynthetic pathway of, 150
Stigmasterol, in grapefruit, 150
Storage, effects on general composition of citrus fruits, 10-13
Stroke, citrus fruits as possible preventatives of, 146
Succinic acid, in citrus fruits, 32
Sucrose, formation in oranges and lemons, 24
Sudachitin
  chemical structure of, 114
  in citrus fruits, 123, 132
Sugars, in citrus fruits, 24
Sukego orangelo, botanical classification of, 3
Sweet lime, composition of, information source for, 7
Sweet orange, *see* Orange
Sweetening agents (artificial), from citrus flavonoids, 118, 148
Synephrine
  biosynthesis of, 55
  in citrus fruits, 55
  structural formula of, 56

T

Tangelo
  botanical classification and varieties of, 3
  composition of,
    information sources for, 6
    maturity and storage effects on, 11
  flavonoids in, 122
  leaf and blossom oil, dimethylstyrene in, 99
  nitrogen bases in, 55
  oil
    properties, information sources on, 102
  seed oil lipids of, 83
Tangeraxanthin
  in citrus fruits, 74
  structure of, 75
Tangeretin
  in anti-inflammatory studies, 147
  chemical structure of, 114
  in citrus fruits, 128, 129, 130, 131, 132, 136
  as fungistat, 134
Tangerine,
  acidity of, 31
  amino acids in, 50-51, 52
  botanical classification of, 2
  flavonoids in, 130-131
  general composition of
    climatic factors and, 20
    information sources on, 6
    tree nutrient status and, 14
  leaf and blossom oil, dimethylstyrene in, 99
  nitrogen bases in, 55
  peel oil, compounds in, 95-96, 97, 100
  pigments in, 74-75
  products, debittering of, 162
  seed oil lipids of, 82
Tangerine juice
  amino acids in, 50-51
  pigments in, 68, 78-79
  volatiles of, 95
Tangors
  amino acids in, 52
  botanical classification and varieties of, 3
  flavonoids of, 130
  nitrogen bases in, 55
  peel oil, compounds in, 96
  pigments of, 68
  seed oil lipids of, 83
Tartaric acid, in citrus fruits, 32
Taxonomy of citrus plants
  coumarin identification in, 110-111

flavonoid use in, 129, 134-135, 166
limonoid use in, 166
Temperature, effect on fruit composition 20-22
Temple tangor
amino acids in, 52
botanical classification of, 3
flavonoids of, 130
nitrogen bases in, 55
pigments of, 68
seed oil lipids of, 83
Terpenes
in citrus fruits, 85, 86, 91, 92, 95, 96, 97-99
oxygenated, in oil analysis, 103
Terpinenes, in citrus volatile flavors, 86, 92, 96, 97, 98, 100
Terpenoids, in citrus volatile flavors, 86
Terpinenols, in citrus volatile flavors, 86, 93, 98
Terpineol, in citrus volatile flavors, 86, 93, 94, 95, 98, 99
Terpinolene, in citrus volatile flavors, 86, 97
Terpinyl esters, in citrus volatile flavors, 88
Tetradecanal, in citrus volatile flavors, 87
Tetradecane, in citrus volatile flavors, 86, 91, 98
Tetrahydrogeraniol, in citrus volatile flavors, 86
Tetramethoxyflavone, in citrus fruits, 130
Tetramethylscutellarein
chemical structure of, 114
in orange peel juice, 129
Texas, information sources on citrus fruits of, 7
Thiamine, in citrus fruits, 40
Thin-layer chromatography
of amino acids, 54
of limonoids, 163-164
Threonine, in citrus fruits, 51, 52
α-Thujene, in citrus volatile flavors, 86
Thymol, in citrus volatile flavors, 86, 93, 95, 96
Thymyl methyl ether, in citrus volatile flavors, 87
Tocopherols, as antioxidants in citrus juice, 144
Tricarboxylic acid cycle enzymes, in citrus fruits, 64-66
2,4,5-Trichlorophenoxyacetic acid, effect on fruit composition, 19
2,4,5-Trichlorophenoxypropionic acid, effect on fruit composition, 19
Tridecanal, in citrus volatile flavor, 87
Tridecane, in citrus volatile flavors, 86
Trifoliate orange
botanical classification of, 2
coumarins in, 111
flavonoids in, 116, 132
limonin in, 158
oil of
compounds in, 99-100
properties, information sources of, 102
2,4,6-Trihydroxy-4′,-methoxychalcone, possible occurrence in citrus fruits, 124
α,α,p-Trimethylbenzyl alcohol, in citrus volatile flavors, 86, 95
α,α,p-Trimethylbenzyl methyl ether, in citrus volatile flavors, 87
2,6,6-Trimethyl-2-vinyltetrahydropyran, in citrus volatile flavors, 87
Triphasia trifolia, carotenoid ketones in, 77
Trisaccharides
in grapefruit, 125
in orange peel, 127
Triterpenoids and derivatives in citrus fruits, 149-164
Trollein, in citrus fruits, 71
Trollichrome, in citrus fruits, 71
Trolliflor, in citrus fruits, 71
Trollixanthin, in citrus fruits, 71
Troyer citrange
botanical classification of, 3
nitrogen bases in, 55
Tunisian sweet lemon, acid biosynthesis in, 35-36
Tyramine
in citrus fruits, 54
structural formula of, 56
Tyrosine, in citrus fruits, 51

U

Ultraviolet spectroscopy
of citrus oils, 101
of coumarins, 110
in flavonoid analysis, 120-121
Umbelliferone, in citrus fruits, 108, 111, 123, 137
Undecanal, in citrus volatile flavors, 87, 92
Undecane, in citrus volatile flavors, 86
Undecanoic acid, in citrus volatile flavors, 88

Undecanol, in citrus volatile flavors, 87
Undecyl acetate, in citrus volatile flavors, 88

V

Valencene, in citrus volatile flavors, 86, 91, 101
Valencia oranges
  acids in, 33-35
    juice, 32
  amino acids in. 50-51
  composition of
    climatic factors and, 20-21
    fruit size and, 13, 14
    general composition of, information source-for, 7
    maturity and storage effects on, 10, 11
    sprays and, 17, 20
    tree nutrient status and, 15-18
  enzymes in, 60, 62
  fatty acids in seed oil of, 82
  flavonoids of, 126, 136
  granulation in, 25
  inorganic constituents of, 43, 45, 57
  limonin variation in, 157-159
  pectin in, 24-25
    distribution of, 26
  pigments in, 67, 68, 70-72
  regreening effects on fruit, 11
  vitamins in, 39-40
  volatiles of, 90, 91, 95
    in peel oil, 92-93, 94
Valenciachrome, in citrus fruits, 72
Valenciaxanthin, in citrus fruits, 72
Valine, in citrus fruits, 51
Vanillin, in colorimetric method for oxygenated terpenes of citrus oils, 103
Vanillin-piperidine reagent, for citrus oil analysis, 103
Vasopressor compounds, in citrus compounds, 55
Violaxanthin, in citrus fruits, 69, 71, 75
Vitamins in citrus fruits, 39-42
  as adulteration index, 41-42
  analytical aspects, 41-42
  ascorbic acid, 39-41
  table of, 40
Vitamin C, *see* Ascorbic acid
Vitamin P (bioflavonoids), 146
  hesperidin determination in, 120
Vitexin
  chemical structure of, 114
  in citrus fruits,
    as *C*-glycosides, 124
Volatile flavoring constituents, 85-103
  analytical aspects, 101, 103
  biogenetic aspects, 100-101
  in grapefruit, 96-97
  in lemons, 97-98
  in limes, 99
  in mandarins, 95-96
  in Natsudaidai, 99
  in oranges, 90-95
  table of, 86-88
  in trifoliate oranges, 99-100

W

Wax of citrus fruits, valencene in, 91

X

Xanthophyll pigments, in citrus fruits, 68
Xanthotoxol, in trifoliate orange seeds, 111
Xylose, in orange juice (by pentose equivalent), 30

Z

Zapoterin, occurrence of, 156
β-Zeacarotene, in citrus fruits, 70, 77
Zeaxanthin, in citrus fruits, 71
Zinc
  effect on fruit yield, 17
  in juice, from foliar sprays, 47

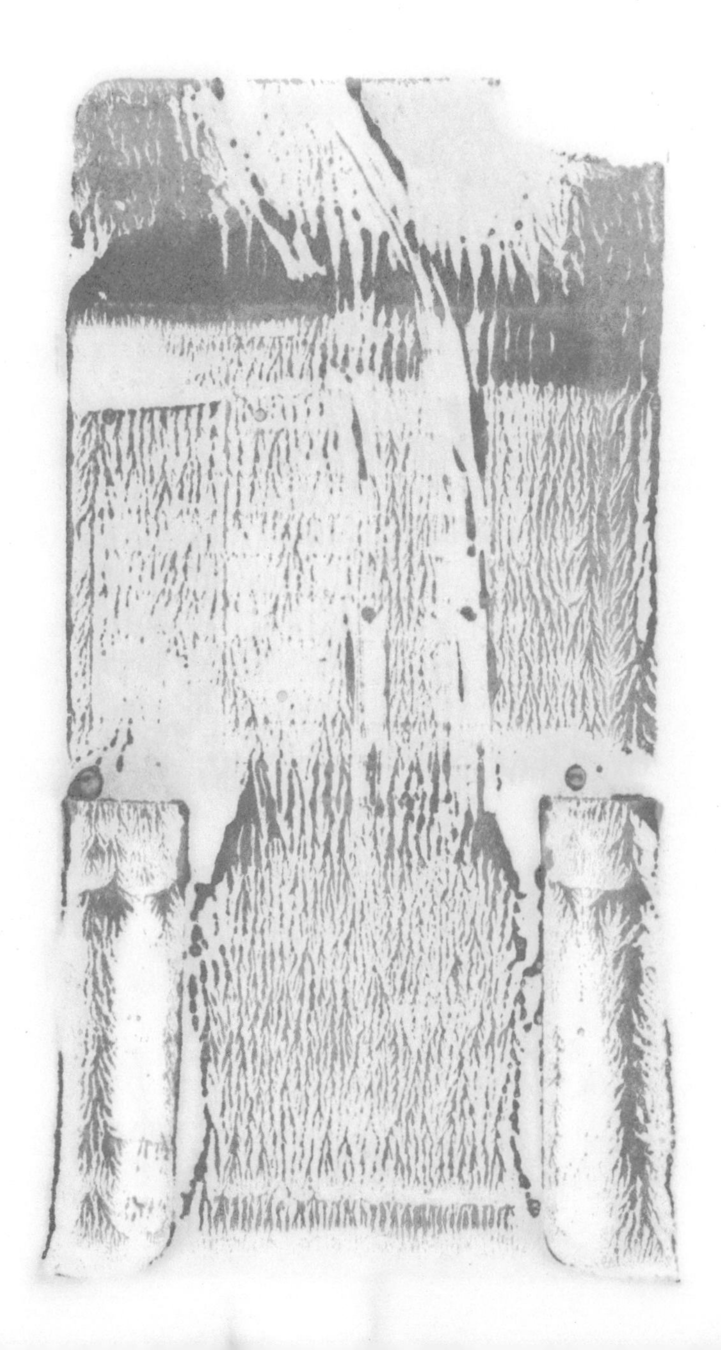